**FINITE ELEMENT ANALYSIS
IN FLUID DYNAMICS**

McGRAW-HILL
INTERNATIONAL
BOOK COMPANY
New York
St. Louis
San Francisco
Beirut
Bogotá
Auckland
Düsseldorf
Johannesburg
Lisbon
London
Lucerne
Madrid
Mexico
Montreal
New Delhi
Panama
Paris
San Juan
São Paulo
Singapore
Sydney
Tokyo
Toronto

T. J. CHUNG

Professor of Mechanical Engineering
The University of Alabama in Huntsville

Finite Element Analysis in Fluid Dynamics

This book was set in Times New Roman 327

British Library Cataloging in Publication Data

Chung, T J
  Finite element analysis in fluid dynamics.
  1. Fluid dynamics   2. Finite element method
  I. Title
  532′.05′01515353   TA357   78-30125

ISBN 0–07–010830–7

**FINITE ELEMENT ANALYSIS
IN FLUID DYNAMICS**

1 2 3 4 MHMH 8 0 7 9 8

Printed and bound in the United States of America

To my family

CONTENTS

PREFACE

This book is an outgrowth of a series of lectures presented to the graduate students at The University of Alabama in Huntsville over a period of years. It begins with mathematical preliminaries and develops into a finite element analysis for solving partial differential equations of boundary and initial value problems in fluid dynamics. The book is intended for not only the uninitiated student but also for the scientist and engineer in practice.

The initial and boundary value problems in continuous media have been the subject of intensive study for centuries. Although many analytical and numerical methods of solution have been developed extensively, there still remains unsolved a considerable number of important problems. The electronic computer in modern times led to the new ideas in the approximation theory. The finite element method with the use of a computer has rapidly become one of the most powerful tools in solving the complex problems of continuous media in general.

The primary objective here is to learn how to *solve* practical problems in fluid dynamics. The beginner, upon thorough comprehension of the solution procedures in some selected subjects, will then find the rest of the topics self-explanatory, in which details are no longer needed and thus only the essence is presented. The mathematical properties of finite elements undoubtedly constitute an important part of our study. Thus the secondary objective is to provide a mathematical treatment with which the analyst can establish validity of his calculations.

A review of the historical background of the finite element method is given in Chapter 1. An elementary account of functional analysis, which is the foundation for the finite element error estimates, is also presented. However, a study of this subject is not a prerequisite for understanding the rest of this book if one just wishes to learn how to solve a problem. Following discussions of variational principles and weighted residual methods, we demonstrate simple one-dimensional finite element solutions for the benefit of the uninitiated student. Chapter 2 discusses various types of finite elements grouped into one-, two-, and three-dimensional and axisymmetric geometries, along with local and global interpolation functions and dual spaces. Assembly of local equations into a global form, imposition of boundary conditions, and solution of nonlinear equations and time-dependent problems are presented in Chapter 3. Error estimates for linear problems are also included.

The topics of fluid dynamics begin with Chapter 4 where basic fluid dynamics equations are reviewed. Incompressible and compressible flows are covered in Chapters 5 and 6, respectively. Here solutions of the Laplace equation for two-dimensional domain and Stokesian equation for axisymmetric geometry are highlighted by complete details using triangular and isoparametric elements. Then various alternative formulations of steady and unsteady incompressible flow problems follow, including the velocity, pressure, stream function, and vorticity as variables. Error analyses from the concept of Sobolev spaces are discussed where appropriate. Problems of free surface, eigenvalues and eigenfunctions for wave motion, turbulent boundary layers, three-dimensional flow, boundary singularities, and some discussions of finite elements versus finite differences are also included. In the case of a compressible flow with temperature and density as additional variables, the methodology as used for incompressible flow can be applied, and thus unnecessary repetitions are avoided. Both viscous and inviscid compressible flows are discussed. Some recent developments of transonic aerodynamics are summarized. Finally, selected miscellaneous topics such as diffusion, magneto-hydrodynamics, and rarefied gas dynamics are presented in Chapter 7.

The finite element method is rapidly expanding in its scope of applications. In the meantime, commitments of the mathematicians, much to the surprise of the engineers, have provided new meaning, momentum, and confidence to this new field of research—the finite element analysis. It is for the unification of our knowledge sought by both engineers and mathematicians and for the spirit of teamwork that the present book is intended.

In the writing of this book, I am indebted to a countless number of authors of pioneering works from which I have freely drawn some of the materials and viewpoints. Professor J. Tinsley Oden reviewed the manuscript and offered invaluable suggestions for improvement. I wish to express my deepest appreciation to him. Thanks are also due to my former and present students who assisted in the solution of example problems. Among them are Dr. J. K. Lee, Messrs. C. G. Hooks, J. N. Chiou, and R. H. Rush. I owe a particular debt of gratitude to Mrs. Barbara Sweeney, who provided excellent service in computer programming. My thanks are further extended to Professor S. T. Wu for reviewing a portion of the manuscript, and to Professors J. J. Brainerd and C. C. Shih, among other colleagues, who shared useful discussions with me.

Some of the results presented in this book were obtained from the research under contract with the U.S. Air Force Office of Scientific Research during 1974, and subsequently with the U.S. Army Research Office since 1975. Their support is gratefully acknowledged.

The entire manuscript was typed by Mrs. Vivian M. Patterson, who, with enthusiasm and unfailing patience, volunteered the arduous task of proofreading and retyping. To her I am truly grateful.

University of Alabama, Huntsville T. J. Chung
March 1977

INTRODUCTION

1-1 GENERAL

The finite element method is an approximate method of solving differential equations of boundary and/or initial value problems in engineering and mathematical physics. In this method, a continuum is divided into many small elements of convenient shapes—triangular, quadrilateral, etc. Choosing suitable points called "nodes" within the elements, the variable in the differential equation is written as a linear combination of appropriately selected interpolation functions and the values of the variable or its various derivatives specified at the nodes. Using variational principles or weighted residual methods, the governing differential equations are transformed into "finite element equations" governing all isolated elements. These local elements are finally collected together to form a global system of differential or algebraic equations with proper boundary and/or initial conditions imposed. The nodal values of the variable are determined from this system of equations.

The finite element method was originally developed by aircraft structural engineers in the 1950's to analyze large systems of structural elements in the aircraft. Turner, Clough, Martin, and Topp [1956] presented the first paper on the subject, followed by Clough [1960] and Argyris [1963], among others. Application of the finite element method to nonstructural problems such as fluid flows and electromagnetism was initiated by Zienkiewicz [1965], and applications to a wide class of problems of interest in nonlinear mechanics was contributed by Oden [1972].

The close relationship of finite element analysis to the classical variational concept of the Rayleigh–Ritz method [Rayleigh, 1877; Ritz, 1909] or the weighted residual methods modeled after the well-known method of Galerkin [1915] has established the finite element method as an important branch of approximation theory. In recent years various authors have contributed to the development of the mathematical theory of finite elements. Among them are Babuska and Aziz [1972], Ciarlet and Raviart [1972], Aubin [1972], Strang and Fix [1972], and Oden and Reddy [1976]. These recent developments have been greatly influenced by the pioneering works of Lions and Magenes [1972, English translation].

Today various theories of fluid behavior are available which encompass virtually any type of phenomena of much immediate practical interest. However, there remains a surprising number of unsolved important problems in fluid dynamics due to difficulties encountered in most of the conventional analytical and numerical methods. In fluid dynamics, a choice of Eulerian coordinates renders the resulting governing equations nonlinear in general, and thus analytically difficult to solve. The most widely used numerical method of overcoming these difficulties has been the method of finite differences [Richtmyer and Morton, 1967; Roache, 1972] in which the partial derivatives in the governing equations are replaced by finite difference quotients. Another numerical method in limited use is the particle-in-cell method [Evans and Harlow, 1957], in which a system of cells is constructed so as to define the position of fluid particles in terms of these cells, and each cell is characterized by a set of variables describing the mean components of velocity, internal energy, density, and pressure in the cell. Among other popular methods are the variational methods and methods of weighted residuals [Finlayson, 1972]. Variational principles are used in the Rayleigh–Ritz method. Unfortunately, variational principles often cannot be found in some engineering problems, particularly when the differential equations are not self-adjoint. Weighted residuals are applied in the methods of Galerkin, least squares, and collocation. The method of weighted residuals utilizes a concept of orthogonal projection of a residual of a differential equation onto a subspace spanned by certain weighting functions. In the finite element method, we may use either variational principles when they exist, or weighted residuals through approximations. In finite element applications to fluid dynamics, the Galerkin method is often considered the most convenient tool for formulating finite element models since it requires no variational principles. The least squares method requires higher order interpolation functions in general, even if the physical behavior may be adequately described by linear or lower order functions. For these reasons, our discussions in this book are centered around the Galerkin method, although the finite element formulations via the methods of variational principles and least squares are demonstrated to a limited extent.

In the following sections of this chapter, we discuss basic mathematical preliminaries and notations. The reader is assumed to have been exposed to the vector analysis and matrix algebra. Tensor equations or index notations are used throughout the text, although no extensive knowledge of tensor algebra is required of the reader. All necessary mathematical preliminaries are presented in sufficient detail for the benefit of the beginner and for those who may require review. Brief discussions of functional analysis including Sobolev spaces, which is essential in error estimates and convergence, are also presented. Subsequently, we discuss the concepts of variational principles and weighted residuals as used in the classical approximate methods of analysis such as the Rayleigh–Ritz and Galerkin methods, respectively. The relationships of these classical concepts with the finite element theory are clarified. At the end of this chapter, simple example problems are solved to demonstrate the basic idea of finite element approximation for the beginner; more general discussions of finite element analysis are taken up in subsequent chapters.

1-2 MATHEMATICAL PRELIMINARIES

1-2-1 Vector and Index Notations

Some of the basic relations in mechanics may be written conveniently in vector notations. Let us begin with commonly used vector expressions in cartesian coordinates. Consider the vectors $\mathbf{A} = A_i \mathbf{i}_i$ and $\mathbf{B} = B_i \mathbf{i}_i$ with an angle between them being α. Here A_i and B_i are the components of the vectors $\mathbf{A}$ and $\mathbf{B}$, respectively, and $\mathbf{i}_i$ are the unit vectors with $i = 1, 2, 3$. The dot product and cross product are given by

$$\mathbf{A} \cdot \mathbf{B} = AB \cos \alpha = A_i \mathbf{i}_i \cdot B_j \mathbf{i}_j = A_i B_j \delta_{ij} = A_i B_i \qquad (1\text{-}1a)$$

$$|\mathbf{A} \times \mathbf{B}| = AB \sin \alpha = |A_i \mathbf{i}_i \times B_j \mathbf{i}_j| = |A_i B_j \varepsilon_{ijk} \mathbf{i}_k| \qquad (1\text{-}1b)$$

where the repeated indices imply summing; δ_{ij} is the Kronecker delta having the property $\delta_{ij} = 1$ for $i = j$ and $\delta_{ij} = 0$ for $i \neq j$; and ε_{ijk} is the permutation symbol defined as

$$\varepsilon_{123} = \varepsilon_{231} = \varepsilon_{312} = 1 \qquad \text{(clockwise permutation)}$$

$$\varepsilon_{132} = \varepsilon_{213} = \varepsilon_{321} = -1 \qquad \text{(counterclockwise permutation)}$$

with all other $\varepsilon_{ijk} = 0$ for any two or more indices being repeated. The permutation symbol and Kronecker delta are related by

$$\varepsilon_{ijk} \varepsilon_{mnp} = \delta_{im} \delta_{jn} \delta_{kp} + \delta_{in} \delta_{jp} \delta_{km} + \delta_{ip} \delta_{jm} \delta_{kn}$$

$$- \delta_{im} \delta_{jp} \delta_{nk} - \delta_{in} \delta_{jm} \delta_{pk} - \delta_{ip} \delta_{jn} \delta_{mp} \qquad (1\text{-}1c)$$

The continued products are

$$\mathbf{A} \times (\mathbf{B} \times \mathbf{C}) = A_i B_j C_k \varepsilon_{ijk} \qquad (1\text{-}1d)$$

$$\mathbf{A} \times (\mathbf{B} \times \mathbf{C}) = (\mathbf{A} \cdot \mathbf{C})\mathbf{B} - (\mathbf{A} \cdot \mathbf{B})\mathbf{C} = A_i B_j C_k \varepsilon_{jkl} \varepsilon_{ilm} \mathbf{i}_m = A_i (C_i B_j - B_i C_j) \mathbf{i}_j \quad (1\text{-}1e)$$

We introduce the vector differential operator, or simply del operator, $\mathbf{V}$, defined as

$$\mathbf{V} = \mathbf{i}_i \frac{\partial}{\partial x_i} \tag{1-2}$$

where $\mathbf{i}_i$ are the components of the unit vector. Then

$$\mathbf{V}E = \mathbf{i}_i \frac{\partial E}{\partial x_i} = E_{,i}\mathbf{i}_i$$

where E is a scalar and the comma denotes the partial derivative. The divergence and curl are given by

$$\text{div } \mathbf{A} = \mathbf{V} \cdot \mathbf{A} = \frac{\partial A_i}{\partial x_i} = A_{i,i} \tag{1-3}$$

$$\text{curl } \mathbf{A} = \mathbf{V} \times \mathbf{A} = \varepsilon_{ijk} A_{j,i}\mathbf{i}_k \tag{1-4}$$

In view of (1-1) through (1-4) we can show that

$$\mathbf{V}(\mathbf{A} \cdot \mathbf{B}) = \mathbf{A} \times (\mathbf{V} \times \mathbf{B}) + (\mathbf{A} \cdot \mathbf{V})\mathbf{B} + \mathbf{B} \times (\mathbf{V} \times \mathbf{A}) + (\mathbf{B} \cdot \mathbf{V})\mathbf{A} \tag{1-5}$$

Let us consider the velocity vector $\mathbf{V} = V_i\mathbf{i}_i$. From (1-5), we may write

$$\mathbf{V}(\mathbf{V} \cdot \mathbf{V}) = 2\mathbf{V} \times (\mathbf{V} \times \mathbf{V}) + 2(\mathbf{V} \cdot \mathbf{V})\mathbf{V} \tag{1-6}$$

in which

$$\mathbf{V} \times \mathbf{V} = \varepsilon_{ijk} V_{j,i}\mathbf{i}_k = \omega_k\mathbf{i}_k = \omega \tag{1-7}$$

where ω_k are the components of vorticity vector

$$\omega = \omega_1\mathbf{i}_1 + \omega_2\mathbf{i}_2 + \omega_3\mathbf{i}_3 = (V_{3,2} - V_{2,3})\mathbf{i}_1 + (V_{1,3} - V_{3,1})\mathbf{i}_2 + (V_{2,1} - V_{1,2})\mathbf{i}_3 \tag{1-8}$$

For a two-dimensional problem, we have

$$\omega_3 = V_{2,1} - V_{1,2} \tag{1-9}$$

The acceleration vector may be written in the form

$$\mathbf{a} = \frac{\partial \mathbf{V}}{\partial t} + (\mathbf{V} \cdot \mathbf{V})\mathbf{V} \tag{1-10a}$$

Substituting the relationship (1-6) in (1-10a) yields

$$\mathbf{a} = \frac{\partial \mathbf{V}}{\partial t} + \mathbf{V}\left(\frac{V^2}{2}\right) - \mathbf{V} \times \omega \tag{1-10b}$$

in which $V^2 = \mathbf{V} \cdot \mathbf{V} = V_j V_j$. Using index notion, we have

$$a_k = \dot{V}_k + (\tfrac{1}{2}V_j V_j)_{,k} - \varepsilon_{ijk} V_i \omega_j$$

or

$$a_k = \dot{V}_k + (\tfrac{1}{2}V_j V_j)_{,k} - \varepsilon_{ijk}\varepsilon_{mnj} V_i V_{n,m} \tag{1-10c}$$

We have shown here that all vector equations may be written in terms of cartesian components of the vectors using index notations. The components a_i of the acceleration vector **a** may be referred to as a tensor of the first order. The scalar is the zero order tensor. The second order tensor, such as the stress tensor σ_{ij} or σ^{ij}, has two indices whereas the tensor of material constants is of fourth order E_{ijkl} or E^{ijkl}. It is seen that the number of indices determine the order of a tensor. If a vector or tensor is referred to a cartesian basis, we refer to their components as cartesian. For curvilinear coordinates and/or nonorthogonal coordinates, the resulting tensor equations are noncartesian leading to the covariant and contravariant components of a vector. Details of this subject are discussed in Sec. 4-1 [see Sokolnikoff, 1964].

1-2-2 Matrix and Index Notations

In continuum mechanics, many of the physical laws are expressed by differential equations which may then be transformed into tensor equations in a local or global form. The earlier development of the finite element analysis was made using the matrix equations because they appeared straightforward and convenient when dealing with linear problems. However, as many nonlinear problems were considered, the inadequacy of matrix notations became apparent. The power of tensor analysis in mechanics in general is well known [see Sec. 4-1; also, Sokolnikoff, 1967].

Consider a matrix equation of the form

$$\mathbf{Au} = \mathbf{f} \qquad (1\text{-}11)$$

where **A** represents a matrix of the size $m \times n$. A matrix with $m \neq n$ is called the rectangular matrix whereas that with $m = n$ is referred to as the square matrix. The matrix with $m = 1$ and $n > 1$ is called the row matrix whereas that with $m > 1$ and $n = 1$, the column matrix. Suppose that **A**, **u**, and **f** are of $m \times m$, $m \times 1$, and $m \times 1$, respectively, and we write (1-11) in the form

$$\begin{bmatrix} A_{11} A_{12} \cdots A_{1m} \\ A_{21} A_{22} \cdots A_{2m} \\ \vdots \qquad \vdots \\ A_{m1} A_{m2} \cdots A_{mm} \end{bmatrix} \begin{bmatrix} u_1 \\ u_2 \\ \vdots \\ u_m \end{bmatrix} = \begin{bmatrix} f_1 \\ f_2 \\ \vdots \\ f_m \end{bmatrix} \qquad (1\text{-}12)$$

If we premultiply both sides of (1-11) by $\mathbf{u}^T$ with T denoting a transpose, then

$$\underset{1 \times m}{\mathbf{u}^T} \; \underset{m \times m}{\mathbf{A}} \; \underset{m \times 1}{\mathbf{u}} = \underset{1 \times m}{\mathbf{u}^T} \; \underset{m \times 1}{\mathbf{f}} \qquad (1\text{-}13)$$

It is seen that (1-13) becomes a matrix of size 1×1, a scalar quantity, an invariant, or a form of energy which has a physical significance in continuum mechanics. Had (1-12) been postmultiplied by $\mathbf{u}^T$, the resulting matrix would be of $m \times m$

$$\underset{m \times m}{\mathbf{A}} \; \underset{m \times 1}{\mathbf{u}} \; \underset{1 \times m}{\mathbf{u}^T} = \underset{m \times 1}{\mathbf{f}} \; \underset{1 \times m}{\mathbf{u}^T} \qquad (1\text{-}14)$$

If we choose to use index notations instead of matrix notations, we write (1-11) in the form

$$A_{ij}u_j = f_i \tag{1-15a}$$

or

$$u_j A_{ij} = f_i \tag{1-15b}$$

The repeated indices imply summing according to the ranges of indices. A repeated index is often called a dummy index and all indices not repeated are referred to as free indices. In the case of (1-15), i is free index and j is dummy index. If there is only one free index, then the number of equations is equal to the maximum range of the free index. If there is more than one free index, then the total number of equations is equal to the products of the maximum ranges of all free indices.

Suppose that in (1-15), $i = 1, 2, \ldots, m$ and $j = 1, 2, \ldots, n$. Then there result m equations obtained by summing the products of the jth columns of A_{ij} and corresponding jth rows of u_j. It is obvious that the free indices of both sides of Eq. (1-15) must match. If $i, j = 1, 2, \ldots, m$ in (1-15), the expanded form of (1-15) becomes

$$
\begin{aligned}
A_{11}u_1 + A_{12}u_2 + \cdots + A_{1m} &= f_1 \\
A_{21}u_1 + A_{22}u_2 + \cdots + A_{2m} &= f_2 \\
&\vdots \\
A_{m1}u_1 + A_{m2}u_2 + \cdots + A_{mm} &= f_m
\end{aligned}
\tag{1-16}
$$

These equations are identical to the matrix form (1-12). If index notations are used for (1-13), we write

$$A_{ij}u_j u_i = f_i u_i \tag{1-17}$$

Note that the choice index i on u_i is equivalent to premultiplication of $\mathbf{u}^T$ in (1-13) and the index i is also repeated here. This will leave no free index and thus expansion of (1-17) results in a single equation

$$
\begin{aligned}
&A_{11}u_1 u_1 + A_{12}u_2 u_1 + \cdots + A_{1m}u_m u_1 \\
&+ A_{21}u_1 u_2 + A_{22}u_2 u_2 + \cdots + A_{2m}u_m u_2 \\
&+ \cdots \\
&+ A_{m1}u_1 u_m + A_{m2}u_2 u_m + \cdots + A_{mm}u_m u_m \\
&= f_1 u_1 + f_2 u_2 + \qquad \cdots + f_m u_m \qquad (m \text{ not to be summed})
\end{aligned}
\tag{1-18}
$$

which is identical to the expansion of the matrix equation (1-13). Likewise, Eq. (1-14) with index notation takes the form

$$A_{ij}u_j u_k = f_i u_k \tag{1-19}$$

It is clear that the free indices are i and k on both sides and thus the total number of equations is equal to $m \times m$. The expansion of (1-19) is of the form

$$B_{ik} = C_{ik}$$

or

$$
\begin{bmatrix}
B_{11}B_{12} \cdots B_{1m} \\
B_{21}B_{22} \cdots B_{2m} \\
\vdots \quad \vdots \qquad \vdots \\
B_{m1}B_{m2} \cdots B_{mm}
\end{bmatrix}
=
\begin{bmatrix}
C_{11}C_{12} \cdots C_{1m} \\
C_{21}C_{22} \cdots C_{2m} \\
\vdots \quad \vdots \qquad \vdots \\
C_{m1}C_{m2} \cdots C_{mm}
\end{bmatrix}
$$

where

$$B_{11} = A_{11}u_1u_1 + A_{12}u_2u_1 + \cdots + A_{1m}u_mu_1 = C_{11} = f_1u_1$$

$$B_{12} = A_{11}u_1u_2 + A_{12}u_2u_2 + \cdots + A_{1m}u_mu_2 = C_{12} = f_1u_2$$

$$\vdots$$

$$B_{mm} = A_{m1}u_1u_m + A_{m2}u_2u_m + \cdots + A_{mm}u_mu_m = C_{mm} = f_mu_m$$

$$(m \text{ not to be summed})$$

The results shown above are identical to those obtained by expanding the matrix equation (1-14).

Consider now a special case of a matrix equation of the form

$$\underset{6\times 3}{\mathbf{B}^T} \; \underset{3\times 3}{\mathbf{E}} \; \underset{3\times 6}{\mathbf{B}} \; \underset{6\times 1}{\mathbf{u}} = \underset{6\times 1}{\mathbf{F}} \qquad (1\text{-}20)$$

The corresponding index equation may be obtained by introducing indices $\alpha, \beta = 1, 2, \ldots, 6$ and $i, j = 1, 2, 3$.

$$B_{i\alpha}E_{ij}B_{j\beta}u_\beta = E_{ij}B_{i\alpha}B_{j\beta}u_\beta = F_\alpha \qquad (1\text{-}21)$$

with the free index α on both sides. Expanding (1-21) yields

$$E_{11}B_{11}B_{11}u_1 + E_{11}B_{11}B_{12}u_2 + \cdots + E_{11}B_{11}B_{16}u_6$$

$$+ E_{12}B_{11}B_{21}u_1 + E_{12}B_{11}B_{22}u_2 + \cdots + E_{12}B_{11}B_{26}u_6$$

$$+ E_{13}B_{11}B_{31}u_1 + E_{13}B_{11}B_{32}u_2 + \cdots + E_{13}B_{11}B_{36}u_6$$

$$\vdots \qquad\qquad \vdots \qquad\qquad \vdots \qquad\qquad \vdots$$

$$+ E_{33}B_{31}B_{31}u_1 + E_{33}B_{31}B_{32}u_2 + \cdots + E_{33}B_{31}B_{36}u_6 = F_1$$

$$\vdots \qquad\qquad\qquad\qquad\qquad\qquad \vdots$$

To obtain the first equation from the matrix equation (1-20), one must write out completely all components of each matrix, but these matrix multiplications may become quite cumbersome.

Let us now consider a fourth order symmetric tensor E_{pqrs} with $p, q, r, s = 1, 2$

$$
E_{pqrs} =
\begin{bmatrix}
E_{1111} & E_{1122} & 0 \\
E_{2211} & E_{2222} & 0 \\
0 & 0 & E_{1212}
\end{bmatrix}
$$

with all other E_{pqrs} being zero. Thus (1-21) may be written in an alternative form

$$E_{pqrs}B^l_{M\,rs}B^k_{N\,pq}u^M_l = F^k_N \qquad \text{or} \qquad E_{pqrs}B_{M\,rsl}B_{N\,pqk}u_{Ml} = F_{Nk} \qquad (1\text{-}22)$$

with $k, l = 1, 2$ and $M, N = 1, 2, 3$. The position of indices whether subscripted or superscripted is immaterial, unlike the covariance and contravariance of the tensor algebra.† In (1-22) we note that the free indices are N and k with their maximum ranges being 3 and 2, thus leading to 6 equations. It can be easily verified that expansion of (1-22) leads to identical results as obtained from (1-21).

As a last example, let us consider a matrix equation

$$\underset{m\,\times\,m}{\mathbf{A}}\ \underset{m\,\times\,1}{\mathbf{u}}\ \underset{1\,\times\,m}{\mathbf{I}}\ \underset{m\,\times\,1}{\mathbf{u}} = \underset{m\,\times\,1}{\mathbf{f}}\ \underset{1\,\times\,m}{\mathbf{I}}\ \underset{m\,\times\,1}{\mathbf{u}} \qquad (1\text{-}23)$$

where $\mathbf{I}$ is the $1 \times m$ identity row matrix. The equivalent tensor equation is of the form

$$A^r_{ij}u_j\delta^k_r u_k = f^r_i\delta^k_r u_k \qquad (1\text{-}24)$$

where $\delta^k_r = \delta_{rk}$ is the Kronecker delta having the property as discussed in Sec. 1-2-1

and

$$A^r_{ij}\delta^k_r = A^k_{ij}, \qquad f^r_i\delta^k_r = f^k_i$$

Thus

$$A^k_{ij}u_j u_k = f^k_i u_k \qquad (1\text{-}25)$$

Expanding (1-25) yields

$$A^1_{11}u_1u_1 + A^1_{12}u_2u_1 + \cdots + A^1_{1m}u_mu_1 + A^2_{11}u_1u_2 + A^2_{12}u_2u_2 + \cdots A^2_{1m}u_mu_2$$

$$+ \cdots + A^m_{11}u_1u_m + A^m_{12}u_2u_m + \cdots + A^m_{1m}u_mu_m = f^1_1u_1 + f^2_1u_2 + \cdots + f^m_1u_m$$

$$A^1_{21}u_1u_1 + A^1_{22}u_2u_1 + \cdots + A^m_{2m}u_mu_m = f^1_2u_1 + f^2_2u_2 + \cdots + f^m_2u_m$$

$$\vdots$$

$$A^1_{m1}u_1u_1 + A^1_{m2}u_2u_2 + \cdots + A^m_{mm}u_mu_m = f^1_mu_1 + f^2_mu_m + \cdots f^m_mu_m$$

$$(m \text{ not to be summed})$$

Here it should be noted that the superscripts on A^k_{ij} and f^k_i are kept for summing, and the components of A^k_{ij} and f^k_i are identified by the subscripts. The advantage of using the index notation in the finite element method becomes more apparent in later chapters, particularly when nonlinear problems are dealt with. Moreover, in computer programming, these indices are the determining factors for efficient bookkeeping.

1-2-3 Green–Gauss Theorem

One of the most important concepts in mechanics of continua is the relationship connecting the surface integral for the flux of a vector field with the volume integral

---

† See Sec. 4-1 for details.

$\theta - theta \quad \sigma = signal \quad \int_\Gamma V \cdot n \, d\Gamma = \int_\Omega \nabla \cdot V \, d\Omega$

$\quad V \cdot n \, d\Gamma$

of its divergence. This relationship is known as the Green–Gauss theorem or divergence theorem and is given by _gamma_ _omega._

$$\int_\Gamma V \cdot n \, d\Gamma = \int_\Omega \nabla \cdot V \, d\Omega \tag{1-26}$$

where **V** is a vector representing the flux, velocity, displacement, etc.; **n** is a vector normal to the surface; and $d\Gamma$ and $d\Omega$ represent the infinitesimal surface area and interior volume, respectively. Substituting $V = V_i i_i$, $n = n_i i_i$, and (1-2) into (1-26) yields

$$\int_\Gamma V_i n_i \, d\Gamma = \int_\Omega V_{i,i} \, d\Omega \tag{1-27}$$

If **V** is multiplied by a scalar α, it follows that

$$\int_\Gamma \alpha V \cdot n \, d\Gamma = \int_\Omega \nabla \cdot (\alpha V) \, d\Omega = \int_\Omega \alpha \nabla \cdot V \, d\Omega + \int_\Omega V \cdot \nabla \alpha \, d\Omega \tag{1-28a}$$

or

$$\int_\Gamma \alpha V_i n_i \, d\Gamma = \int_\Omega (\alpha V_i)_{,i} \, d\Omega = \int_\Omega \alpha V_{i,i} \, d\Omega + \int_\Omega V_i \alpha_{,i} \, d\Omega \tag{1-28b}$$

It is interesting to note that (1-28b) is related to an integration by parts in the form

$\int a \, db \qquad ab - \int b \, da$

$$\int_\Omega \alpha V_{i,i} \, d\Omega = \int_\Gamma \alpha V_i n_i \, d\Gamma - \int_\Omega V_i \alpha_{,i} \, d\Omega \tag{1-29}$$

For example, we may write (1-29) in the form

$$\int_\Omega \alpha \left(\frac{\partial V_1}{\partial x_1} + \frac{\partial V_2}{\partial x_2} + \frac{\partial V_3}{\partial x_3} \right) dx_1 \, dx_2 \, dx_3$$

$$= \int_\Gamma \alpha V_1 \, dx_2 \, dx_3 + \int_\Gamma \alpha V_2 \, dx_1 \, dx_3 + \int_\Gamma \alpha V_3 \, dx_1 \, dx_2$$

$$- \int_\Omega \left(V_1 \frac{\partial \alpha}{\partial x_1} + V_2 \frac{\partial \alpha}{\partial x_2} + V_3 \frac{\partial \alpha}{\partial x_3} \right) dx_1 \, dx_2 \, dx_3 \tag{1-30}$$

On inclined boundary surfaces, we have $dx_2 \, dx_3 = d\Gamma_1 = n_1 \, d\Gamma$, $dx_1 \, dx_3 = d\Gamma_2 = n_2 \, d\Gamma$, and $dx_1 \, dx_2 = d\Gamma_3 = n_3 \, d\Gamma$ where $d\Gamma$ is the inclined surface area and n_1, n_2, and n_3 are the direction cosines denoting the components of the vector **n** normal to the surface as shown in Fig. 1-1. Therefore, (1-30) may be rewritten as

$$\int_\Omega \alpha (V_{1,1} + V_{2,2} + V_{3,3}) \, d\Omega$$

$$= \int_\Gamma \alpha (V_1 n_1 + V_2 n_2 + V_3 n_3) \, d\Gamma - \int_\Omega (V_1 \alpha_{,1} + V_2 \alpha_{,2} + V_3 \alpha_{,3}) \, d\Omega \tag{1-31}$$

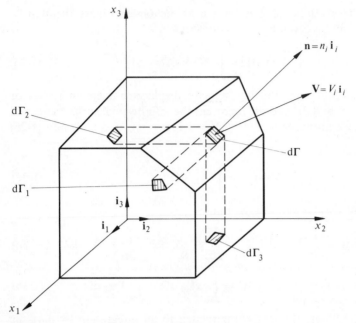

Figure 1-1 Vector normal to inclined surface—$d\Gamma_1$, $d\Gamma_2$, $d\Gamma_3$, are projections of $d\Gamma$.

This is identical to (1-29) expanded into components. It is concluded that integration by parts of the form (1-29) is equivalent to an application of the Green–Gauss theorem (1-28).

1-2-4 Functional Analysis

This section is not a prerequisite to the later chapters dealing with formulations and solutions of finite element equations. Functional analysis is needed only when error estimates are desired. Thus, the reader primarily interested in the finite element solutions of practical problems may postpone this section and Sec. 3-4 on mathematical properties of finite elements until the rest of the book is comprehended.

In this section we introduce a brief account on functional analysis. Definitions of various terminologies, distributions, Sobolev spaces, Sobolev embedding theorems, duals of Hilbert spaces, and trace theorems are presented.†

Basic definitions The usual set theoretics and notations are used here. For example, $A \in B$ is read "A belongs to the set B"; $\forall$ means "for every"; $A \subset B$ indicates that

---

† Consult standard texts for detailed treatments of functional analysis such as Taylor [1958] or Kolmogorov and Fomin [1957, 1961]. For materials on distributions, Sobolev spaces, and trace theorems we draw heavily from Oden and Reddy [1976]. See Aubin [1972] and Lions and Magenes [1972] for mathematics of boundary value problems.

A is a subset of B; and the *union, intersection,* and *difference* of two sets A and B are denoted as $A \cup B$, $A \cap B$, and $A - B$, respectively.

Two sets A and B are said to be *disjoint* if their intersection is the *null set,* i.e., if $A \cap B = \varnothing$. The cartesian product of two sets A and B is a set $A \times B$ of ordered pairs (a, b), that is $A \times B = \{(a, b): a \in A, b \in B\}$.

Let $A \subset B$; then we denote $b \in B$ as an *accumulation point* of A if every neighborhood of b contains a point of A different from b. The set consisting of A and all its accumulation point is called the *closure* of A. If A and B are two sets such that the closure of A contains the set B, then A is said to be *dense* in B.

Let the set of real numbers be denoted by **R**, and the set of positive real numbers by $\mathbf{R}^+$. The cartesian product $\mathbf{R}^2$ represents the *real euclidian plane,* and $\mathbf{R}^n$ denotes the set of n-tuples $(x_1, x_2, \ldots, x_n)$ where $x_i \in \mathbf{R}$ $(i = 1, 2, \ldots, n)$ which we call the *n-dimensional real euclidian space.* The set of complex numbers also occurs frequently so we denote it by **C**.

If a unique element of T can be associated with each element of S, such a process is referred to as a *mapping* of S into T and we write $f(x): S \rightarrow T$ where the element $f(x)$ is called the image of x under the mapping. If every element of T is the image of some element of S, i.e., in which $f(S) = T$, we say that f maps S *onto* T. This is called a *surjective mapping.* It is possible that any element of the set T could be the image of more than one element of S. If each element of $f(S) \subset T$ is the image of one and only one element of S, we refer to such mapping as *one-to-one* or *injective mapping.* A mapping which is both surjective and injective is called *bijective.*

Suppose that S is a set with elements $x, y, \ldots$ which admits addition and multiplication by a complex scalar, i.e., for each $x, y \in S$, there exists an element $(x + y) \in S$ and for each $x \in S$ and each $x \in \mathbf{C}$, there exists an element $\alpha x \in S$. The set is called a *linear (vector) space* if the following properties hold:

$$(\alpha\beta)x = \alpha(\beta y) \qquad \forall\, \alpha, \beta \in \mathbf{C}, \qquad \forall\, x \in S \qquad (1\text{-}32a)$$

$$(\alpha + \beta)x = \alpha x + \beta x \qquad \forall\, \alpha, \beta \in \mathbf{C}, \qquad \forall\, x \in S \qquad (1\text{-}32b)$$

$$\alpha(x + y) = \alpha x + \alpha y \qquad \forall\, \alpha \in \mathbf{C}, \text{'} \qquad \forall\, x, y \in S \qquad (1\text{-}32c)$$

$$1x = x \qquad \forall\, x \in S \qquad (1\text{-}32d)$$

A subset M of S is called a *vector subspace* of S if, for each $\alpha \in \mathbf{C}$, $\alpha M \subset M$ and if

$$M + M = \{x + y: x \in M, y \in M\} \subset M$$

If A is any subset of S, the set of all finite linear combinations

$$\sum_i \alpha_i x_i \qquad \alpha_i \in \mathbf{C} \qquad x_i = A \qquad (1\text{-}33)$$

is called the vector subspace spanned by A. Moreover, A subset A of S is said to be *linearly independent* if

$$\sum_{1 \leq i \leq n} \alpha_i x_i = 0 \Rightarrow \alpha_1 = \alpha_2 = \cdots \alpha_n = 0 \qquad (1\text{-}34)$$

for every choice of the positive integer n and of the vectors $x_1, x_2, \ldots, x_n$ in S. A linearly independent subset spanning S is called a *base* of S.

A mapping $f: V_1 \to V_2$ of one vector space V_1 onto another vector space V_2 is called a linear mapping if, and only if, for all elements x, y of V_1 and all $\alpha \in \mathbf{K}$ ($\mathbf{K} = \mathbf{R}, \mathbf{C}$)

$$f(x + y) = f(x) + f(y) \tag{1-35a}$$

$$f(\alpha x) = \alpha f(x) \tag{1-35b}$$

$$f(\alpha x + \beta y) = \alpha f(x) + \beta f(y) \qquad \forall\, x, y \in V \qquad \forall\, \alpha, \beta \in \mathbf{K} \tag{1-35c}$$

If such a mapping is bijective, we say that V_1 and V_2 are isomorphic.

A *gauge* (or a *sublinear functional*) on a vector space V is a mapping $g: V \to \mathbf{R}$ such that

$$g(x + y) \le g(x) + g(y) \qquad g(\alpha x) = \alpha g(x)$$

for all $x, y \in V$ and α an arbitrary positive real scalar.

If $g: V \to \mathbf{R}$ is a gauge on a real vector space V and U is a subspace of V, then if f is linear on U and

$$f(x) \le g(x) \qquad x \in U$$

there is a linear mapping on V such that

$$F(x) = f(x) \qquad x \in U$$

$$F(x) \le g(x) \qquad x \in U$$

This is known as the *Hahn–Banach Theorem.*

A *seminorm* is defined as a mapping $p: V \to \mathbf{R}^+$ such that

$$p(x + y) \le p(x) + p(y) \qquad p(\alpha x) = |\alpha|\, p(x)$$

for all $x, y \in V$ and all $\alpha \in \mathbf{K}$. For example, we can define a seminorm in $\mathbf{R}^n$ by

$$p(x) = \max_{1 \le j \le n} |x_j| \tag{1-36}$$

A seminorm which satisfies the additional condition that

$$p(x) > 0 \qquad x \ne 0$$

is called a *norm*. A norm is usually denoted by $\|x\|$ rather than by $p(x)$.

If S is a set of elements, any function $d: S^2 \to \mathbf{R}^+$ is called a *metric* if, for each pair of elements x and y of S, it satisfies the following axioms:

$$d(x, y) \ge 0 \qquad d(x, x) = 0 \qquad \text{and} \qquad d(x, y) \in 0 \Rightarrow x \ne y; \tag{1-37a}$$

$$d(x, y) = d(y, x); \tag{1-37b}$$

$$d(x, z) \le d(x, y) + d(y, z) \tag{1-37c}$$

Here the function $d(x, y)$ may be considered as the distance between the points in the set S. Thus (1-37c) is known as the *triangle inequality*. Any set on which a metric can be defined is called a *metric space*.

A class **E** of subsets of a set E which has the property that (1) E and $\phi \in$ **E**, (2) the union of any number of elements of **E** is in **E**, and (3) the interaction of any two elements of **E** is in **E** is called a *topology* on E. The pair $(E, \mathbf{E})$ is called a topological space and the elements of the topology **E** are called open sets. A point $a_0 \in E$ is an interior point of a set $F \subset E$ if there exists an open set containing a_0 which is contained in F. The set of all interior points of a set F is called the interior of F. A set is closed if its complement is open.

Let (X, d_1) and (Y, d_2) be metric spaces. A transformation T from X into Y is called a *contraction* if there exists a number $\lambda \in (0, 1)$ such that

$$d_2(T(x), T(y)) \le \lambda d_1(x, y) \qquad \forall \, x, y \in X$$

A contraction T transforms each pair of points (x, y) into a pair $(T(x), T(y))$ whose members are closer together. From this definition, we arrive at the *fixed-point theorem* stated as follows: *Let T be a contraction of a complete metric space (X, d) into itself:*

$$d(T(x), T(y)) \le \lambda d(x, y) \qquad \lambda \in (0, 1)$$

Then T has a unique fixed point $\bar{x}$, i.e., there is a unique x such that $T(x) = x$.

A linear space in which a norm has been defined is called a *normed linear space*. In a normed linear space, a sequence of elements $\{x_n\} = x_1, x_2, \ldots, x_n$ which has the property that

$$\lim_{m, n \to \infty} \| x_m - x_n \| = 0 \tag{1-38}$$

is called a *Cauchy sequence*. If there exists x_0 such that $\lim_{n \to \infty} \| x_n - x_0 \| = 0$, we say that the sequence $\{x_n\}$ converges to x_0. If every Cauchy sequence in a normed linear space V converges to a point in the space, the space is said to be *complete*. We define a *covering* (or cover) of a set S as a family of open sets such that S is contained in their union. Then a topological space S is said to be *compact* if, and only if, every open covering of S contains a finite subcovering. A complete normed linear space is called a *Banach space*. As examples of complete normed linear spaces, we have

1. The space $\mathbf{R}^n$ $\qquad \| x \| = (x_1^2 + x_2^2 + \cdots + x_n^2)^{1/2}$

2. The space l_1 $\qquad \displaystyle\sum_{n=1}^{\infty} | x_n | < \infty \qquad \| x \| = \sum_{n=1}^{\infty} | x_n |$

3. The space l_2 $\qquad \displaystyle\sum_{n=1}^{\infty} | x_n |^2 < \infty \qquad \| x \| = \left[\sum_{n=1}^{\infty} | x_n |^2 \right]^{1/2}$

4. The space l_p $\qquad \displaystyle\sum_{n=1}^{\infty} | x_n |^p < \infty \qquad \| x \| = \left[\sum_{n=1}^{\infty} | x_n |^p \right]^{1/p}, | < p < \infty |$

5. The space l_∞ $\qquad \displaystyle\sup_n | x_n | < \infty \qquad \| x \| = \sup_n | x_n |$

Here sup denotes the supremum, i.e., the least upper bound.

Let G be a linear operator (mapping or function) and c be a positive real number such that

$$\| Gx \| \le c \| x \| \tag{1-39}$$

Then the operator G is said to be *bounded*. An operator G is called a *functional* if its range space is a set of real numbers.

A vector space V is called an *inner product space* if there is a scalar-valued function on $V \times V$ having properties

$$(x, y_1) + (x, y_2) = (x, y_1 + y_2)$$

$$(\alpha x, y) = \alpha(x, y)$$

$$(x, y) = (y, x)$$

$$(x, x) \ge 0; \quad (x, x) = 0 \quad \text{only if } x = 0$$

If (x, y) is an inner product on a vector space V, then the function $\| x \| = \sqrt{(x, x)}$ is a norm on V, and for all x and y in V

$$|(x, y)| \le \| x \| \; \| y \|$$

This is known as the *Schwarz inequality*.

An inner product space which is complete is called a *Hilbert space*. It is an infinite-dimensional Banach space in which an inner product is defined and which is complete with respect to the norm

$$\| x \| = \sqrt{(x, x)}$$

We may define a scalar product between two spaces V and V^* defined over the same field by mapping pairs (x, x^*), where $x \in V$ and $x^* \in V^*$, into real numbers. Here the spaces V and V^* are called *dual spaces*, often known as *conjugate spaces*.

Another result which is of great importance in the theory of Hilbert spaces is the *Riesz representation theorem* stated as follows: *if a is an arbitrary point of a Hilbert space H, the mapping $f: H \to K$, defined by $f(x) = (x, a)$ is a continuous linear functional on H with norm $\| a \|$.* Conversely, in a Hilbert space $\hat{H}$, corresponding to every continuous linear functional f on $\hat{H}$, there exists a point $a \in \hat{H}$ such that $f(x) = (x, a)$ for all $x \in \hat{H}$. Also, $\| f \| = \| a \|$.

In analysis, a generalization of the integral considered by Lebèsgue overcomes the limitation of the Riemann integral. To this end, consider an integral $\int f(x) \, dx$ representing the area under the curve $y = f(x)$. The Riemann integral can be approximated by the sum

$$s = y_1(x_1 - x_0) + y_2(x_2 - x_1) + \cdots + y_n(x_n - x_{n-1})$$

Clearly the Riemann integral does not exist if $f(x)$ oscillates too violently. The decisive idea in the Lebèsgue integral is the notion of *measure*. The measure of an open interval $a < x < b$ is simply the length a–b. If a set consists of a finite collection of such intervals, the measure is the sum of the lengths. The

Lebèsgue integral is then approximated by the sum

$$s = y_1 m(e_1) + y_2 m(e_2) + \cdots + y_n m(e_n) \tag{1-40}$$

where $m(e_1)$, etc., denote the measure of the sets e_i, where $[a, b] = \cup_{i=1}^{n} e_i$. Riemann's definition breaks down if $f(x)$ does not remain close to y_k whereas Lebèsgue's definition cannot break down because $f(x)$ is automatically close to y_k throughout the set e_k.

In many applications, we are concerned with the special case in which the Hilbert space H is the space of square integrable functions on Ω. Then the inner product of two elements u and $v \in H$ is given by

$$(u, v) = \int_{\Omega} uv \, d\Omega \tag{1-41}$$

where integration in the Lebèsgue sense is implied; u and v are orthogonal if $(u, v) = 0$, and

$$(u, u) < \infty \qquad \forall u \in H$$

The space of all equivalence classes of real-valued (or complex-valued) Lebèsgue-measurable functions u, such that $|u|^p$ is Lebèsgue integrable, $1 \le p \le \infty$, is a Banach space denoted $L_p(\Omega)$, and equipped with the norm

$$\| u \|_{L_p(\Omega)} = \left(\int_{\Omega} |u|^p \, dx \right)^{1/p} \tag{1-42}$$

where $dx = mes(dx_1 dx_2 \cdots dx_n)$.

Linear (differential) operator Consider a linear boundary value problem

$$Au = f \qquad \text{in } \Omega \tag{1-43a}$$

$$B_r u = g_r \qquad \text{on } \Gamma \tag{1-43b}$$

where A and B_r are linear differential operators in the domain Ω and on the boundaries Γ. Consider the $2m$th order operator in the form

$$A = a_1 \frac{d^{2m}}{dx^{2m}} + a_2 \frac{d^{2m-1}}{dx^{2m-1}} + \cdots + a_n$$

If the coefficients are zero except for $a_1 = 1$, then

$$\frac{d^{2m}u}{dx^{2m}} = f \qquad \text{in } \Omega \qquad \text{with } A = \frac{d^{2m}}{dx^{2m}}$$

Let us consider another function v and construct an inner product of Au with v within a domain $0 < x < 1$. Then

$$(Au, v) = \int_0^1 (Au)v \, dx$$

Integrating by parts yields

$$\int_0^1 (Au)v\,dx = \frac{d^{2m-1}u}{dx^{2m-1}}\,v\,\Big|_0^1 - \int_0^1 \frac{d^{2m-1}u}{dx^{2m-1}}\frac{dv}{dx}\,dx$$

$$= \frac{d^{2m-1}u}{dx^{2m-1}}\,v\,\Big|_0^1 - \frac{d^{2m-2}u}{dx^{2m-2}}\frac{dv}{dx}\,\Big|_0^1 + \int_0^1 \frac{d^{2m-2}u}{dx^{2m-2}}\frac{d^2v}{dx^2}\,dx$$

$$= \frac{d^{2m-1}}{dx^{2m-1}}\,v\,\Big|_0^1 - \cdots + \cdots - u\,\frac{d^{2m-1}v}{dx^{2m-1}}\,\Big|_0^1 + \int_0^1 u\,\frac{d^{2m}v}{dx^{2m}}\,dx$$

This may be put in the form

$$(Au, v) = (u, A^*v) + [B_r^{(E)}(u, v) + B_r^{(N)}(u, v)]_0^1 \tag{1-44}$$

with

$$B_r^{(E)}(u, v) = \sum_{r=0}^{m-1} (-1)^{r+1} G_r u F_r v \tag{1-45a}$$

$$B_r^{(N)}(u, v) = \sum_{r=m}^{2m-1} (-1)^{r+1} F_r^* u G_r^* v \tag{1-45b}$$

where $G_r u = g_r^{(E)}$ and $F_r^* u = g_r^{(N)}$ are called the essential (Dirichlet) and natural (Neumann) boundary conditions, respectively.

$$G_r u = \left[\frac{d^0}{dx^0}, \frac{d}{dx}, \frac{d^2}{dx^2}, \ldots, \frac{d^{m-1}}{dx^{m-1}}\right]u \tag{1-46a}$$

$$F_r^* u = \left[\frac{d^m}{dx^m}, \frac{d^{m+1}}{dx^{m+1}}, \frac{d^{m+2}}{dx^{m+2}}, \ldots, \frac{d^{2m-1}}{dx^{2m-1}}\right]u \tag{1-46b}$$

Here G_r and F_r^* are the boundary operators. The expression (1-44) is known as Green's formula. In two- or three-dimensional problems, the Green's formula takes the form

$$(Au, v) = \int_\Omega uA^*v\,d\Omega + \int_\Gamma [B_r^{(E)}(u, v) + B_r^{(N)}(u, v)]\,d\Gamma \tag{1-47}$$

It should be noted that for the $2m$th equation, we have $A^* = A$, $G_r^*(r = 2m - 1, 2m - 2, \ldots, m) = G_r(r = 0, 1, \ldots, m)$ and $F_r^*(r = 2m - 1, 2m - 2, \ldots, m) = F_r(r = 0, 1, \ldots, m)$. Equations with these conditions are referred to as self-adjoint; and the linear differential operator A is known as a self-adjoint operator. Moreover, the conditions $A^* = A$ and $v = u$ result in symmetric positive definite properties for the inner product (Au, u). If $A^* \neq A$, then A is a nonself-adjoint operator, resulting in nonself-adjoint equations.

Example 1-1 Let $A = \alpha(d^2/dx^2) + \beta$, $0 < x < 1$, where α and β are constants. Then

$$(Au, v) = \int_0^1 \left(\alpha\frac{d^2u}{dx^2} + \beta u\right)v\,dx$$

$$= \alpha \frac{du}{dx} v \Big|_0^1 - \int_0^1 \alpha \frac{du}{dx} \frac{dv}{dx} \, dx + \int_0^1 \beta uv \, dx$$

$$= \alpha \left(\frac{du}{dx} v - u \frac{dv}{dx} \right)_0^1 + \int_0^1 u \left(\alpha \frac{d^2v}{dx^2} + \beta v \right) dx$$

Note that $m = 1$ here, and the results above can be written in the form (1-44),

$$(Au, v) = (u, A^*v) - \alpha(G_0 u F_0 v - F_1^* u G_1^* v)_0^1$$

where

$$A^* = \alpha \frac{d^2}{dx^2} + \beta \qquad G_0 = 1 \qquad F_0 = \frac{d}{dx} \qquad F_1 = \frac{d}{dx} \qquad G_1^* = 1$$

Thus we have $A^* = A$ and $G_1^* = G_0$, and the operator A is the self-adjoint operator. Here $G_0 u = u$ and $F_1 u = du/dx$ are the Dirichlet and Neumann boundary conditions, respectively.

Example 1-2 Let

$$A = a_{ij} \frac{\partial^4}{\partial x_i^2 \, \partial x_j^2} \text{ in } \Omega, \ (i,j = 1, 2), \text{ and } a_{ij} \text{ are constants}$$

$$A = a_{11} \frac{\partial^4}{\partial x_1^4} + a_{12} \frac{\partial^4}{\partial x_1^2 \, \partial x_2^2} + a_{21} \frac{\partial^4}{\partial x_2^2 \, \partial x_1^2} + a_{22} \frac{\partial^4}{\partial x_2^4}$$

Constructing the inner product (Au, v) and integrating by parts four times, we arrive at the form identical to (1-47) with $A^* = A$ and

$$G_0 u F_0 v = - \int_\Gamma u \left[a_{11} \frac{\partial^3 v}{\partial x_1^3} \, dy + a_{12} \frac{\partial^3 v}{\partial x_1^2 \, \partial x_2} \, dx + a_{21} \frac{\partial^3 v}{\partial x_1 \, \partial x_2^2} \, dy + a_{22} \frac{\partial^3 v}{\partial x_2^3} \, dx \right]$$

$$\vdots$$

$$F_3^* u G_3^* v = \int_\Gamma v \left[a_{11} \frac{\partial^3 u}{\partial x_1^3} \, dy + a_{12} \frac{\partial^3 u}{\partial x_1 \, \partial x_2^2} \, dy + a_{21} \frac{\partial^3 v}{\partial x_1^2 \, \partial x_2} \, dx + a_{22} \frac{\partial^3 v}{\partial x_2^3} \, dx \right]$$

Note that for $G_r^*(r = 2m - 1, 2m - 2, \ldots, m) = G_r(r = 0, 1, \ldots, m)$, it is necessary that $a_{ij} = a_{ji}$, which confirms the symmetry of the self-adjoint operator.

The partial differential equation (1-43) is seen to be equivalent to the integral relation (1-47). We then say that u is a solution *in the weak sense* of the original equation if it satisfies this integral relation for all functions v of the class considered.

Distributions In mathematical physics problems, we encounter the so-called Dirac delta function and the idea of a *weak solution* of partial differential equation, neither of which cannot be handled by the ordinary theory of functions. The theory of *distributions* is then motivated by an attempt to generalize the concept of a function sufficient to embrace the Dirac delta function and weak solutions (Schwartz, 1966). A distribution is often called a *generalized function*.

Let Ω be an open set in $\mathbf{R}^n$. By the *support* of a function ϕ defined on Ω—denoted by supp ϕ—we mean the closure of the set $\{x: x \in \Omega, \phi(x) \neq 0\}$. Thus supp ϕ is defined as the smallest relatively closed subset of Ω outside of which ϕ vanishes. We define $C_0^k(\Omega)$ as the set of all functions whose partial derivatives up to and including those of order k are continuous in Ω, and which have compact supports contained in Ω. The space $C_0^\infty(\Omega)$ is called the *space of test functions*, the functions themselves being called *test functions*.

A class of *test functions* on $\mathbf{R}$ denoted as $C_0^\infty(\mathbf{R})$ is a special class of real-valued functions defined on the real line $\mathbf{R} = (-\infty, \infty)$, which has the properties: (1) Each $\phi(x)$ in $\mathscr{F}(\mathbf{R})$ is infinitely differentiable, and (2) every sequence of test functions $\phi_n(x)$ must vanish outside some common interval of $\mathbf{R}$ and $\phi_n(x)$ and all its derivatives with respect to x must approach zero uniformly as $n \to \infty$

$$\lim_{n \to \infty} \left| \frac{d^r \phi_n(x)}{dx^r} \right| = 0 \qquad \forall x \in \mathbf{R} \qquad r \geq 0$$

Consider now a function $f(\mathbf{R})$ as a collection of weighted values over $\mathbf{R}$ and $\mathscr{F}(\mathbf{R})$ as a class of smooth accessory functions $\phi(x)$ defined on $\mathbf{R}$. Then the scalar product of $f(\mathbf{R})$ and $\mathscr{F}(\mathbf{R})$ is defined by the Lebèsgue integral

$$(f, \phi) = \int_{-\infty}^{\infty} f\phi \, dx \tag{1-48}$$

With the role of the test functions ϕ defined as above, we examine a *functional q* on the space of test functions, which simply implies a mapping of the form

$$q(\alpha\phi_1 + \beta\phi_2) = \alpha q(\phi_1) + \beta q(\phi_2) \qquad \forall \phi_1, \phi_2 \in \mathscr{F}(\mathbf{R})$$

A linear functional on $\mathscr{F}(\mathbf{R})$ denoted by $C_0^\infty(\mathbf{R})$ is called a distribution on $\mathbf{R}$ if, and only if, for each compact subset K of $\mathbf{R}$ there exists an integer m and a constant C_k such that

$$|q(\phi)| \leq C_k \sup_{\substack{k \leq m \\ x \in K}} \left| \frac{d^k \phi(x)}{dx^k} \right| \qquad \forall \phi \in \mathscr{F}(K) \tag{1-49}$$

We define the dual space $\mathscr{F}(\mathbf{R})'$ of the space of test functions as the space of distributions. Thus, if $q \in \mathscr{F}(\mathbf{R})'$, we may write

$$q(\phi) \equiv (q, \phi) \qquad \phi \in \mathscr{F}(\mathbf{R})$$

where the symbol $(\cdot, \cdot)$ represents a bilinear map of $\mathscr{F}(\mathbf{R})' \times \mathscr{F}(\mathbf{R})$ into $\mathbf{R}$, called a *duality pairing*.† Furthermore, we have

$$(\alpha q_1, \phi) + (\beta q_2, \phi) = (\alpha q_1 + \beta q_2, \phi)$$

---

† Some authors use the symbol $\langle \cdot, \cdot \rangle$ for scalar product or duality pairing. In this book, we use the symbol $(\cdot, \cdot)$ which also denotes the inner product. The reader may distinguish them by noting that both elements in the inner product belong to the same space, whereas they belong to different spaces in the event of duality pairing.

and

$$\lim_{n \to \infty} (q_n, \phi) = (q, \phi)$$

We now examine the derivatives of distributions q. We consider first duality pairing (f', ϕ) with $f' = df/dx$ being locally integrable. Assuming that the test functions ϕ have finite support, i.e., $\phi(-\infty) = \phi(\infty) = 0$, we have

$$(f', \phi) = \int_{-\infty}^{\infty} f'(x)\phi(x)\, dx = -\int_{-\infty}^{\infty} f(x)\phi'(x) + f\phi \Big|_{-\infty}^{\infty} = -(f, \phi')$$

Thus, the *distributional derivative* of q is the functional $p = q'$ defined by

$$(p, \phi) = -(q, \phi') \qquad \forall\, \phi \in \mathcal{F}(\mathbf{R})$$

Here q' is also a distribution and $|q'(\phi)|$ must be bounded as follows:

$$|q'(\phi)| \le \mathop{C_k}_{\substack{k+1 \le m \\ x \in K}} \left| \frac{d^{k+1}\phi(x)}{dx^{k+1}} \right| \qquad \forall\, \phi \in \mathcal{F}(\mathbf{R}) \tag{1-50}$$

It follows that the kth derivative $q^{(k)} = d^k q/dx^k$ is defined as

$$(q^k, \phi) = (-1)^k \left(q, \frac{d^k \phi}{dx^k} \right) \qquad \forall\, \phi \in \mathcal{F}(\mathbf{R}) \tag{1-51}$$

Example 1-3 Consider a generalized derivative of the form (f', ϕ) where the piecewise continuous function f is as shown in Fig. 1-2.

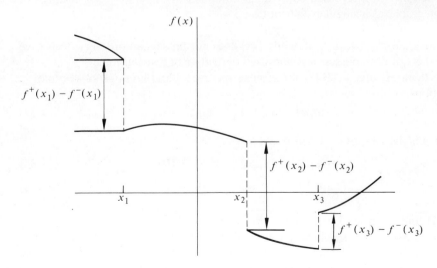

Figure 1-2 Piecewise continuous function.

Integrating by parts, Oden and Reddy (1976) have shown that

$$f', \phi) = -(f, \phi')$$

$$= \int_{-\infty}^{x_1} \frac{df}{dx} \phi \, dx - f(x)\phi(x)\Big|_{-\infty}^{x_1} + \int_{x_1}^{x_2} \frac{df}{dx} \phi(x) - f(x)\phi(x)\Big|_{x_1}^{x_2} \cdots$$

$$+ \int_{x_n}^{\infty} \frac{df}{dx} \phi \, dx - f(x)\phi(x)\Big|_{x_n}^{\infty}$$

$$= \int_{-\infty}^{\infty} \frac{df}{dx} \phi \, dx + \phi(x_1)[f^+(x_1) - f^-(x_1)]$$

$$+ \cdots + \phi(x_n)[f^+(x_n) - f^-(x_n)]$$

$$= \left(\frac{df}{dx}, \phi\right) + \sum_{i=1}^{n} [\![f(x_i)]\!] \phi(x_i)$$

$$= \left(\frac{df}{dx}, \phi\right) + \sum_{i=1}^{n} [\![f(x_i)]\!] (\delta(x - x_i), \phi(x))$$

$$= \left(\left\{ \frac{df}{dx} + \sum_{i=1}^{n} [\![f(x_i)]\!] \delta(x - x_i) \right\}, \phi\right)$$

where $[\![f(x_i)]\!] = f^+(x_i) - f^-(x_i)$ and $\delta(x - x_i)$ is the Dirac delta function. Note that f' is given by

$$f' = \frac{df}{dx} + \sum_{i=1}^{n} [\![f(x_i)]\!] \delta(x - x_i)$$

Here it is seen that the generalized derivative of a piecewise continuous function $f(x)$ is represented by delta functions. We refer to (δ, ϕ), locally not integrable, as a singular distribution.

This example reveals that, while $f(x)$ does not possess an ordinary derivative everywhere, it does possess a generalized derivative of distributions.

As demonstrated in (1-47), for a linear operator A and homogenous boundary conditions,

$$(Au, \phi) = (u, \overset{*}{A}\phi) \qquad \forall \, \phi \in \mathscr{F}(\mathbf{R}) \tag{1-52}$$

where $\overset{*}{A}$ is the adjoint of A. Similarly,

$$(Aq, \phi) = (f, \phi) \qquad \forall \, \phi \in \mathscr{F}(\mathbf{R}) \tag{1-53}$$

or

$$(q, \overset{*}{A}\phi) = (f, \phi) \qquad \forall \, \phi \in \mathscr{F}(\mathbf{R}) \tag{1-54}$$

Here A and q act as distributions. We note that differentiability required of q must be weakened if f is locally integrable whereas it must be defined distributionally if f is a singular distribution. From these observations, we define three classes of solution $q(x)$ of the differential equation of the form

$$Aq(x) = f(x) \tag{1-55}$$

1. If $q(x)$ is sufficiently differentiable to satisfy (1-55), $q(x)$ is called a *classical solution*.
2. If $q(x)$ is only locally integrable but not sufficiently differentiable to satisfy (1-55), $q(x)$ is referred to as a *strong solution*.
3. If $q(x)$ is a singular distribution satisfying (1-54), q is called a *weak solution*.

Since we expect to deal with multidimensional problems, we introduce here useful notations. Let the set of ordered n-tuples of nonnegative integers be given by

$$\boldsymbol{\alpha} = (\alpha_1, \alpha_2, \ldots, \alpha_n)$$

Also, we define

$$|\boldsymbol{\alpha}| \equiv \alpha_1 + \alpha_2 + \cdots + \alpha_n \qquad \boldsymbol{\alpha}! = \alpha_1! \, \alpha_2! \cdots \alpha_n!$$

$$\mathbf{x} = (x_1, x_2, \ldots, x_n) \in \mathbf{R}^n$$

$$|\mathbf{x}| = \left(\sum_{i=1}^{n} |x_i|^2 \right)^{1/2}$$

$$\mathbf{x}^{\alpha} = x_1^{\alpha_1} x_1^{\alpha_2} \cdots x_n^{\alpha_n}$$

and

$$D^{\alpha} u(\mathbf{x}) = \frac{\partial^{|\alpha|} u(x)}{\partial x_1^{\alpha_1} \, \partial x_2^{\alpha_2} \cdots \partial x_n^{\alpha_n}} = \left(\frac{\partial^{\alpha_1}}{\partial x_1^{\alpha_1}} \right) \left(\frac{\partial^{\alpha_2}}{\partial x_2^{\alpha_2}} \right) \cdots \left(\frac{\partial^{\alpha_n}}{\partial x_n^{\alpha_n}} \right) u(\mathbf{x}) \qquad (1\text{-}56)$$

together with the duality pairing

$$(u, \phi) = \int_{-\infty}^{\infty} \int_{-\infty}^{\infty} \cdots \int_{-\infty}^{\infty} u(\mathbf{x}) \phi(\mathbf{x}) \, dx_1 \, dx_2 \cdots dx_n \equiv \int_{\mathbf{R}^n} u\phi \, d\mathbf{x} \qquad (1\text{-}57)$$

Now the αth generalized partial derivative of the distribution q is given by

$$(p, \phi) = (-1)^{|\alpha|} (q, D^{\alpha}\phi) \qquad \forall \, \phi \in \mathscr{F}(\mathbf{R}^n) \qquad (1\text{-}58)$$

for the continuous linear functional $p = D^{\alpha} q$.

An additional topic in the theory of distributions includes the concepts of weak and strong derivatives on Banach spaces. Recall that the space of all equivalence classes of real-valued (or complex-valued) Lebèsgue-measurable functions u on an open, bounded domain $\Omega \subset \mathbf{R}^n$ was denoted as $L_p(\Omega)$, $1 \leq p \leq \infty$, with the norm given by (1-42)

$$\| u \|_{L_p(\Omega)} = \left(\int_{\Omega} |u|^p \, dx \right)^{1/p} \qquad (1\text{-}59)$$

Let the dual of $L_p(\Omega)$ be $L_q(\Omega)$ where

$$\frac{1}{p} + \frac{1}{q} = 1$$

then for $u \in L_p(\Omega)$ and $v \in L_q(\Omega)$ we have

$$\int uv \, dx \le \| u \|_{L_p(\Omega)} \| v \|_{L_q(\Omega)} \tag{1-60}$$

This is known as Hölder's inequality. Now let a sequence $\{u_k\}$ of functions be *strongly convergent* in $L_p(\Omega)$ to the function $u \in L_p(\Omega)$. Then it follows that

$$\lim_{k \to \infty} \| u_k - u \|_{L_p(\Omega)} \equiv \lim_{k \to \infty} \left(\int_\Omega |u_k - u|^p \, dx \right)^{1/p} = 0$$

Therefore, a sequence $\{u_k\}$ is said to be weakly convergent to $u \in L_p(\Omega)$ if for an an arbitrary bounded linear functional, $l(u_k) = \int_\Omega vu \, dx \in L_q(\Omega)$, $v \in L_q(\Omega)$

$$\lim_{k \to \infty} l(u_k) = l(u)$$

or

$$\lim_{k \to \infty} \int_\Omega u_k v \, dx = \int_\Omega uv \, dx$$

With these definitions, let $\{\phi_k\}$ denote a bounded sequence of m times continuously differentiable functions with compact support in $\Omega \subset \mathbf{R}^n$ which are strongly convergent to $u \in L_p(\Omega)$

$$\lim_{k \to \infty} \| \phi_k - u \|_{L_p(\Omega)} = 0 \tag{1-61}$$

Then u is said to have *strong L_p derivatives* up to order m if there exist functions $v^\alpha \in L_p(\Omega)$ such that

$$\lim_{k \to \infty} \| D^\alpha \phi_k - v^\alpha \|_{L_p(\Omega)} = 0 \qquad \forall \, |\alpha| \le m$$

On the contrary, a function u, locally integrable on $\Omega \subset \mathbf{R}^n$ is said to have a weak derivative w^α of order α if a locally integrable function w^α exists such that

$$\int_\Omega w^\alpha \phi \, dx = (-1)^{|\alpha|} \int_\Omega u D^\alpha \phi \, dx \qquad \forall \, \phi \in \mathscr{F}(\mathbf{R})$$

Here w^α is the αth generalized derivative of the regular distribution generated by u.

Sobolev spaces Sobolev spaces [Sobolev, 1950, 1963 (translation), and Adams, 1975] are the generalization of $L_p(\Omega)$ spaces so that all weak derivatives of functions $u(\mathbf{x})$ are included in $L_p(\Omega)$ whose norm is defined by (1-59), implying a Banach space. If all weak partial derivatives $u(\mathbf{x})$ of order $\le m$, m being an integer ≥ 0, are in $L_p(\Omega)$, then $u(\mathbf{x})$ is said to belong to a Sobolev space denoted as $W_p^m(\Omega)$,

$$W_p^m(\Omega) = \{u : D^\alpha u \in L_p(\Omega) \qquad \forall \, \alpha \text{ such that } |\alpha| \le m\}$$

For $m = 0$, we have $W_p^0(\Omega) = L_p(\Omega)$. The Sobolev norm is defined as

$$\| u \|_{W_p^m(\Omega)} = \left(\int_\Omega \sum_{|\alpha| \le m} | D^\alpha u |^p \, dx \right)^{1/p} = \left(\sum_{|\alpha| \le m} \| D^\alpha u \|_{L_p(\Omega)}^p \right)^{1/p} \qquad (1\text{-}62)$$

Example 1-4 Consider the spaces: L_2, W_2^1, and W_3^2 with an open interval on the real line $\Omega = (x_1, x_2) \subset \mathbf{R}$. The associated norms are

$$\| u \|_{L_2(x_1, x_2)} = \left[\int_{x_1}^{x_2} u^2 \, dx \right]^{1/2} < \infty$$

$$\| u \|_{W_2^1(x_1, x_2)} = \left\{ \int_{x_1}^{x_2} \left[u^2 + \left(\frac{du}{dx} \right)^2 \right] dx \right\}^{1/2} < \infty$$

$$\| u \|_{W_3^2(x_1, x_2)} = \left[\int_{x_1}^{x_2} \left(|u|^3 + \left| \frac{du}{dx} \right|^3 + \left| \frac{d^2 u}{dx^2} \right|^3 \right) dx \right]^{1/3} < \infty$$

If the domain is $\Omega = (x_1, x_2) \times (y_1, y_2) \subset \mathbf{R}^2$, then the Sobolev norm W_3^2 is

$$\| u \|_{W_3^2(\Omega)} = \left[\int_{y_1}^{y_2} \int_{x_1}^{x_2} \left(|u|^3 + \left| \frac{\partial u}{\partial x} \right|^3 + \left| \frac{\partial u}{\partial y} \right|^3 + \left| \frac{\partial^2 u}{\partial x^2} \right|^3 + \left| \frac{\partial^2 u}{\partial x \, \partial y} \right|^2 \right. \right.$$

$$\left. \left. + \left| \frac{\partial^2 u}{\partial y^2} \right|^3 \right) dx \, dy \right]^{1/3}$$

With these preliminaries, we list here the basic properties of the Sobolev spaces:

1. $W_p^m(\Omega)$ is complete with respect to the norm $\| \cdot \|_{W_p^m}$ of (1-62). To see this, we examine a Cauchy sequence in $W_p^m(\Omega)$

$$\lim_{j,k \to \infty} \| u_j - u_k \|_{W_p^m(\Omega)}^p = \lim_{j,k \to \infty} \| D^\alpha u_j - D^\alpha u_k \|_{L_p(\Omega)}^p = 0$$

Since each term in this series is nonnegative

$$\lim_{j,k \to \infty} \| D^\alpha u_j - D^\alpha u_k \|_{L_p(\Omega)} = 0$$

This indicates that for each index α such that $|\alpha| \le m$, the sequence $\{ D^\alpha u_k \}$ is a Cauchy sequence in $L_p(\Omega)$.

2. Let Ω be the union of a countable collection of open sets Ω_k. Then $W_p^m(\Omega)$ is the completion (closure) with respect to the norm of (1-62) of the space $C^m(\Omega)$ of functions with continuous derivatives of all orders $\le m$, and of the space $\hat{C}^\infty(\Omega)$ of infinitely differentiable functions with finite norms of (1-62).

To illustrate the foregoing definitions, we consider the Sobolev spaces W_2^1, W_2^2, W_1^2, and W_1^4 represented by piecewise functions $u \in W_2^1(a, b)$, $v \in W_2^2(a, b)$, $\hat{u} \in W_1^2(a, b)$, and $\hat{v} \in W_1^4(a, b)$, respectively. Note that differentiations lower the

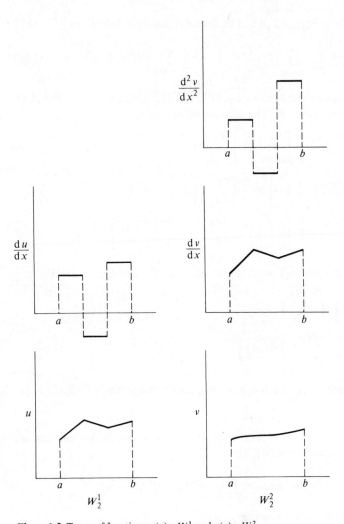

Figure 1-3 Types of functions $u(x) \in W_2^1$ and $v(x) \in W_2^2$.

order of the space (Fig. 1-3). If u belongs to W_2^2, then du/dx belongs to W_2^1 with both spaces in $L_2(a, b)$. Similarly, $d^4\hat{v}/dx^4$ is of the form of $d^2\hat{u}/dx^2$ in $L_1(a, b)$ as shown in Fig. 1-4. These observations indicate

$$W_2^2(a, b) \subset W_2^1(a, b)$$

and

$$W_1^4(a, b) \subset W_1^2(a, b)$$

Some of the most important properties of the Sobolev spaces are: their ability to generate smooth functions and a given space $W_p^m(\Omega)$ can be mapped into (i.e., embedded in) other spaces. These properties are known as the Sobolev embedding theorems.

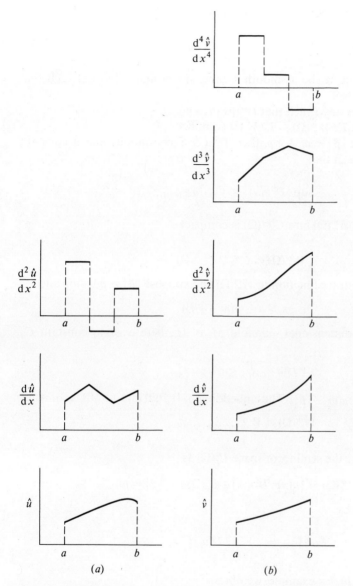

Figure 1-4 Types of functions $\hat{u}(x) \in W_1^2$ and $\hat{v}(x) \in W_1^4$. (a) space W_1^2; (b) space W_1^4.

Sobolev embedding theorems The notion of Sobolev embedding theorems is related to the degree of smoothness that can be expected of functions in certain Sobolev spaces. Namely, a given space $W_p^m(\Omega)$ can be mapped continuously into other spaces. These properties may be summarized as follows:

1. Let $u(x) \in W_p^m$, Ω being a bounded domain in $\mathbf{R}^n$ with $mp > n$. Then there exists

a constant $C > 0$, independent of u such that

$$\sup_{x \in \Omega} |u(x)| \leq C \, \| u \|_{W_p^m(\Omega)} \tag{1-63}$$

2. If $p > 1$, $mp \leq n$, let M_s be a smooth s-dimensional manifold of Ω such that $n - mp \leq s \leq n$. Then
   (a) $u(x) \in L_l(M_s)$ for any l such that $l < ps/(n - mp)$.
   (b) the embedding $E: W_p^m(\Omega) \to L_l(M_s)$ is compact.
3. If $mp > n$ and $0 < |\beta| < m - n/p$, then $D^\beta u(x)$ is continuous and a constant C_0 can be found such that

$$\sup_{x \in \Omega} |D^\beta u(x)| \leq C_0 \, \| u \|_{W_p^m(\Omega)}$$

The imbedding of $W_p^m(\Omega)$ into $\bar{C}^{|\beta|}(\Omega)$ is compact

$$W_p^m(\Omega) \subset \bar{C}^{m - (n/p) - 1}(\Omega)$$

Let $u(x)$ be an arbitrary function in $W_p^m(\Omega)$ and suppose $|\beta| \geq m - n/p$ and

$$s > n - p(m - |\beta|)$$

Then, on every s-dimensional manifold M_s of Ω, there exists a constant C_1 such that

$$\| D^\beta u \|_{L_l(M_s)} \leq C_1 \, \| u \|_{W_p^m(\Omega)}$$

where $l < sp/[n - p(m - |\beta|)]$. The embedding of $W_p^m(\Omega)$ into $L_l(M_s)$ is compact,

$$W_p^m(\Omega) \subset W_{np/[n - p(m - |\beta|)]}^{|\beta|}(\Omega)$$

Finally, we define the seminorm space $L_p^m(\Omega)$ as

$$L_p^m(\Omega) = \{u(x): D^\alpha u(x) \in L_p(\Omega) \quad |\alpha| = m\}$$

with the norm given by

$$\| u \|_{L_p^m(\Omega)}^p = \sum_{|\alpha| = m} \| D^\alpha u \|_{L_p(\Omega)}^p$$

If we let $u(x) \in C^2(a, b)$, for example, we obtain

$$\| u \|_{W_2^2(a,b)}^2 = \int_a^b \left[u^2 + \left(\frac{du}{dx} \right)^2 + \left(\frac{d^2 u}{dx^2} \right)^2 \right]^2 dx$$

and

$$\| u \|_{L_2^2(a,b)}^2 = \int_a^b \left(\frac{d^2 u}{dx^2} \right)^2 dx$$

These definitions, together with some of the properties to be discussed in Hilbert spaces below, constitute the foundations for the finite element theory.

Duals of Hilbert spaces In most linear boundary value problems, we encounter $W_2^m(\Omega)$ spaces. These spaces are Hilbert spaces $H^m(\Omega)$ endowed with special kinds of inner products,

$$(u, v)_{H^m(\Omega)} = \sum_{|\alpha| \le m} (D^\alpha u, D^\alpha v)_{L_2(\Omega)} = \sum_{|\alpha| \le m} \int_\Omega D^\alpha u D^\alpha v \, dx$$

and the associated norm

$$\| u \|_{H^m(\Omega)} = \left(\sum_{|\alpha| < m} \| D^\alpha u \|_{L_2(\Omega)}^2 \right)^{1/2} = [(u, u)_{H^m(\Omega)}]^{1/2} < \infty$$

where
$$H^m(\Omega) = \{u : D^\alpha u \in L_2(\Omega) \quad \forall \, \alpha \text{ such that } |\alpha| \le m\}$$

with $D^\alpha u$ denoting the weak αth derivative of u.

The dual of a Hilbert space H is the space of continuous linear functionals on H, and denoted H'

$$H' = \mathscr{L}(H, \mathbf{R})$$

If l is a continuous linear functional on H, we define

$$l(u) = (l, u) \qquad l \in H'; \; u \in H$$

The bilinear form (l, u) represents the duality pairing on the product space $H' \times H$, with H' supplied with the norm

$$\| l \|_{H'} = \sup_{\substack{u \in H \\ u \ne 0}} \frac{|(l, u)|}{\| u \|_H} \tag{1-64}$$

and
$$|(l, u)| \le \| l \|_{H'} \| u \|_H$$

In connection with the dual spaces, we introduce the function $\hat{u}(\mathbf{y})$ as the Fourier transform of $u(\mathbf{x}) \in L_2(\mathbf{R}^n)$ given by

$$\hat{u}(\mathbf{y}) = F(u) = \frac{1}{(2\pi)^{n/2}} \int_{\mathbf{R}_n} u(\mathbf{x}) \exp(-i\mathbf{x}\mathbf{y}) \, dx$$

with its inverse

$$\hat{u}(\mathbf{x}) = F^{-1}(\hat{u}) = \frac{1}{(2\pi)^{n/2}} \int_{\mathbf{R}_n} \hat{u}(\mathbf{y}) \exp(i\mathbf{x}\mathbf{y}) \, dy$$

Fourier transforms of distribution can also be defined for the space $\mathscr{S}(\mathbf{R}^n)$

$$\mathscr{S}(\mathbf{R}^n) = \{\phi : \mathbf{x}^\beta D^\alpha \phi \in L_2(\mathbf{R}^n)\}$$

Thus, Fourier transforms of a continuous linear functional q on $\mathscr{S}(\mathbf{R}^n)$, called a tempered distribution, are then given by

$$(\hat{q}, \phi) = (q, \hat{\phi})$$

where ϕ is the test function. Clearly, if $u \in L_2(\mathbf{R}^n)$, it generates a *tempered distribution* q and the Fourier transform $\hat{u}$ of u coincides with the distribution q. It follows that, for any tempered distribution $q \in \mathscr{S}(\mathbf{R}^n)$, we have

$$F(D_x^\alpha q) = (iy)^\alpha F(q)$$
$$D_y^\alpha F(q) = F((-ix)^\alpha q)$$

Additionally

$$\| u \|_{L_2(\mathbf{R}^n)} = \| F(u) \|_{L_2(\mathbf{R}^n)}$$

which is known as Plancherel's equality.

Let the dual of $H^s(\mathbf{R}^n)$ be denoted by $H^{-s}(\mathbf{R}^n)$. The Sobolev space $H^{-s}(\mathbf{R}^n)$, $s > 0$ is defined by

$$H^{-s}(\mathbf{R}^n) = \left(H^s(\mathbf{R}^n)\right)'$$

where $H^s(\mathbf{R}^n)$ is the Hilbert space

$$H^s(\mathbf{R}^n) = \{u : u \in \mathscr{S}(\mathbf{R}^n) \qquad (1 + |\mathbf{y}|^2)^{s/2} \hat{u} \in L_2(\mathbf{R}^n)\}$$

with the norm

$$\| u \|_{H^s(\mathbf{R}^n)} = \| (1 + |\mathbf{y}|^2)^{s/2} \hat{u} \|_{L^2(\mathbf{R}^n)} \qquad |\mathbf{y}|^2 = y_1^2 + y_2^2 + \cdots + y_n^2$$

The process of $H^m(\mathbf{R})$ onto $H^{-m}(\mathbf{R})$ is called the canonical isometry given by a differential operator K such that

$$Ku = \sum_{j=0}^{m} (-1)^j D^{2j} u$$

with the distribution q in $H^{-m}(\mathbf{R})$

$$q = \sum_{j=0}^{m} D^j q_j$$

These observations point to the negative Sobolev spaces $H^{-m}(\mathbf{R})$ which can be obtained by considering those functions $u(x)$ whose derivatives of various orders are in $L_2(\mathbf{R}) = H^0(\mathbf{R})$. For example, let $u(x)$ be given by

$$D^2 u(x) = \tilde{h}(x)$$

such that

$$D^{-1} D^2 u = Du = \int \tilde{h}(x) \, dx$$

$$D^{-1} Du = u = \int \int \tilde{h}(x) \, dx \, dx$$

Note that the functions $D^2 u$, Du, and u as shown in Fig. 1-4b are in $L_2(\mathbf{R})$. If we consider a function $v(\mathbf{x})$ less smooth than $u(\mathbf{x})$ (discontinuous), we obtain $\tilde{h}(x)$ by integrating $v(x)$

$$D^{-1} v = \tilde{h}(x)$$

and $D^{-2}v$ becomes piecewise linear. Thus the function v is in $H^{-1}(\mathbf{R})$ whereas $D^{-1}v$ is in $L_2(\mathbf{R})$. This implies that

$$D^{-m}v = \tilde{h}(x) \qquad w \in H^{-m}(\mathbf{R})$$

$$\tilde{h} = \int \int \cdots \int v(x)\, dx\, dx \cdots dx \qquad (m \text{ times})$$

If
$$D^m w = \tilde{h}(x)$$

then w is in $H^m(\mathbf{R})$. In view of (1-64), we note that

$$\| v \|_{H^{-s}(\mathbf{R}^n)} = \sup_{u \in H^s(\mathbf{R}^n)} \frac{|(v, u)|}{\| u \|_{H^s(\mathbf{R}^n)}}$$

Furthermore,

$$\mathscr{F}(\mathbf{R}^n) \subset H^s(\mathbf{R}^n) \subset H^0(\mathbf{R}^n) \subset H^{-s}(\mathbf{R}^n) \subset (\mathscr{F}(\mathbf{R}^n))'$$

indicating that all inclusions describe a dense embedding (or continuous injection) of one space into another with the pivot space being $H^0(\mathbf{R}^n) = L_2(\mathbf{R}^n)$.

Let us, for example, consider the Sobolev space $H_0^1(-1, 1)$. Here a Dirac measure δ has the property $\delta(\phi) = \phi(0), \forall \phi \in C_0^\infty(\mathbf{R})$. From the Sobolev embedding theorem, $v(\mathbf{x})$ at $x = 0$ is uniquely defined since each $u(\mathbf{x}) \in H^1(-1, 1)$ is equivalent to a continuous function when $m > n/p$ (here $m = 1$, $n = 1$, $p = 2$). Also, from (1-63)

$$\sup_{\mathbf{x} \in (-1, 1)} |v(\mathbf{x})| \equiv C \| v \|_{H^1(-1,1)}$$

and
$$\| \delta \|_{H^{-1}(-1,1)} = \sup_{v \in H_0^1(-1,1)} \frac{|v(0)|}{\| v \|_{H^1(-1,1)}} < C \qquad v \neq 0$$

This indicates that the delta distribution exists and is in $H^{-1}(-1, 1)$. Once again from the Sobolev embedding theorem, the smallest space to which δ can belong is determined as

$$v(\mathbf{x}) \in H_0^\mu(-1, 1)$$

with $\mu > n/2 = \frac{1}{2}$. Thus, setting $\mu = \frac{1}{2} + \varepsilon$ where $\varepsilon > 0$, we have

$$\delta \in H^{-\mu}(-1, 1) = H^{-1/2 - \varepsilon}(-1, 1)$$

The trace theorem We now come to the problems of boundary conditions. The well-known trace theorem will be discussed. Consider the transition from $H^m(\mathbf{R}^n)$ to $H^m(\mathbf{R}_+^n)$, where $\mathbf{R}_+^n$ is the half space,

$$\mathbf{R}_+^n = \{x; x = (x_1, x_2, \ldots, x_{n-1}, x_n) \in \mathbf{R}^n: x_n > 0\}$$

Let $0 \leq j \leq m$ and let the trace operators γ_j be defined such that

$$(\gamma_j u)(x) = \left. \frac{\partial^j u(x, t)}{\partial n^j} \right|_{t=0}$$

If $u \in H^m(\mathbf{R}_+^2)$, it would be possible to extend γ_j to a continuous operator from $H^m(\mathbf{R}_+^2)$ into some boundary space $H^\mu(\Gamma) = H^\mu(\mathbf{R}_x)$. Here we are interested in determining the largest μ that admits a continuous mapping from $H^m(\mathbf{R}_+^2)$ into $H^\mu(\mathbf{R}_x)$. Let $v(x, t) \in \mathscr{S}_t(\mathbf{R})$. Then $\partial^j v(x, t)/\partial t^j \in \mathscr{S}_t(\mathbf{R})$, and $D_t^j v(x, t)$ can be used as a test function for tempered distributions. Since the Fourier transform of the distribution 1 is $\sqrt{2\pi}\,\delta$, we have

$$(\hat{1}, \phi) = (1, \hat{\phi}) = \int_{-\infty}^{\infty} \hat{\phi}(s)\, ds = \sqrt{2\pi}\,\phi(0)$$

Thus

$$\sqrt{2\pi}\, D_t^j v(x, 0) = \int_{-\infty}^{\infty} s^j \bar{v}(x, s)\, ds$$

where $\bar{v}(x, s) = F_t v(x, t)$ is the Fourier transform of $v(x, t)$ with respect to t. Thus

$$\| \gamma_j v \|^2_{\dot{H}^\mu(\mathbf{R}_x)} = \int_{\mathbf{R}_x} (1 + |y|^2)^\mu \frac{1}{2\pi} \left(\int_{-\infty}^{\infty} s^j \hat{v}(y, s)\, ds \right)^2 dy$$

Using Schwartz' inequality, Oden and Reddy (1976) have shown that

$$\int_{-\infty}^{\infty} s^j \hat{v}(y, s)\, ds \leq \int_{-\infty}^{\infty} |s^j \hat{v}(y, s)|\, ds$$

$$= \int_{-\infty}^{\infty} |s^j \hat{v}|\, Q^{m/2} Q^{-m/2}\, ds \leq \left[\int_{-\infty}^{\infty} |\hat{v}|^2 Q^m\, ds \right]^{1/2}$$

$$\times \left[\int_{-\infty}^{\infty} Q^{-m} s^{2j}\, ds \right]^{1/2}$$

with $Q = 1 + |y|^2 + s^2$. Let $p = s(1 + |y|^2)^{-1/2}$. Then

$$\int_{-\infty}^{\infty} Q^{-m} s^{2j}\, ds = c(1 + |y|^2)^{j - m + 1/2}$$

where

$$c = \int_{-\infty}^{\infty} (1 + |p|^2)^{-m} p^{2j}\, dp > 0 \qquad j \leq m - 1$$

Therefore we obtain

$$\| \gamma_j v \|^2_{\dot{H}^\mu(\mathbf{R}_x)} \leq \frac{c}{2\pi} \int_{-\infty}^{\infty} (1 + |y|^2)^{\mu - m + j + 1/2} \int_{-\infty}^{\infty} Q^m |\bar{v}|^2\, ds\, dy$$

If $\mu - m + j + \frac{1}{2} \leq 0$, then $(1 + |y|^2)^{\mu - m + j + 1/2} \leq 1$. Thus, for any $\mu \leq m - j - \frac{1}{2}$, we have

$$\| \gamma_j v \|^2_{\dot{H}^\mu(\mathbf{R}_x)} \leq \bar{c}^2 \int_{-\infty}^{\infty} \int_{-\infty}^{\infty} Q^m |v|^2\, ds\, dy$$

$$= \bar{c}^2 \| v \|^2_{\dot{H}^m(\mathbf{R}^2)}$$

where $\bar{c}^2 = \sqrt{c/2\pi}$ and the largest μ for which this inequality holds is $\mu = m - j - \frac{1}{2}$. Now for $u \in H^m(\mathbf{R}^2_+)$, we have

$$\| \gamma_j u \|_{H^{m-j-1/2}(\mathbf{R}_x)} \leq C \| u \|_{H^m(\mathbf{R}^2_+)} \tag{1-65}$$

It is concluded that the trace operators γ_j can be extended to continuous operators from $H^m(\mathbf{R}^n)$ into $H^{m-j-1/2}(\mathbf{R}^{n-1})$, with $0 \leq j \leq m - 1$.

Remarks We have summarized here the important theorems which have a direct bearing on the subject of error estimates of the finite element analysis. However, as mentioned earlier, the rigorous study of this section and Sec. 3-4 may be delayed until the reader has studied the finite element computational techniques covered in the rest of the book.

1-3 VARIATIONAL METHODS

1-3-1 General

Physical laws can often be deduced from concise mathematical principles to the effect that certain integrals attain extreme values. The problems concerned with the determination of extreme values of integrals whose integrands contain unknown functions belong to the calculus of variations. The simplest of such problems concerns the determination of an unknown function $y = y(x)$ for which the integral

$$I = \int_{x_0}^{x_1} F(x, y, y') \, dx \tag{1-66}$$

between two points $p_0(x_0, y_0)$ and $p_1(x_1, y_1)$ is a minimum. Here the prime denotes a derivative with respect to x.

We shall suppose that $F(x, y, y')$, viewed as a function of its arguments x, y, and y', has continuous partial derivatives of the second order, and we assume that there is a curve $y = y(x)$ with a continuously turning tangent that minimizes the integral. We then choose the competing family of curves as follows: Let $y = \eta(x)$ be any function with continuous second derivatives which vanishes at the end points of the interval (x_0, x_1). Then

$$\eta(x_0) = 0 \qquad \eta(x_1) = 0 \tag{1-67}$$

If α is a small parameter,

$$\bar{y}(x) = y(x) + \alpha\eta(x) \tag{1-68}$$

represents a family of curves passing through (x_0, y_0) and (x_1, y_1). This situation is shown in Fig. 1-5. Substituting (1-68) into (1-66) leads to the integral as a function of α

$$I(\alpha) = \int_{x_0}^{x_1} F[x, y(x) + \alpha\eta(x), y'(x) + \alpha\eta'(x)] \, dx \tag{1-69}$$

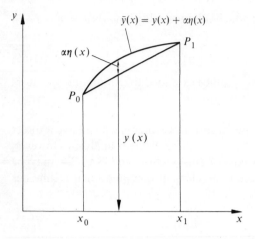

Figure 1-5 Minimization of integral in one dimension.

For $\alpha = 0$, (1-68) yields $y(x) = \bar{y}(x)$, and since $y = y(x)$ minimizes the integral, we conclude that $I(\alpha)$ must have a minimum for $\alpha = 0$. A necessary condition for this is to have the variation of the integral $I(\alpha)$, denoted by $\delta I(\alpha)$, set equal to zero. Thus,

$$\delta I(\alpha) = \frac{\partial I}{\partial \alpha} \, \delta\alpha = 0 \tag{1-70}$$

Here the differentiation is performed under the integral sign in (1-69). The symbol δ denotes a variational operator functioning similarly as differential operator. Thus we obtain

$$\delta I(\alpha) = \int_{x_0}^{x_1} \frac{\partial}{\partial \alpha} F(x, \bar{y}, \bar{y}') \, \delta\alpha \, dx$$

where

$$\bar{y} = y(x) + \alpha\eta(x) \qquad \bar{y}' = y'(x) + \alpha\eta'(x)$$

and

$$\frac{\partial}{\partial \alpha} F(x, \bar{y}, \bar{y}') = \frac{\partial F}{\partial \bar{y}} \frac{\partial \bar{y}}{\partial \alpha} + \frac{\partial F}{\partial \bar{y}'} \frac{\partial \bar{y}'}{\partial \alpha} = \frac{\partial F}{\partial \bar{y}} \eta(x) + \frac{\partial F}{\partial \bar{y}'} \eta'(x)$$

so that

$$\delta I(\alpha) = \int_{x_0}^{x_1} \left[\frac{\partial F}{\partial y} \eta(x) + \frac{\partial F}{\partial y'} \eta'(x) \right] \delta\alpha \, dx$$

On setting $\delta I(0) = 0$, we get

$$\int_{x_0}^{x_1} \left[\frac{\partial F}{\partial y} \eta(x) + \frac{\partial F}{\partial y'} \eta'(x) \right] \delta\alpha \, dx = 0 \tag{1-71}$$

The second term in the integral (1-71) can be integrated by parts to yield

$$\int_{x_0}^{x_1} \frac{\partial F}{\partial y'} \eta'(x) \, dx = \frac{\partial F}{\partial y'} \eta(x) \Big|_{x_0}^{x_1} - \int_{x_0}^{x_1} \eta(x) \frac{d}{dx}\left(\frac{\partial F}{\partial y'}\right) dx = -\int_{x_0}^{x_1} \eta(x) \frac{d}{dx}\left(\frac{\partial F}{\partial y'}\right) dx$$

Here it is seen that $\eta(x)$ vanishes at the end points, x_0 and x_1. Accordingly, (1-71) becomes

$$\int_{x_0}^{x_1} \eta(x) \left[\frac{\partial F}{\partial y} - \frac{d}{dx}\left(\frac{\partial F}{\partial y'}\right)\right] \delta\alpha \, dx = 0 \tag{1-72}$$

where $\eta(x)$ is an arbitrary function other than zero within the domain bounded by the end points x_0 and x_1. Since $\delta\alpha$ is also arbitrary, the vanishing of the integral in (1-72) is assured only if

$$\frac{\partial F}{\partial y} - \frac{d}{dx}\left(\frac{\partial F}{\partial y'}\right) = 0 \tag{1-73}$$

This is called the Euler–Lagrange equation. On carrying out the differentiation in (1-73), we get

$$\frac{\partial F}{\partial y} - \frac{\partial^2 F}{\partial x \, \partial y'} - \frac{\partial^2 F}{\partial y \, \partial y'} y' - \frac{\partial^2 F}{\partial y'^2} y'' = 0 \tag{1-74}$$

Similar operations when performed on the integral

$$I = \int_{x_0}^{x_1} F(x, y, y', y'', \ldots, y^n) \, dx$$

yield the Euler–Lagrange equation of the form

$$\frac{\partial F}{\partial y} - \frac{d}{dx}\left\{\frac{\partial F}{\partial y'}\right\} + \frac{d^2}{dx^2}\left\{\frac{\partial F}{\partial y''}\right\} - \cdots (-1)^n \frac{d^n}{dx^n} \frac{\partial F}{\partial y^{(n)}} = 0 \tag{1-75}$$

The foregoing discussion can also be generalized to the problems of minimizing the double integral (see Fig. 1-6),

$$I(u) = \iint_\Omega F\left(x, y, u, \frac{\partial u}{\partial x}, \frac{\partial u}{\partial y}\right) dx \, dy \tag{1-76}$$

in which the competing functions $u(x, y)$ assume on the boundary C of the region Ω preassigned continuous values $u = \phi(s)$. The Euler–Lagrange equation corresponding to (1-76) may be derived from a procedure similar to (1-66) through (1-73). The result is obtained in the form

$$\frac{\partial F}{\partial u} - \frac{\partial}{\partial x}\left\{\frac{\partial F}{\partial(\partial u/\partial x)}\right\} - \frac{\partial}{\partial y}\left\{\frac{\partial F}{\partial(\partial u/\partial y)}\right\} = 0 \tag{1-77}$$

For example, we may assume in (1-76)

$$F\left(x, y, u, \frac{\partial u}{\partial x}, \frac{\partial u}{\partial y}\right) = \frac{1}{2}\left\{\left(\frac{\partial u}{\partial x}\right)^2 + \left(\frac{\partial u}{\partial y}\right)^2\right\} - f(x, y)u \tag{1-78}$$

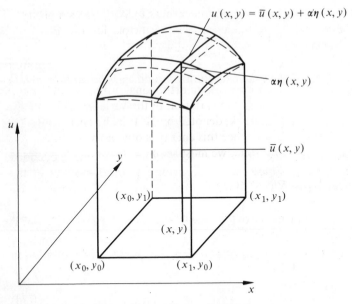

Figure 1-6 Minimization of integral in two dimensions.

where $f(x, y)$ is an unknown function. Then the substitution of this function in (1-77) gives

$$- f(x, y) - \frac{\partial}{\partial x}\left(\frac{\partial u}{\partial x}\right) - \frac{\partial}{\partial y}\left(\frac{\partial u}{\partial y}\right) = 0$$

or

$$- \frac{\partial^2 u}{\partial x^2} - \frac{\partial^2 u}{\partial y^2} = f(x, y)$$

Using the del operator, we write

$$- \nabla^2 u = f(x, y) \tag{1-79}$$

The expression (1-79) is known as the Poisson equation. The integral (1-76) with the function of the type (1-78) is called the "variational principle" for Poisson's equation. The function F, the integrand of the variational principle, is called the "variational functional."

The variational method, one of the most powerful methods of solution for engineering problems, begins with finding a functional or the variational principle of the type (1-76). A minimization of such a functional provides solution of the unknowns. The procedure for obtaining such a solution, referred to as the Rayleigh–Ritz method, is discussed below.

To qualify for the variational functional, a substitution of such a functional into the corresponding Euler–Lagrange equation must recover the governing differential equation, as demonstrated in (1-77), (1-78), and (1-79). Here a question arises: How does one obtain a variational principle for a given differential equation in the first place? To answer this question, we proceed as follows. For simplicity, let us assume that we want to find the variational principle for the Poisson equation

$$-\nabla^2 u - f(x, y) = 0 \tag{1-80}$$

We construct an inner product which represents an orthogonal projection of (1-80) onto a weighting function given by δu or the variation of the variable. This inner product may be considered as virtual work, denoted by δI. It is the energy which accounts for deviation or variation δu. Since this energy is obviously fictitious, or a deviation from true energy in the system, we must set $\delta I = 0$, or

$$\delta I = \int \int \left(-\nabla^2 u - f(x, y) \right) \delta u \, dx \, dy = 0 \tag{1-81}$$

Here I denotes the variational principle of the type (1-76). Using the Green–Gauss theorem or integrating by parts, we obtain

$$\delta I = -\int u_{,i} n_i \delta u \, ds + \int \int u_{,i} \delta u_{,i} \, dx \, dy - \int \int f(x, y) \delta u \, dx \, dy$$

$$= \delta \left[-\int u_{,i} n_i u \, ds + \frac{1}{2} \int \int u_{,i} u_{,i} \, dx \, dy - \int \int f(x, y) u \, dx \, dy \right] = 0$$

or

$$\delta I = \delta \left\{ -\int u \frac{\partial u}{\partial x} dy - \int u \frac{\partial u}{\partial y} dx + \frac{1}{2} \int \int \left[\left(\frac{\partial u}{\partial x} \right)^2 + \left(\frac{\partial u}{\partial y} \right)^2 - 2f(x, y) u \right] dx \, dy \right\} = 0$$

$$\tag{1-82a}$$

with $dy = n_1 \, ds = \cos (n, x) \, ds$, $dx = n_2 \, ds = \cos (n, y) \, ds$, $ds = (dx^2 + dy^2)^{1/2}$, as shown in Fig. 1-7. Note that the gradients $u_{,i} n_i$ are prescribed at boundaries as

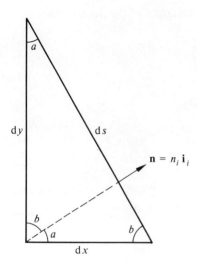

Figure 1-7 Inclined surface. $a = \cos^{-1} (dy/ds)$; $b = \cos^{-1} (dx/ds)$.

constants. Clearly, from (1-82a), the variational principle is identified as

$$I = \int\int \frac{1}{2}\left[\left(\frac{\partial u}{\partial x}\right)^2 + \left(\frac{\partial u}{\partial y}\right)^2 - 2f(x,y)u\right] dx\,dy \qquad (1\text{-}82b)$$

Here it should be noted that the boundary terms in (1-82a) known as natural boundary conditions vanish on free boundaries. Thus the variational functional assumes the form,

$$F\left(x, y, u, \frac{\partial u}{\partial x}, \frac{\partial u}{\partial y}\right) = \frac{1}{2}\left[\left(\frac{\partial u}{\partial x}\right)^2 + \left(\frac{\partial u}{\partial y}\right)^2\right] - f(x,y)u \qquad (1\text{-}82c)$$

This is the variational functional identical to (1-78) derived for Poisson's equation. We remark that the integrals given by (1-82a) and (1-82b) are referred to as the minimum and stationary conditions of energy, respectively. Additional discussion of variational principles as applied to finite element analysis will be given in later chapters. For additional details of variational principles, see Mikhlin [1964], Finlayson [1972], and Oden and Reddy [1975].

1-3-2 Rayleigh–Ritz Method

Any continuous medium consists of, theoretically, an infinite number of points at which all pertinent variables are defined. We may consider physical phenomena such as flows of stress, heat, fluid, electricity, etc. These variables may be specified at an infinite number of points in the system. The Rayleigh–Ritz method is an approximate procedure by which such continuous systems are reduced to systems with a finite number of variables. This method was first presented in 1877 by Rayleigh and was refined and extended by Ritz in 1909. It will be shown that the Rayleigh–Ritz method is the direct application of variational principles of the type discussed previously.

Let us consider the general problem of the minimum of the double integral

$$I(u) = \int\int F\left(x, y, u, \frac{\partial u}{\partial x}, \frac{\partial u}{\partial y}\right) dx\,dy \qquad (1\text{-}83)$$

under the boundary condition

$$u = \phi(s) \text{ on the boundary } s_1 \qquad (1\text{-}84)$$

where s_1 is the contour bounding the domain. Now consider a family of functions depending on several parameters:

$$u = \phi(x, y, a_1, a_2, \ldots, a_n) \qquad (1\text{-}85)$$

such that the boundary condition (1-84) is satisfied for all values of the parameters. Substituting (1-85) into (1-83) converts the integral $I(u)$ into a function of the n variables $a_1, a_2, \ldots, a_n$, or

$$I(u) = I(a_1, a_2, \ldots, a_n)$$

Since we are seeking the minimum of this function, the numbers a_k must satisfy the system of equations

$$\frac{\partial I}{\partial a_k} = 0 \qquad (k = 1, 2, \ldots, n) \tag{1-86}$$

Solving for these unknown parameters $a_1, a_2, \ldots, a_n$ and substituting them in (1-85) yields the required approximate solution:

$$\bar{u}(x, y) = \phi(x, y, a_1, a_2, \ldots, a_n) \tag{1-87}$$

If we assume that the solution (1-87) is only an approximation to the true solution $u(x, y)$ for which $I(u) = M$, the value of the minimum, then it can be assured that the sequence of functions $\bar{u}_1, \bar{u}_2, \ldots$ is minimal; that is, the sequence of integrals $I(\bar{u}_1), I(\bar{u}_2) \ldots$ tends to the true minimum:

$$\lim_{n \to \infty} I(\bar{u}_n) = I(u) = M$$

We shall show that the sufficient condition for this is the relative completeness of the system of families

$$u_n(x, y) = \phi_n(x, y, a_1, a_2, \ldots, a_n), \text{ with } n = 1, 2, \ldots \tag{1-88}$$

which consists of the following: whatever may be the function u, continuous together with $\partial u / \partial x$ and $\partial u / \partial y$ and satisfying the boundary condition (1-84), and whatever may be the positive number $\varepsilon > 0$, one can indicate an n and a function of the nth family (1-88),

$$u_n(x, y) = \phi_n(x, y, a_1, a_2, \ldots, a_n)$$

such that the following inequality will be valid everywhere in the region Ω:

$$\left| u_n - u \right| < \varepsilon, \qquad \left| \frac{\partial u_n}{\partial x} - \frac{\partial u}{\partial x} \right| < \varepsilon, \qquad \left| \frac{\partial u_n}{\partial y} - \frac{\partial u}{\partial y} \right| < \varepsilon \tag{1-89}$$

That is, any admissible function in which each successive family contains all the functions of the one preceding together with its partial derivatives may be approximated as closely as one pleases by means of functions of the given families. For example, let us consider the Poisson equation

$$-\frac{\partial^2 u}{\partial x^2} - \frac{\partial^2 u}{\partial y^2} - f(x, y) = 0$$

whose solution is assumed to be in the form

$$u(x, y) = \sum_{i=1}^{n} C_i \phi_i(x, y) \tag{1-90}$$

Here C_i are the constants to be determined and $\phi_i(x, y)$ are the appropriately chosen functions satisfying the global boundary conditions for the problem under consideration. Now the integral to be minimized or the variational principle for the Poisson equation is found in the form

$$I = \frac{1}{2} \int \int \left[\left(\frac{\partial u}{\partial x} \right)^2 + \left(\frac{\partial u}{\partial y} \right)^2 - 2f(x, y)u \right] dx \, dy \tag{1-91}$$

under the condition $u = \phi(s)$ on the boundary s_1. The variation of the integral (1-91) in view of (1-90) assumes the form

$$\delta I = \frac{\partial I}{\partial C_i} \delta C_i = \int \int \left[\frac{\partial u}{\partial x} \frac{\partial \phi_i}{\partial x} + \frac{\partial u}{\partial y} \frac{\partial \phi_i}{\partial y} - f(x, y)\phi_i \right] dx \, dy \, \delta C_i = 0 \tag{1-92}$$

Integrating by parts or using the Green–Gauss theorem, we have

$$\delta I = \left\{ \int \left(\frac{\partial u}{\partial x} \phi_i \, dy + \frac{\partial u}{\partial y} \phi_i \, dx \right) - \int \int \left[\frac{\partial^2 u}{\partial x^2} + \frac{\partial^2 u}{\partial y^2} + f(x, y) \right] \phi_i \, dx \, dy \right\} \delta C_i$$

$$= \left\{ \int \int \left(\frac{\partial u}{\partial x} \phi_i \cos(n, x) \, ds + \frac{\partial u}{\partial y} \phi_i \cos(n, y) \, ds \right) \right.$$

$$\left. - \int \int \left[\frac{\partial^2 u}{\partial x^2} + \frac{\partial^2 u}{\partial y^2} + f(x, y) \right] \phi_i \, dx \, dy \right\} \delta C_i \tag{1-93}$$

Since the integrand in the bracket of the last integral term is zero by definition of the governing differential equation, we obtain the following by setting (1-92) equal to (1-93):

$$\int \int \left(\frac{\partial u}{\partial x} \frac{\partial \phi_i}{\partial x} + \frac{\partial u}{\partial y} \frac{\partial \phi_i}{\partial y} - f(x, y)\phi_i \right) dx \, dy$$

$$= \int \left(\frac{\partial u}{\partial x} \phi_i \cos(n, x) + \frac{\partial u}{\partial y} \phi_i \cos(n, y) \right) ds \tag{1-94}$$

Replacing u in (1-94) by (1-90) results in n linear simultaneous algebraic equations from which the coefficients C_i are to be calculated. The desired solution is obtained finally by substituting these unknown coefficients into (1-90).

The developments presented can be summarized more rigorously in terms of notations given in Sec. 1-2-4. Equation (1-80) may be written as

$$-Au(x, y) = f(x, y) \qquad (x, y) \in \Omega \subset \mathbf{R}^2$$
$$u(x, y) = u_0 \qquad (x, y) \in \Gamma \tag{1-95a}$$

where the linear operator A is given by

$$A = \frac{\partial^2}{\partial x^2} + \frac{\partial^2}{\partial y^2}$$

If f belongs to $H^0(\Omega)$, then the solution exists in $H^2(\Omega)$. Consider a function v such that $Av \in H^0(\Omega)$. The variational boundary value problem corresponding to

(1-95a) becomes

$$\int_\Omega \left(\frac{\partial u}{\partial x} \frac{\partial v}{\partial x} + \frac{\partial u}{\partial y} \frac{\partial v}{\partial y} \right) dx\, dy = \int_\Omega fv\, dx\, dy \qquad \forall\, v \in H_0^1(\Omega) \qquad (1\text{-}95b)$$

Note that the second derivatives of (1-95a) are now replaced by the first derivatives in (1-95b), suggesting that weaker conditions may be required of the solution. This weak solution exists in $H_0^1(\Omega)$. Since $H^2(\Omega)$ is densely embedded in $H^1(\Omega)$, two solutions required by (1-95a) and (1-95b) are equivalent. This implies that $u \in H^2(\Omega) \cap H_0^1(\Omega)$. Here $H_0^1(\Omega)$ corresponds to the space of admissible variations.

Let us consider a quadratic functional of the type (1-95b) written in a general form

$$I(v) = B(v, v) - 2(f, v)$$

and for $\alpha \in \mathbf{R}$

$$I(u + \alpha v) = I(u) + 2\alpha[B(u, v) - (f, v)] + \alpha^2 B(v, v)$$

The first variation $\delta I(u, v)$ vanishes at the point $u \in H_0^1(\Omega)$

$$\delta I(u, v) = \lim_{\alpha \to 0} \frac{1}{\alpha} [I(u + \alpha v) - I(u)] = 2[B(u, v) - (f, v)] = 0$$

or
$$B(u, v) = (f, v) \qquad \forall\, v \in H_0^1(\Omega)$$

This is identical to (1-95b). If $f \in H^0(\Omega)$, then

$$(Au, v)_{H^0(\Omega)} = (f, v)_{H^0(\Omega)}$$

and
$$B(u, v) = (f, v)_{H^0(\Omega)} \qquad \forall\, v \in H_0^1(\Omega)$$

To determine stationary conditions we proceed with

$$I(v) = B(v, v) - 2(f, v) = \sum_{i,j=1}^{G} A_{ij} C_i C_j - 2 \sum_{i=1}^{G} f_i C_i$$

where $v = C_i \phi_i$ with C_i to be chosen such that

$$\frac{\partial I(v)}{\partial C_i} = 0, \qquad 1 \le i \le G$$

This becomes identical to (1-94) with the exception that the Neumann boundary conditions are set equal to zero.

There are two limitations in the Rayleigh–Ritz method: First, the variational principle, although derivable in the manner of (1-81) for self-adjoint equations (even order derivatives with constant coefficients) may not exist in many fluid mechanics problems such as in nonself-adjoint equations (odd order derivatives). Secondly, it is difficult if not impossible to find the function ϕ_i satisfying the global boundary conditions for the domain with complicated geometries.

1-4 METHOD OF WEIGHTED RESIDUALS

The basic idea of the methods of weighted residual is to obtain an approximate solution to a differential equation of the form

$$- Au - f = 0 \text{ in } \Omega \tag{1-96a}$$

subject to the boundary conditions

$$B_r u = g_r \quad \text{on } \Gamma \tag{1-96b}$$

by introducing a set of functions in the form

$$\hat{u} = \sum_{i=1}^{n} C_i \phi_i \tag{1-97}$$

where C_i are constants and ϕ_i are the linearly independent functions chosen such that all global boundary conditions are satisfied. Since (1-97) is an approximate function, we note that, if substituted into (1-96a), it will not satisfy exactly the governing differential equation. Thus we set (1-81) equated to an error or residual ε

$$- A\hat{u} - f = \varepsilon \tag{1-98}$$

Now let us introduce a set of weighting functions $w_i (i = 1, 2, \ldots, N)$ and construct an inner product (ε, w_i). We now set this inner product equal to zero

$$(\varepsilon, w_i) = 0 \tag{1-99}$$

which is equivalent to forcing the error of the approximate differential equation (1-98) to be zero in an average sense.

There are various ways to choose the weighting functions w_i, leading to (1) Galerkin method, (2) least squares method, (3) method of moments, and (4) collocation method. In the Galerkin method, the weighting function w_i is the test function which is made equal to the trial (basis) function. Thus

$$(\varepsilon, \phi_i) = 0 \tag{1-100}$$

The least squares method is to set the weighting function equal to the residual itself and minimize the inner product (square of the error) with respect to every constant coefficient C_i

$$\frac{\partial}{\partial C_i} (\varepsilon, \varepsilon) = 0 \tag{1-101}$$

In the method of moments, the weighting functions are chosen from any set of linearly independent functions such as $1, x, x^2, x^3, \ldots$ for one-dimensional problems so that

$$(\varepsilon, x^i) = 0 \quad i = 0, 1, 2, \ldots \tag{1-102}$$

The collocation method is to choose a set of points in the domain as collocation points, the weighting functions being the displaced Dirac delta function

$$w_i = \delta(x - x_i)$$

which has the property that

$$\left(\varepsilon, \delta(x - x_i)\right) = \varepsilon\big|_{x_i} \tag{1-103}$$

Thus the residual is zero at N specified collocation points x_i. The subject of classical weighted residual methods has an excellent treatment in the book of Finlayson [1972].

The Galerkin and least squares methods are well adapted to finite element applications whereas the methods of moments and collocation do not lend themselves to direct applications to the finite element method. In what follows, therefore, we further elaborate the Galerkin and least squares method.

1-4-1 Galerkin Method

The Galerkin method [Galerkin, 1915] is an orthogonal projection of the residual ε to a set of linearly independent complete functions ϕ_i as implied by the inner-product (1-100). In order for $\hat{u}(x, y)$ to be the exact solution of the given equation, it is necessary for ε to be identically equal to zero; and this requirement, if ε is considered to be continuous, is equivalent to the requirement of the orthogonality of the expression ε to all the functions of the system $\phi_i(x, y)$ $(i = 1, 2, \dots n)$. However, having at our disposal only n constants $C_1, C_2, \dots, C_n$, we can satisfy only n conditions of orthogonality. Based on these considerations, we arrive at the system of equations

$$\int_\Omega \varepsilon \phi_i \, d\Omega = -\int_\Omega \left[A\left(\sum_{j=1}^{n} C_j \phi_j \right) + f \right] \phi_i \, d\Omega = 0$$

which serves for the determination of the coefficients C_i. On finding the C_i from this system and substituting them in the expression for $\hat{u}$, we obtain the required approximate solution.

To illustrate, let us consider the Poisson equation

$$-\frac{\partial^2 u}{\partial x^2} - \frac{\partial^2 u}{\partial y^2} - f(x, y) = 0$$

Let the approximate solution $\hat{u}$ be given by (1-97). Then the residual is defined as

$$\varepsilon = -\frac{\partial^2 \hat{u}}{\partial x^2} - \frac{\partial^2 \hat{u}}{\partial y^2} - f(x, y) \tag{1-104}$$

We then write the Galerkin integral in the form

$$(\varepsilon, \phi_i) = \int\int \varepsilon \phi_i \, dx \, dy = -\int\int \left(\frac{\partial^2 u}{\partial x^2} + \frac{\partial^2 u}{\partial y^2} + f(x, y) \right) \phi_i \, dx \, dy = 0 \tag{1-105}$$

Integrating (1-105) by parts or using the Green–Gauss theorem,

$$-\int \left(\frac{\partial u}{\partial x}\phi_i\,dy + \frac{\partial u}{\partial y}\phi_i\,dx\right) + \int\int \left(\frac{\partial u}{\partial x}\frac{\partial \phi_i}{\partial x} + \frac{\partial u}{\partial y}\frac{\partial \phi_i}{\partial y} - f(x,y)\phi_i\right)dx\,dy$$

$$= -\int \left(\frac{\partial u}{\partial x}\phi_i\cos(n,x)\,ds + \frac{\partial u}{\partial y}\phi_i\cos(n,y)\,ds\right)$$

$$+ \int\int \left(\frac{\partial u}{\partial x}\frac{\partial \phi_i}{\partial x} + \frac{\partial u}{\partial y}\frac{\partial \phi_i}{\partial y} - f(x,y)\phi_i\right)dx\,dy = 0$$

This is identical to the expression (1-94).

To summarize our discussions, let us suppose that we wish to find $u \in U$ such that

$$B(u,v) = l(v) \qquad \forall\, v \in V \tag{1-106a}$$

where l is a continuous linear functional on V. If $U_h \subset U$ and $V_h \subset V$ are two finite-dimensional subspaces of U and V, respectively, then the Galerkin approximation of the solution u of (1-106) is represented by

$$B(\hat{u},\hat{v}) = l(\hat{v}) \qquad \forall\, \hat{v} \in V_h \tag{1-106b}$$

Denoting

$$\hat{u} = C_i\phi_i \qquad \hat{v} = \overset{*}{C}_i\psi_i \tag{1-107a, b}$$

where C_i and $\overset{*}{C}_i$ are arbitrary constants, we obtain

$$B(C_j\phi_j, \overset{*}{C}_i\psi_i) = l(\overset{*}{C}_i\psi_i)$$

or

$$[B(\psi_i,\phi_j)C_j - l(\psi_i)]\overset{*}{C}_i = 0$$

This leads to

$$B(\psi_i,\phi_j)C_j = l(\psi_i)$$

Solving for C_j and substituting into (1-107a) yields

$$\hat{u} = [B(\psi_j,\phi_i)]^{-1}l(\psi_j)\phi_i$$

The subspace U_h is often referred to as the space of trial functions, and subspace V_h as the space of test functions. If $U = V$ and $U_h = V_h$, then we denote

$$B(\phi_i,\phi_j) = B_{ij}, \qquad l(\phi_i) = l_i$$

where B_{ij} is symmetric. Thus,

$$u = B_{ij}^{-1}l_j\phi_i$$

Consider now the bilinear form

$$B(u,\hat{v}) = l(\hat{v})$$

Subtracting the above from (1-106b) gives

$$B(u - \hat{u}, v) = 0$$

This is known as an orthogonality condition. The error $e = u - \hat{u}$ is "orthogonal" to the subspace V_h with respect to the bilinear map $B(\cdot, \cdot)$. Thus,

$$B(u - \hat{v}, u - \hat{v}) = B(e + (\hat{u} - \hat{v}), e + (\hat{u} - \hat{v})) = B(e, e) + B(\hat{u} - \hat{v}, \hat{u} - \hat{v})$$

Clearly, of all the functions $\hat{v}$, the closest to the true solution u is the Galerkin approximation $\hat{u}$. This is called the best approximation.

It should be noted that for all engineering problems for which the variational functional exists, the Galerkin integral (1-105) gives results identical to those one would obtain using the Rayleigh–Ritz method. Furthermore, it is once again emphasized that when the variational functional does not exist for some particular problem, the Galerkin method is unfailingly applicable. However, the determination of the function ϕ_i in (1-97) still poses great difficulties for complicated geometries and boundary conditions.

1-4-2 Least Squares Method

As in the Galerkin method, the least squares method [Mikhlin, 1964] requires no variational principle. The idea is to minimize the mean square error, which is accomplished by setting equal to zero the following integral:

$$\int \int \varepsilon^2 \, dx \, dy \qquad (1\text{-}108)$$

where we once again define ε to be the residual of the differential equation. We shall determine the coefficients C_i in (1-90) from the condition that (1-108) is a minimum. Thus,

$$\frac{\partial}{\partial C_i} \left[\int \int \varepsilon^2 \, dx \, dy \right] = 0$$

or

$$2 \int \int \varepsilon \frac{\partial \varepsilon}{\partial C_i} \, dx \, dy = 0 \qquad (1\text{-}109)$$

Note that (1-109) is similar to (1-105) in that the weighting function ϕ_i is replaced by $2 \, \partial\varepsilon/\partial C_i$ in (1-109). It is also seen that the least squares method involves higher derivatives which will, in general, lead to a better convergence than the Rayleigh–Ritz method or the Galerkin method, but it has the disadvantage of requiring higher order weighting functions.

For finite element applications of the least squares method, see Akin [1973] and Lynn [1974, 1976], and Lynn and Bapana [1976].

1-5 FINITE ELEMENT METHOD IN ONE-DIMENSIONAL PROBLEMS

1-5-1 Global and Local Finite Element Models

For simplicity, let us consider a one-dimensional problem as depicted in Fig. 1-8 with the domain of study $\Omega(0 < x < 1)$ and its boundaries Γ at $x = 0$ and $x = L$.

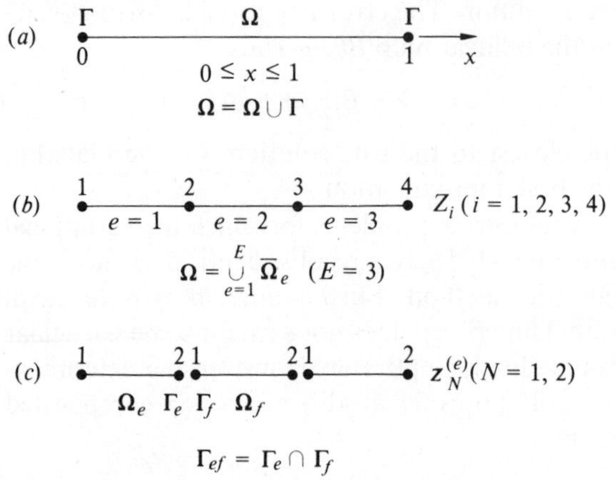

Figure 1-8 Finite element models. (a) Domain of study, (b) global finite element model, and (c) local finite element model.

Let the entire domain be divided into subdomains called *finite elements*, say three elements in this example. The joints of the elements are called *nodes*. The finite element model $\bar{\Omega}$ is expressed as the union of the domain and its boundaries

$$\bar{\Omega} = \Omega \cup \Gamma \qquad (1\text{-}110a)$$

We now isolate all elements from the global domain. Each local element $\bar{\Omega}_e$ is now identified as

$$\bar{\Omega}_e = \Omega_e \cup \Gamma_e$$

The boundaries of this element and the neighboring element are the intersection

$$\Gamma_e \cap \Gamma_f = \Gamma_{ef}$$

Thus the connected finite element model (1-110a) is the union of all elements

$$\bar{\Omega} = \Omega \cup \Gamma = \bigcup_{e=1}^{E=3} \bar{\Omega}_e \qquad (1\text{-}110b)$$

where E is the total number of elements.

The global nodes of connected model $\bar{\Omega}$ and the local nodes of isolated elements are identified by Z_i ($i = 1, 2, 3, 4$) and $z_N^{(e)}$ ($N = 1, 2; e = 1, 2, 3$). They are related as follows: $z_1^{(1)} = Z_1, z_2^{(1)} = Z_2, z_1^{(2)} = Z_2, z_2^{(2)} = Z_3, z_1^{(3)} = Z_3$, and $z_2^{(3)} = Z_4$. Writing these relations in matrix form yields

$$\begin{bmatrix} z_1^{(1)} \\ z_2^{(1)} \end{bmatrix} = \begin{bmatrix} 1 & 0 & 0 & 0 \\ 0 & 1 & 0 & 0 \end{bmatrix} \begin{bmatrix} Z_1 \\ Z_2 \\ Z_3 \\ Z_4 \end{bmatrix}, \qquad \begin{bmatrix} z_1^{(2)} \\ z_2^{(2)} \end{bmatrix} = \begin{bmatrix} 0 & 1 & 0 & 0 \\ 0 & 0 & 1 & 0 \end{bmatrix} \begin{bmatrix} Z_1 \\ Z_2 \\ Z_3 \\ Z_4 \end{bmatrix},$$

$$\begin{bmatrix} z_1^{(3)} \\ z_2^{(3)} \end{bmatrix} = \begin{bmatrix} 0 & 0 & 1 & 0 \\ 0 & 0 & 0 & 1 \end{bmatrix} \begin{bmatrix} Z_1 \\ Z_2 \\ Z_3 \\ Z_4 \end{bmatrix}$$

Using index notations, we may express the above in the form

$$z_N^{(e)} = \Delta_{Ni}^{(e)} Z_i \tag{1-111}$$

where $\Delta_{Ni}^{(e)}$ is called the Boolean matrix having the property

$$\Delta_{Ni}^{(e)} = \begin{cases} 1, \text{ if the local node } N \text{ coincides with the global node } i \\ 0, \text{ otherwise} \end{cases} \tag{1-112}$$

Similarly, we may write

$$Z_i = \Delta_{Ni}^{(e)} z_N^{(e)} \tag{1-113}$$

where $\Delta_{Ni}^{(e)}$ is seen to be a transpose of $\Delta_{Ni}^{(e)}$ of (1-111). Inserting (1-111) into (1-113) yields

$$Z_i = \Delta_{Ni}^{(e)} \Delta_{Nj}^{(e)} Z_j$$

from which we obtain the relation

$$\Delta_{Ni}^{(e)} \Delta_{Nj}^{(e)} = \delta_{ij}$$

where δ_{ij} is the Kronecker delta. Likewise, substituting (1-113) into (1-111), we get

$$z_N^{(e)} = \Delta_{Ni}^{(e)} \Delta_{Mi}^{(e)} z_M$$

Once again we have

$$\Delta_{Ni}^{(e)} \Delta_{Mi}^{(e)} = \delta_{NM}$$

For example, take element 1, the above relationship is verified as

$$\begin{bmatrix} 1 & 0 & 0 & 0 \\ 0 & 1 & 0 & 0 \end{bmatrix} \begin{bmatrix} 1 & 0 \\ 0 & 1 \\ 0 & 0 \\ 0 & 0 \end{bmatrix} = \begin{bmatrix} 1 & 0 \\ 0 & 1 \end{bmatrix}$$

1-5-2 Local and Global Interpolation Functions

The concepts of variational and weighted residual methods are used in the finite element method, but the basic difference is that the approximating functions are constructed first in the subdomain (local elements), thus avoiding the difficult task of requiring satisfaction of global boundary conditions.

Suppose that the variable u is approximated within a local element e as a linear function $u^{(e)}$

$$u^{(e)} = a_1 + a_2 x \tag{1-114}$$

Writing (1-114) at $x = 0$ and $x = h$, we obtain

$$u_{x=0}^{(e)} = u_1 = a_1$$

$$u_{x=h}^{(e)} = u_2 = a_1 + a_2 h$$

Solving for the constants a_1 and a_2 and substituting into (1-114) yield

$$u^{(e)} = \Phi_N^{(e)} u_N^{(e)} \qquad (N = 1, 2) \tag{1-115}$$

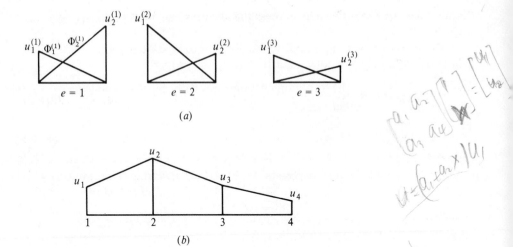

(a)

(b)

Figure 1-9 Interpolation functions. (a) Local interpolation functions and local nodal values $(u^{(e)} = \Phi_N^{(e)} u_N)$; (b) global interpolation functions and global nodal values

$$\left(u = \sum_{e=1}^{3} \Phi_N^{(e)} \Delta_{Ni}^{(e)} u_i \right)$$

where $\Phi_N^{(e)}$ are called interpolation functions,

$$\Phi_1^{(e)} = 1 - \frac{x}{h}, \qquad \Phi_2^{(e)} = \frac{x}{h} \qquad (1\text{-}116)$$

Schematically, these functions are shown in Fig. 1-9a. They have the properties

$$0 \le \Phi_N^{(e)} \le 1, \qquad \sum_{N=1}^{r} \Phi_N^{(e)} = 1, \qquad \Phi_N^{(e)}(z_M) = \delta_{NM} \qquad (1\text{-}117)$$

with r the total number of elemental nodes ($r = 2$ here).

The local nodal values $u_N^{(e)}$ can be related to the global nodal values u_i similarly as (1-111),

$$u_N^{(e)} = \Delta_{Ni}^{(e)} u_i \qquad (1\text{-}118)$$

Thus the global function u can be written as the sum of all local element contributions.

$$u = \sum_{e=1}^{E} u^{(e)} = \sum_{e=1}^{E} \Phi_N^{(e)} u_N^{(e)} = \sum_{e=1}^{E} \Phi_N^{(e)} \Delta_{Ni}^{(e)} u_i$$

or

$$\boxed{u = \Phi_i u_i} \qquad (1\text{-}119)$$

where Φ_i is called the global interpolation function

$$\Phi_i = \sum_{e=1}^{E} \Phi_N^{(e)} \Delta_{Ni}^{(e)} \qquad \text{with} \qquad \Phi_i(Z_j) = \delta_{ij}$$

In view of (1-119) and (1-118), the expanded form of (1-119) appears as shown in Fig. 1-9b.

If the second order interpolation function is used, we write

$$u^{(e)} = a_1 + a_2 x + a_3 x^2$$

This requires installation of an additional node, preferably at midpoint within the element. It can be shown that[†]

$$u^{(e)} = \Phi_N^{(e)} u_N^{(e)} \qquad (N = 1, 2, 3) \tag{1-120}$$

where

$$\Phi_1 = 1 - 3\frac{x}{h} + 2\left(\frac{x}{h}\right)^2 \qquad \Phi_2 = 4\left[\frac{x}{h} - \left(\frac{x}{h}\right)^2\right] \qquad \Phi_3 = 2\left(\frac{x}{h}\right)^2 - \frac{x}{h}$$

It is interesting to note that the relations (1-117) still hold with $r = 3$ in this case. With these quadratic functions, the curve describing u will be parabolic in Fig. 1-9b.

1-5-3 Derivation of Finite Element Equations by Variational Method (Rayleigh–Ritz Method)

Consider a differential equation of the form

$$\frac{d^2 u}{dx^2} - \alpha^2 u = f(x) \qquad 0 < x < 1 \tag{1-121}$$

subject to boundary conditions. As demonstrated in Sec. 1-3, the variational principle corresponding to (1-121) is

$$I = \int_0^1 \frac{1}{2}\left[\left(\frac{du}{dx}\right)^2 + \alpha^2 u^2 + 2fu\right] dx \tag{1-122}$$

Substituting (1-119) into (1-122) yields

$$I = \left[\frac{1}{2}\int_0^1 \left(\frac{\partial\Phi_i}{\partial x}\frac{\partial\Phi_j}{\partial x} + \alpha^2\Phi_i\Phi_j\right) dx\right] u_i u_j + \left(\int_0^1 f\Phi_i\, dx\right) u_i$$

From (1-120), the global integral above can be shifted to the sum of the local integrals

$$I = \left[\sum_{e=1}^{E} \frac{1}{2}\int_0^h \left(\frac{\partial\Phi_N^{(e)}}{\partial x}\frac{\partial\Phi_M^{(e)}}{\partial x} + \alpha^2\Phi_N^{(e)}\Phi_M^{(e)}\right) dx\right]\Delta_{Ni}^{(e)}\Delta_{Mj}^{(e)} u_i u_j$$

$$+ \left[\sum_{e=1}^{E}\int_0^h f\Phi_N^{(e)}\, dx\right]\Delta_{Ni}^{(e)} u_i$$

or

$$I = \tfrac{1}{2} A_{ij} u_i u_j + F_i u_i \tag{1-123}$$

[†] An alternative method with the origin set at the center is shown in Sec. 2-2-1.

where A_{ij} and F_i are the global coefficient matrix and global input vector, respectively,

$$A_{ij} = \int_0^1 \left(\frac{\partial \Phi_i}{\partial x} \frac{\partial \Phi_j}{\partial x} + \alpha^2 \Phi_i \Phi_j \right) dx = \sum_{e=1}^E A_{NM}^{(e)} \Delta_{Ni}^{(e)} \Delta_{Mj}^{(e)} \qquad (1\text{-}124)$$

$$F_i = - \int_0^1 f\Phi_i \, dx = - \sum_{e=1}^E F_N^{(e)} \Delta_{Ni}^{(e)} \qquad (1\text{-}125)$$

Here $A_{NM}^{(e)}$ represents the local element coefficient matrix

$$A_{NM}^{(e)} = \int_0^h \left(\frac{\partial \Phi_N^{(e)}}{\partial x} \frac{\partial \Phi_M^{(e)}}{\partial x} + \alpha^2 \Phi_N^{(e)} \Phi_M^{(e)} \right) dx \qquad (1\text{-}126)$$

and $F_N^{(e)}$ is the local input vector

$$F_N^{(e)} = - \int_0^h f\Phi_N^{(e)} \, dx \qquad (1\text{-}127)$$

Note also that from (1-123), (1-124), and the relation $u_i = \Delta_{Ni}^{(e)} u_N^{(e)}$, we may write

$$I = \sum_{e=1}^E I^{(e)}$$

where $I^{(e)}$ is the variational principle corresponding to a local element

$$I^{(e)} = \tfrac{1}{2} A_{NM}^{(e)} u_N^{(e)} u_M^{(e)} + F_N^{(e)} u_N^{(e)} \qquad (1\text{-}128)$$

Minimizing $I^{(e)}$ with respect to every $u_N^{(e)}$ yields

$$\delta I^{(e)} = \frac{\partial I^{(e)}}{\partial u_N} \delta u_N = 0$$

Since δu_N is arbitrary, we must have

$$\frac{\partial I^{(e)}}{\partial u_N} = 0$$

This gives

$$A_{NM}^{(e)} u_M^{(e)} = F_N^{(e)} \qquad (1\text{-}129)$$

The expression (1-129) is known as the local finite element equations.

The solution is obtained by minimizing the global variational principle I (1-123) with respect to every global nodal value u_i

$$\delta I = \frac{\partial I}{\partial u_i} \delta u_i = 0$$

Thus, using similar arguments as in the local equations

$$A_{ij} u_j = F_i \qquad (1\text{-}130)$$

Here A_{ij} is the $n \times n$ positive-definite symmetric square matrix. The equation (1-130) is called the global finite element equations. It may be said that the global equations (1-130) represent the *collection* or *assembly* of local equations (1-129).

A glance at (1-124) and (1-125) indicates that the local element coefficient matrices $A_{NM}^{(e)}$ and the local input vector $F_N^{(e)}$ are assembled according to the Boolean matrices which place the appropriate local nodal contributions to the corresponding global system. We delay discussion of boundary conditions until Sec. 3-2.

Example 1-5 We demonstrate here the calculations of local and global matrices and vectors. Consider the system of three equal elements and four global nodes. From (1-126) and (1-127), we have

$$A_{NM}^{(e)} = \int_0^h \left(\frac{\partial \Phi_N^{(e)}}{\partial x} \frac{\partial \Phi_M^{(e)}}{\partial x} + \alpha^2 \Phi_N^{(e)} \Phi_M^{(e)} \right) dx$$

$$F_N^{(e)} = - \int_0^h f \Phi_N^{(e)} \, dx$$

with $\Phi_1^{(e)} = 1 - (x/h)$ and $\Phi_2^{(e)} = x/h$. Integrating yields

$$A_{NM}^{(e)} = \frac{1}{h} \begin{bmatrix} 1 & -1 \\ -1 & 1 \end{bmatrix} + \frac{\alpha^2 h}{6} \begin{bmatrix} 2 & 1 \\ 1 & 2 \end{bmatrix} = \begin{bmatrix} \dfrac{1}{h} + \dfrac{\alpha^2 h}{3} & \dfrac{\alpha^2 h}{6} - \dfrac{1}{h} \\ \dfrac{\alpha^2 h}{6} - \dfrac{1}{h} & \dfrac{1}{h} + \dfrac{\alpha^2 h}{3} \end{bmatrix}$$

$$F_N^{(e)} = - f \frac{h}{2} \begin{bmatrix} 1 \\ 1 \end{bmatrix}$$

From (1-124), we have

$$A_{ij} = \sum_{e=1}^{E} A_{NM}^{(e)} \Delta_{Ni}^{(e)} \Delta_{Mj}^{(e)} = \begin{bmatrix} A_{11}^{(1)} & A_{12}^{(1)} & & \\ A_{21}^{(1)} & A_{22}^{(1)} + A_{11}^{(2)} & A_{12}^{(2)} & \\ & A_{21}^{(2)} & A_{22}^{(2)} + A_{11}^{(3)} & A_{12}^{(3)} \\ & & A_{21}^{(3)} & A_{22}^{(3)} \end{bmatrix}$$

$$F_i = \sum_{e=1}^{E} F_N^{(e)} \Delta_{Ni}^{(e)} = \begin{bmatrix} F_1^{(1)} \\ F_2^{(1)} + F_1^{(2)} \\ F_2^{(2)} + F_1^{(3)} \\ F_2^{(3)} \end{bmatrix} = f \frac{h}{2} \begin{bmatrix} 1 \\ 2 \\ 2 \\ 1 \end{bmatrix}$$

It is shown here that the global quantities A_{ij} and F_i are obtained as the components of the local contributions of $A_{NM}^{(e)}$ and $F_N^{(e)}$ superimposed (algebraically summed) at the common global nodes.

1-5-4 Derivation of Finite Element Equations by Weighted Residual Method (Galerkin Integral)

Consider again the differential equation (1-121)

$$\frac{d^2u}{dx^2} - \alpha^2 u - f(x) = 0$$

By substituting the approximate function (1-119) into this differential equation, we expect to have committed an error or a residual ε. Thus we may write

$$\frac{d^2u}{dx^2} - \alpha^2 u - f = \varepsilon \tag{1-131}$$

As described in Sec. 1-4-1, we construct an inner product of this residual and the global finite element interpolation function Φ_i

$$(\varepsilon, \Phi_i) = \int_0^1 \left(\frac{d^2u}{dx^2} - \alpha^2 u - f\right)\Phi_i \, dx = 0 \tag{1-132}$$

This is an orthogonal projection of the residual space onto a subspace spanned by Φ_i. Integrating (1-132) by parts yields

$$\frac{du}{dx}\Phi_i\Big|_0^1 - \int_0^1 \left(\frac{du}{dx}\frac{d\Phi_i}{dx} + \alpha^2 u\Phi_i + f\Phi_i\right)dx = 0 \tag{1-133}$$

The boundary term obtained here is the natural boundary condition and requires careful examination. We note that the interpolation function Φ_i in (1-132) does not include the boundary. If a two-dimensional problem were considered, we would have required two types of interpolation functions: one for the interior domain and the other for the boundary surfaces; that is

$$u(x, y) = \Phi_i(x, y)u_i \quad \text{For interior} \tag{1-134}$$

and

$$u(\Gamma) = \overset{*}{\Phi}_k(\Gamma)u_k \quad \text{for boundary} \tag{1-135}$$

where i denotes all interior global nodes in Ω and k denotes all boundary nodes along Γ. Clearly, $\overset{*}{\Phi}(\Gamma)$ is the interpolation function which represents the variation of du/dn along the boundary surface (line) so that the global boundary integral of the type

$$\int_\Gamma \frac{du}{dn}\overset{*}{\Phi}_k(\Gamma)\,d\Gamma = \sum_{e=1}^E \int_0^a \frac{du^{(e)}}{dn}\overset{*}{\Phi}_N^{(e)}\Delta_{Nk}^{(e)}\,d\Gamma \quad (N = \text{boundary element nodes}) \tag{1-136}$$

can be performed as the union of each of the boundary elements. However, in a one-dimensional problem, there exists no boundary surface (line); there are only two boundary points, one at each end of the domain. Returning to the boundary term in (1-133), if du/dx is specified at ends, Φ_i which must be the boundary interpolation $\overset{*}{\Phi}_i$ is simply a unity,

$$\overset{*}{\Phi}_i = \Delta_{Ni}^{(e)} \overset{*}{\Phi}_N^{(e)} = 1, \quad \overset{*}{\Phi}_i(Z_j) = \delta_{ij}, \quad \overset{*}{\Phi}_N^{(e)}(z_M) = \delta_{NM} \tag{1-137}$$

Here i, j and N, M represent the boundary node for the global and local systems with only the boundary element and boundary node being involved. Therefore, we rewrite (1-133) in the form

$$\int_0^1 \left(\frac{du}{dx} \frac{d\Phi_i}{dx} + \alpha^2 u \Phi_i + f \Phi_i \right) dx = \frac{du}{dx} \overset{*}{\Phi}_i \Big|_0^1$$

and

$$A_{ij} u_j = F_i + \frac{du}{dx} \overset{*}{\Phi}_i \Big|_{x=0, x=1} = F_i + \frac{du}{dx} \Big|_{i(x=0, x=1)} \tag{1-138}$$

If the given problem is the Dirichlet type, then we simply have

$$A_{ij} u_j = F_i$$

Treatment of boundary conditions for multidimensional problems will be elaborated in Chap. 3.

Remarks We note that the same results are obtained from either the variational method or the Galerkin method. This is because the weighting functions (test functions) in the Galerkin method are set equal to trial functions (basis or interpolation functions). If the Neumann condition existed in the variational method, we would have started with the variational principle

$$I = \int_0^1 \left[\frac{1}{2} \left(\frac{du}{dx} \right)^2 + \frac{\alpha^2}{2} u^2 + f u \right] dx - \frac{du}{dx} u \Big|_0^1$$

If $du/dx = 0$, then simply the last term vanishes. Once again, we emphasize that the advantage of the weighted residual method is to require no variational principle.

Example 1-6 Consider a second order differential equation in one dimension

$$\frac{d^2 u}{dx^2} - \alpha^2 u = 0 \tag{E1-6a}$$

whose analytical solution is $u = \exp(-\alpha x)$ with $\alpha > 0$ for decay and $\alpha < 0$ for growth. Our objective is to solve (E1-6a) by the finite element method with the following boundary conditions:

$$u(0) = e^0 = 1 \qquad u(2) = e^2 = 7.389 \tag{E1-6b}$$

with $\alpha = -1$ per unit length.

The global finite element equations are derived from either the variational principle or Galerkin weighted residual method in the form (following the discussions in Sec. 1-5-3 or 1-5-4)

$$A_{ij}u_j = F_i \qquad \text{(E1-6c)}$$

where

$$A_{ij} = \sum_{e=1}^{E} A_{NM}^{(e)} \Delta_{Ni}^{(e)} \Delta_{Mj}^{(e)} \qquad \text{(E1-6d)}$$

$$F_i = 0$$

$$A_{NM}^{(e)} = \int_0^h \left(\frac{\partial \Phi_N^{(e)}}{\partial x} \frac{\partial \Phi_M^{(e)}}{\partial x} + \alpha^2 \Phi_N^{(e)} \Phi_M^{(e)} \right) dx \qquad \text{(E1-6e)}$$

Using the linear variation of u within an element, we have $\Phi_1 = 1 - (x/h)$ and $\Phi_2 = x/h$. Thus

$$A_{NM}^{(e)} = \begin{bmatrix} \dfrac{1}{h} + \dfrac{h}{3} & \dfrac{h}{6} - \dfrac{1}{h} \\[2ex] \dfrac{h}{6} - \dfrac{1}{h} & \dfrac{1}{h} + \dfrac{h}{3} \end{bmatrix}$$

Let us assume that the domain is divided into four equal elements with five nodes. Then we have $h = 0.5$ and

$$A_{NM}^{(e)} = \begin{bmatrix} 2.1667 & -1.9167 \\ -1.9167 & 2.1667 \end{bmatrix}$$

The local finite element equations

$$A_{NM}^{(e)} u_M^{(e)} = 0$$

take the form

$$\begin{bmatrix} 2.1667 & -1.9167 \\ -1.9167 & 2.1667 \end{bmatrix} \begin{bmatrix} u_1^{(e)} \\ u_2^{(e)} \end{bmatrix} = \begin{bmatrix} 0 \\ 0 \end{bmatrix} \qquad \text{(E1-6f)}$$

The global finite element equations are obtained from (E1-6c) and (E1-6d) as

$$\begin{bmatrix} 2.1667 & -1.9167 & 0 & 0 & 0 \\ -1.9167 & 4.3334 & -1.9167 & 0 & 0 \\ 0 & -1.9167 & 4.3334 & -1.9167 & 0 \\ 0 & 0 & -1.9167 & 4.3334 & -1.9167 \\ 0 & 0 & 0 & -1.9167 & 2.1667 \end{bmatrix} \begin{bmatrix} u_1 \\ u_2 \\ u_3 \\ u_4 \\ u_5 \end{bmatrix} = \begin{bmatrix} 0 \\ 0 \\ 0 \\ 0 \\ 0 \end{bmatrix} \qquad \text{(E1-6g)}$$

It should be noted that solution of the global finite element equations (E1-6g) would be trivial without boundary conditions imposed. Writing out the second, third, and fourth equations of (E1-6g) with the imposed boundary conditions $u_1 = 1$ and $u_5 = 7.389$, we obtain the following equations:

$$u_1 = 1$$

$$-1.9167(1) + 4.3334u_2 - 1.9167u_3 = 0$$

$$-1.9167u_2 + 4.3334u_3 - 1.9167u_4 = 0$$

$$-1.9167u_3 + 4.3334u_4 - 1.9167(7.389) = 0$$

$$u_5 = 7.389$$

Rewriting the above in matrix form yields

$$
\begin{bmatrix}
1 & 0 & 0 & 0 & 0 \\
0 & 4.3334 & -1.9167 & 0 & 0 \\
0 & -1.9167 & 4.3334 & -1.9167 & 0 \\
0 & 0 & -1.9167 & 4.3334 & 0 \\
0 & 0 & 0 & 0 & 1
\end{bmatrix}
\begin{bmatrix}
u_1 \\ u_2 \\ u_3 \\ u_4 \\ u_5
\end{bmatrix}
=
\begin{bmatrix}
1 \\ 1.9167 \\ 0 \\ 14.1625 \\ 7.389
\end{bmatrix}
\qquad \text{(E1-6h)}
$$

From (E1-6h) we obtain $u_1 = 1$, $u_2 = 1.6348$, $u_3 = 2.6961$, $u_4 = 4.4607$, and $u_5 = 7.389$. The exact solution gives $u_1 = 1$, $u_2 = 1.648$, $u_3 = 2.718$, $u_4 = 4.481$, and $u_5 = 7.389$. The average deviation is less than one percent (Fig. 1-10c). In the solution (E1-6h), since u_1 and u_5 are known, the first and fifth equations may be discarded so that

$$
\begin{bmatrix}
4.3334 & -1.9167 & 0 \\
-1.9167 & 4.3334 & -1.9167 \\
0 & -1.9167 & 4.3334
\end{bmatrix}
\begin{bmatrix}
u_2 \\ u_3 \\ u_4
\end{bmatrix}
=
\begin{bmatrix}
1.9167 \\ 0 \\ 14.1625
\end{bmatrix}
\qquad \text{(E1-6i)}
$$

Note that the solution of (E1-6i) does not and should not satisfy the unrestrained equations (E1-6g) which are not modified for the boundary conditions. Other methods of imposing boundary conditions are discussed in detail in Sec. 3-2.

It is interesting to note that if only two elements are used, we have the global equations of the form

$$
\frac{1}{6}
\begin{bmatrix}
8 & -5 & 0 \\
-5 & 16 & -5 \\
0 & -5 & 8
\end{bmatrix}
\begin{bmatrix}
u_1 \\ u_2 \\ u_3
\end{bmatrix}
=
\begin{bmatrix}
0 \\ 0 \\ 0
\end{bmatrix}
$$

with u_2 now corresponding to u_3 in the four-element system. Modification for boundary conditions yields a single equation

$$\tfrac{1}{6}[(-5)(1) + 16u_2 - 5(7.389)] = 0$$

Solving for u_2 gives $u_2 = 2.6215$ instead of 2.6961 obtained for the four element system, which is a further deviation from the exact solution of 2.718. It should be noted that the inferior result is due to not only the coarser mesh but, more importantly, to the assumption of linear interpolation functions which contradicts the exponential function governing the exact solution.

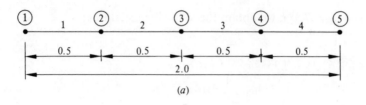

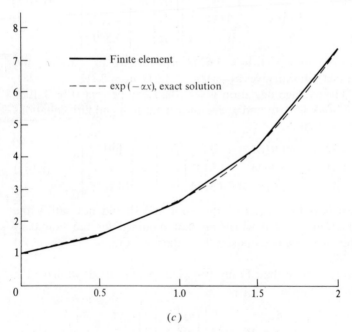

Figure 1-10 Solution of Ex. 1-6. (*a*) One-dimensional domain with four finite elements and five nodes; (*b*) typical element; (*c*) comparison of finite element solution with exact solution.

Example 1-7 Using the Galerkin method, solve the differential equation of the form

$$\frac{d^2u}{dx^2} + u + x = 0 \qquad 0 < x < 1 \qquad \text{(E1-7a)}$$

subject to the boundary conditions

$$u(0) = 0 \qquad \text{(E1-7b)}$$

$$\frac{du(1)}{dx} = 0 \tag{E1-7c}$$

The Galerkin finite element equation takes the form

$$\int_0^1 \left(\frac{d^2u}{dx^2} + u + x\right)\Phi_i \, dx = 0$$

Setting

$$u = \sum_{e=1}^E \Phi_N^{(e)}\Delta_{Ni}^{(e)}u_i = \Phi_i u_i \qquad x = \sum_{e=1}^E \Phi_N^{(e)}\Delta_{Ni}^{(e)}x_i = \Phi_i x_i \tag{E1-7d}$$

and integrating (E1-7d) by parts, we get

$$\sum_{e=1}^E \left[\left(\int_0^h \frac{d\Phi_N^{(e)}}{dx}\frac{d\Phi_M^{(e)}}{dx} \, dx - \int_0^h \Phi_N^{(e)}\Phi_M^{(e)} \, dx\right)u_j\Delta_{Ni}^{(e)}\Delta_{Mj}^{(e)}\right.$$

$$\left. - \left(\int_0^h x\Phi_N \, dx\right)\Delta_{Ni}^{(e)}\right] - \frac{du}{dx}\bigg|_{i(x=1)} = 0 \tag{E1-7e}$$

or

$$B_{ij}u_j = f_i + g_i \tag{E1-7f}$$

where

$$B_{ij} = \sum_{e=1}^E B_{NM}^{(e)}\Delta_{Ni}^{(e)}\Delta_{Mj}^{(e)} \qquad f_i = \sum_{e=1}^E C_{NM}^{(e)}\Delta_{Ni}^{(e)}\Delta_{Mj}^{(e)}x_j \qquad g_i = \frac{du}{dx}\bigg|_{i(x=1)} = 0$$

$$B_{NM}^{(e)} = \int_0^h \left(\frac{d\Phi_N^{(e)}}{dx}\frac{d\Phi_M^{(e)}}{dx} - \Phi_N^{(e)}\Phi_M^{(e)}\right) dx = \frac{1}{h}\begin{bmatrix} 1 & -1 \\ -1 & 1 \end{bmatrix} - \frac{h}{6}\begin{bmatrix} 2 & 1 \\ 1 & 2 \end{bmatrix}$$

$$C_{NM}^{(e)} = \frac{h}{6}\begin{bmatrix} 2 & 1 \\ 1 & 2 \end{bmatrix}$$

The global finite element equation (E1-7e) will now be modified for the essential (Dirichlet) boundary condition as demonstrated in the previous example simply by eliminating the rows and columns corresponding to node 1 for $u(x = 0) = 0$. The natural (Neumann) boundary condition $(du/dx)_{x=1} = 0$ is automatically satisfied by setting $g_i = 0$. Finally, with these boundary conditions, we obtain the linear algebraic equations to solve for the unknowns. For the four-element system, for example, the linear simultaneous equations which satisfy the boundary conditions are

$$\begin{bmatrix} a & b & & \\ b & a & b & \\ & b & a & b \\ & & b & \frac{a}{2} \end{bmatrix}\begin{bmatrix} u_1 \\ u_2 \\ u_3 \\ u_4 \end{bmatrix} = \frac{h}{6}\begin{bmatrix} 1.5 \\ 3.0 \\ 4.5 \\ 2.75 \end{bmatrix}$$

where

$$a = \frac{2(3 - h^2)}{3h} \qquad b = \frac{-(6 + h^2)}{6h} \qquad h = 0.25$$

If the second order interpolation function (1-120) is used, it can be shown that

$$B_{NM}^{(e)} = \frac{1}{3h} \begin{bmatrix} 7 & -8 & 1 \\ -8 & 16 & -8 \\ 1 & -8 & 7 \end{bmatrix} - \frac{h}{30} \begin{bmatrix} 4 & 2 & -1 \\ 2 & 16 & 2 \\ -1 & 2 & 4 \end{bmatrix}$$

$$C_{NM}^{(e)} = \frac{h}{30} \begin{bmatrix} 4 & 2 & -1 \\ 2 & 16 & 2 \\ -1 & 2 & 4 \end{bmatrix}$$

Note that in the assembly procedure, the sum (superposition) of entries of adjacent elements occurs only at the end nodes of each element, not at the midpoint nodes.

$$\begin{bmatrix} a & b & & & & & & & & \\ b & a & b & & & & & & & \\ & b & a & b & & & & & & \\ & & b & a & b & & & & & \\ & & & b & a & b & & & & \\ & & & & b & a & b & & & \\ & & & & & b & a & b & & \\ & & & & & & b & a & b & \\ & & & & & & & b & a & b \\ & & & & & & & & b & \frac{a}{2} \end{bmatrix} \begin{bmatrix} u_2 \\ u_3 \\ u_4 \\ u_5 \\ u_6 \\ u_7 \\ u_8 \\ u_9 \\ u_{10} \\ u_{11} \end{bmatrix} = \begin{bmatrix} 0.6 \\ 1.2 \\ 1.8 \\ 2.4 \\ 3.0 \\ 3.6 \\ 4.2 \\ 4.8 \\ 5.4 \\ 2.9 \end{bmatrix} (0.0167)$$

where $a = 19.9333$, $b = -10.0167$.

Table 1-1 Solutions of Example 1-7

		x 0	$\frac{1}{8}$	$\frac{2}{8}$	$\frac{3}{8}$	$\frac{4}{8}$	$\frac{5}{8}$	$\frac{6}{8}$	$\frac{7}{8}$	1
	4 elem.	0	—	0.2061	—	0.3840	—	0.5072	—	0.5526
Linear	8 elem.	0	0.1055	0.2074	0.3022	0.3865	0.4569	0.5105	0.5444	0.5562
	16 elem.	0	0.1057	0.2078	0.3027	0.3871	0.4577	0.5113	0.5453	0.5571
	2 elem.	0	—	0.2078	—	0.3872	—	0.5114	—	0.5572
Quadratic	4 elem.	0	0.1057	0.2079	0.3029	0.3873	0.4579	0.5116	0.5456	0.5574
	8 elem.	0	0.1057	0.2079	0.3029	0.3873	0.4579	0.5116	0.5456	0.5574
Exact sol. $u = \dfrac{\sin x}{\cos 1} - x$		0	0.1057	0.2079	0.3029	0.3873	0.4579	0.5116	0.5456	0.5574

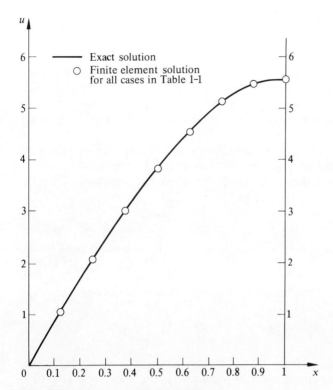

Figure 1-11 Solutions of Ex. 1-7 for all cases.

The results of both linear and quadratic elements for various mesh sizes are summarized in Table 1-1. The convergence of the quadratic elements is conspicuously superior to that of the linear elements, although the results plotted in Fig. 1-11 for all cases do not exhibit the differences for the scale drawn. Had the solution been carried out with more significant digits, it could be shown that the rate of convergence is order of h^{2k} pointwise where k denotes the order of polynomial (Sec. 3-4-4). Thus the error decreases as h^2 and h^4 for the linear and quadratic elements, respectively.

If the exact analytical solution cannot be obtained, then the only means of verification for accuracy is to prove the convergence mathematically and assure that the finite element solutions have been converged with finer mesh sizes. We discuss the problems involved in error estimates and convergence in Sec. 3-4, but some elementary observations into accuracy of the finite element method are presented in the following section.

1-5-5 Accuracy of Finite Element Method

The approximate solution obtained by the finite element method must always be subject to evaluation as to its accuracy and possible errors. There are two primary types of errors: round-off error associated with manipulation of numbers in the computer and discretization error associated with the finite element approximations. The discretization error consists of three factors: (1) order of interpolation functions as related to the physics of the problem under study, (2) size of the element, and (3) shapes or arrangements of the elements. Although these topics are covered in Chap. 3, let us briefly comment on possible errors versus the size of element for the differential equations considered in the previous section.

For the two-element system with three nodes and element sizes given by h, βh, the global finite element equation at node i in terms of nodes $i - 1$ and $i + 1$ for the differential equation (E1-6a) is simply the assembled equation of (E1-6g) corresponding to node i

$$\frac{1}{h}(-u_{i-1} + u_i) + \frac{1}{\beta h}(u_i - u_{i+1}) + \frac{h\alpha^2}{6}(u_{i-1} + 2u_i) + \frac{\beta h\alpha^2}{6}(2u_i + u_{i+1}) = 0$$

$$(1\text{-}139)$$

Note that the standard finite difference for equal mesh sizes may be given by

$$\frac{u_{i+1} - 2u_i + u_{i-1}}{h^2} - \alpha^2 u_i = 0 \qquad (1\text{-}140)$$

If u_i in the second term is replaced by $(u_{i+1} + 4u_i + u_{i-1})/6$ and the mesh sizes for the left and right sides are h and βh, respectively, the expression (1-140) is reduced to (1-139). Writing u_{i-1} and u_{i+1} in the Taylor series about i yields

$$u_{i-1} = u - u'h + \frac{u''}{2!}h^2 - \frac{u'''}{3!}h^3 + \frac{u''''}{4!}h^4 - \cdots$$

$$u_{i+1} = u + u'\beta h + \frac{u''}{2!}(\beta h)^2 + \frac{u'''}{3!}(\beta h)^3 + \frac{u''''}{4!}(\beta h)^4 + \cdots$$

Substituting and setting $u_i = u$, we obtain

$$\frac{d^2 u}{dx^2} - \alpha^2 u + \varepsilon = 0 \qquad (1\text{-}141)$$

with

$$\varepsilon = -\frac{h}{3}(1 - \beta)\frac{d^3 u}{dx^3} + \frac{h^2}{12}\frac{(1 + \beta^3)}{1 + \beta}\frac{d^4 u}{dx^4} + \frac{h\alpha^2}{3}(1 - \beta)\frac{du}{dx}$$

$$- \frac{h^2\alpha^2}{6}\frac{(1 + \beta^3)}{1 + \beta}\frac{d^2 u}{dx^2} + \frac{h^3\alpha^2}{18}\frac{(1 - \beta^4)}{1 + \beta}\frac{d^3 u}{dx^3} - \frac{h^4\alpha^2}{72}\frac{(1 + \beta^5)}{1 + \beta}\frac{d^4 u}{dx^4} + \cdots$$

$$(1\text{-}142)$$

The above differential equation (1-141) is the same as (E1-6a) except for the error terms in ε. We note that the error is 0(h) (order h, the size of element) governed by the lowest order in h in (1-142). If both elements are of the same size ($\beta = 1$), then we have

$$\varepsilon = \frac{h^2}{12}\frac{d^4 u}{dx^4} - \frac{h^2 \alpha^2}{6}\frac{d^2 u}{dx^2} - \frac{h^4 \alpha^2}{72}\frac{d^4 u}{dx^4} + \cdots$$

In this case, the error is $0(h^2)$, giving more rapid rate of convergence. This implies the advantage of using equal size elements in this particular example, although no such general conclusion is valid in two- and three-dimensional elements. In both cases, we are guaranteed of convergence to the exact solution as h tends to zero.

In general, a study of convergence rates via Taylor series expansion of the finite difference operator does not always provide exact conclusions. More rigorous approaches are discussed in Chap. 3.

The question of accuracy in the finite element solution has not been fully explained in reference to the acceptable functions and mesh sizes. The decision as to the size of mesh and the order of approximations warrants a practical judgment. Here a most relevant and important question arises: Do the refined mesh and higher order approximations always improve the accuracy? This question is by no means trivial in two- and three-dimensional problems when a variety of element shapes and various degrees of approximations are considered, particularly in the partial differential equations other than linear elliptic. The mathematical theory of finite elements which delves into this question rightly belongs to a branch of applied mathematics and modern theory of functional analysis. This book is intended for clarification of some of the mathematics of finite elements as well as the study of procedures for obtaining the finite element solutions in various problems in fluid mechanics.

TWO

FINITE ELEMENT INTERPOLATION FUNCTIONS

2-1 GENERAL

It is clear from the discussions in Chap. 1 that the finite element equations are obtained by variational or weighted residual methods. It was shown that the procedures of finite element methods are similar to the Rayleigh–Ritz method or the Galerkin method. However, there are some basic differences in philosophy. In the finite element method, we note that

1. Global functional representations of a variable consist of an assembly of local functional representations. That is, the global integrals over the domain Ω stem from the assembly of integrals over the local domains Ω_e of disjoint finite elements.
2. If the Galerkin method is used, finite element interpolation functions act as weighting functions in the Galerkin integral.

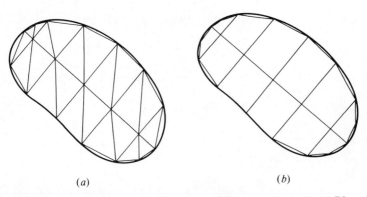

(a)   *(b)*

Figure 2-1 Finite element discretization of two-dimensional domain. (*a*) Discretization by triangular elements; (*b*) discretization by quadrilateral elements.

3. If the Rayleigh–Ritz method is used, the global variational principle is constructed as an assembly of the local variational principle.

The most crucial step in the finite element analysis of a given problem is the choice of adequate interpolation functions. They must meet certain criteria such that convergence to the true solution of the governing differential equation may be attained. Detailed discussions of this subject are given in Chap. 3.

The finite element interpolations are characterized by the shape of the finite element and order of approximations. In general, the choice of a finite element depends on the geometry of the global domain, the degree of accuracy desired in the solution, the ease of integration over the domain, etc.

In Fig. 2-1a, a two-dimensional domain is discretized by a series of triangular elements and quadrilateral elements. It is seen that the global domain consists of many subdomains (the finite elements). The global domain may, of course, be one-, two-, and three-dimensional. The corresponding geometries of finite elements are shown in Fig. 2-2. A one-dimensional element is simply a straight line, a two-

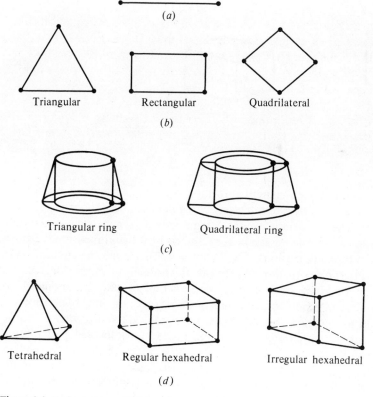

Triangular   Rectangular   Quadrilateral

(b)

Triangular ring   Quadrilateral ring

(c)

Tetrahedral   Regular hexahedral   Irregular hexahedral

(d)

Figure 2-2 Various shapes of finite elements with corner nodes. (a) One-dimensional element; (b) two-dimensional elements; (c) two-dimensional element generated into three-dimensional ring element for axisymmetric geometry; and (d) three-dimensional elements.

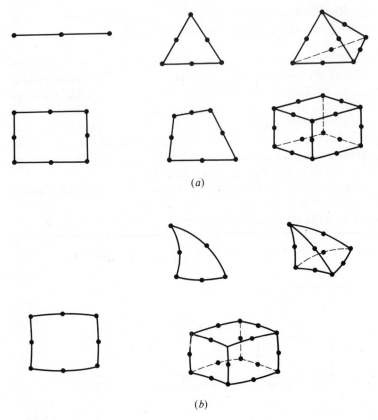

Figure 2-3 Quadratic elements. (*a*) Quadratic elements with straight edges; (*b*) quadratic elements with curved edges.

dimensional element may be triangular, rectangular, or quadrilateral, and a three-dimensional element can be a tetrahedron, a regular hexahedron, or an irregular hexahedron, etc. The three-dimensional domain with axisymmetric geometry and axisymmetric physical behavior can be represented by a two-dimensional element generated into a three-dimensional ring by integration around the circumference. In general, the interpolation functions are the polynomials of various degrees, but often they may be given by products of polynomials with trigonometric or exponential functions. If polynomial expansions are used, the linear variation of a variable within an element can be expressed by the data provided at corner nodes. For quadratic variations, we must add side nodes located preferably at midway between the corner nodes (Fig. 2-3). Cubic variations of the variable or the finite element geometry are represented by two side nodes (Fig. 2-4). Often the sides of the finite element are curvilinear so that additional nodes must be installed along the sides, requiring quadratic or higher degree polynomials. Sometimes a complete expansion of certain degree polynomials may require installation of nodes at various points within the element. However, these interior nodes are

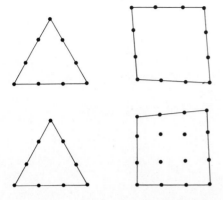

Figure 2-4 Cubic and higher order elements.

undesirable unless absolutely required by a specific physical situation being considered because additional nodes lead to complications in formulation and computation. Types of finite elements may be distinguished by (1) geometries (one-, two-, and three-dimensional); (2) choices of interpolation functions (polynomials, Lagrange or Hermite polynomials); (3) choices of element coordinates (cartesian or natural coordinates); and (4) choices of specified variables at nodes (Lagrange family or Hermite families). These distinctions are discussed below.

2-2 ONE-DIMENSIONAL ELEMENTS

2-2-1 Conventional Elements

The polynomial expansions for a variable u to be approximated in a one-dimensional element may be written as

$$u = a_0 + a_1 x + a_2 x^2 + a_3 x^3 + \cdots \tag{2-1a}$$

or
$$u = a_0 + a_i x^i \tag{2-1b}$$

with $i = 1$ for linear variation, $i = 2$ for quadratic variation, etc., and a_0, a_i are the constants. For a linear variation of u, we need a two-node system with one node at each end (Fig. 2-5). The interpolation functions for this case were derived and used in Sec. 1-5, based on Fig. 2-5a. An alternative method, perhaps the more general approach, is discussed below. Let $\xi = x/l$ with $\xi = 0$ at the center and $\xi = -1$ and $\xi = 1$ at nodes 1 and 2, respectively (Fig. 2-5b). Then (2-1) becomes

$$u = a_0 + a_1 \xi + a_2 \xi^2 + a_3 \xi^3 + \cdots \tag{2-2a}$$

or
$$u = a_0 + a_i \xi^i \tag{2-2b}$$

For a linear element (two-node system), we have

$$u = a_0 + a_1 \xi \tag{2-3}$$

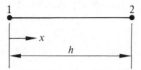

(a) Origin at End Node (Linear Variation)

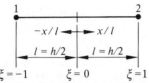

(b) Origin at Center (Linear Variation)

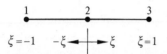

(c) Origin at Center (Quadratic Variation)

Figure 2-5 One-dimensional element. (a) Origin at end node (linear variation); (b) origin at center (linear variation); (c) origin at center (quadratic variation).

As we demonstrated in Sec. 1-5, by writing (2-3) at each node, solving for the constants and substituting them into (2-3), we obtain†

$$u = \Phi_1 u_1 + \Phi_2 u_2 = \Phi_N u_N \tag{2-4}$$

with $N = 1, 2$, and the interpolation functions are

$$\Phi_1 = \tfrac{1}{2}(1 - \xi) \tag{2-5a}$$

$$\Phi_2 = \tfrac{1}{2}(1 + \xi) \tag{2-5b}$$

Likewise, for quadratic approximations in which we require an additional node, preferably at the midside (Fig. 2-5c), we have

$$u = a_0 + a_1 \xi + a_2 \xi^2 \tag{2-6}$$

and writing (2-6) at each node yields

$$u_1 = a_0 - a_1 + a_2 \qquad u_2 = a_0 \qquad u_3 = a_0 + a_1 + a_2$$

Evaluating the constants, we obtain

$$u = \Phi_1 u_1 + \Phi_2 u_2 + \Phi_3 u_3 = \Phi_N u_N \tag{2-7}$$

with $N = 1, 2, 3$ and

$$\Phi_1 = \tfrac{1}{2}\xi(\xi - 1) \qquad \Phi_2 = 1 - \xi^2 \qquad \Phi_3 = \tfrac{1}{2}\xi(\xi + 1) \tag{2-8}$$

† For simplicity, the superscript (e) denoting the local element will be dropped hereafter except when such omission may cause confusion.

It is easily seen that the limits of integration of the interpolation functions as used in Sec. 1-6 should be changed to

$$\int_0^h f(x)\, dx = \int_{-1}^1 g(\xi)\frac{\partial x}{\partial \xi}\, d\xi = l\int_{-1}^1 g(\xi)\, d\xi = \frac{h}{2}\int_{-1}^1 g(\xi)\, d\xi \qquad (2\text{-}9)$$

If the interpolation functions are derived in terms of nondimensionalized spatial variables, then such a normalized system is called a natural coordinate.

2-2-2 Lagrange Polynomial Element

To avoid the inversion of coefficient matrix for higher order approximations, we may use the Lagrange interpolation function L_N of the form

$$\Phi_N = L_N = \prod_{M=1,M\neq N}^n \frac{x - x_M}{x_N - x_M}$$

$$= \frac{(x - x_1)(x - x_2)\cdots(x - x_{N-1})(x - x_{N+1})\cdots(x - x_n)}{(x_N - x_1)(x_N - x_2)\cdots(x_N - x_{N-1})(x_N - x_{N+1})\cdots(x_N - x_n)} \qquad (2\text{-}10)$$

with the symbol Π denoting a product of the binomials of (2-10) over the range of M (see Fig. 2-6). Here the element is divided into equal length segments by the $n = m + 1$ nodes with m and n equal to the order of approximations and the number of nodes in an element, respectively. Let us consider a first order approximation of a variable u such that

$$u = L_N u_N \qquad (N = 1, 2)$$

with
$$L_1 = \frac{x - x_2}{x_1 - x_2} = \frac{x - h}{-h} = 1 - \frac{x}{h}$$

$$L_2 = \frac{x - x_1}{x_2 - x_1} = \frac{x}{h}$$

with $x_1 = 0$ and $x_2 = h$. If the nondimensionalized form $\xi = x/h$ is used, we have

$$L_N = \prod_{M=1,M\neq N}^n \frac{\xi - \xi_M}{\xi_N - \xi_M} \qquad (2\text{-}11)$$

and

$$L_1 = \frac{\xi - \xi_2}{\xi_1 - \xi_2} = 1 - \xi$$

$$L_2 = \frac{\xi - \xi_1}{\xi_2 - \xi_1} = \xi$$

For quadratic approximations, we have $n = m + 1 = 3$ and

$$L_1 = \frac{(\xi - \xi_2)(\xi - \xi_3)}{(\xi_1 - \xi_2)(\xi_1 - \xi_3)} = 2(\xi - \tfrac{1}{2})(\xi - 1)$$

$$L_2 = \frac{(\xi - \xi_1)(\xi - \xi_3)}{(\xi_2 - \xi_1)(\xi_2 - \xi_3)} = -4\xi(\xi - 1)$$

$$L_3 = \frac{(\xi - \xi_1)(\xi - \xi_2)}{(\xi_3 - \xi_1)(\xi_3 - \xi_2)} = 2\xi(\xi - \tfrac{1}{2})$$

If the origin is taken as shown on the right-hand side of Fig. 2-6(c), then we note that

$$L_1 = \tfrac{1}{2}\xi(\xi - 1) \qquad L_2 = 1 - \xi^2 \qquad L_3 = \tfrac{1}{2}\xi(\xi + 1)$$

which are identical to (2-8), the results one would expect to obtain.

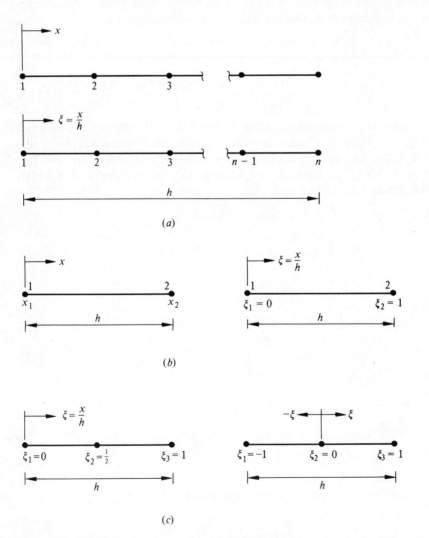

(a)

(b)

(c)

Figure 2-6 Lagrangian element. (a) Lagrangian element coordinates; (b) linear approximation; (c) quadratic approximation.

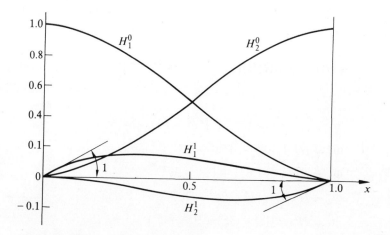

Figure 2-7 Hermite interpolation functions.

2-2-3 Hermite Polynomial Element

If a continuity of the derivative of a variable at common nodes is desired, one efficient way of assuring this continuity is to use the Hermite polynomials. For a one-dimensional element with two end nodes, the Hermite polynomials for a variable u are related by

$$u(\xi) = H_N^0(\xi)u_N + H_N^1(\xi)\left(\frac{\partial u}{\partial \xi}\right)_N \qquad (N = 1, 2) \qquad (2\text{-}12a)$$

or

$$u(\xi) = \Phi_r Q_r \qquad (r = 1, 2, 3, 4) \qquad (2\text{-}12b)$$

where the Hermite polynomials have the properties [see Hildebrand, 1956]

$$H_N^0(\xi_M) = \delta_{NM} \qquad \frac{d}{d\xi} H_N^1(\xi_M) = \delta_{NM}$$

Here $H_N^0(\xi)$ and H_N^1, which are now used as the finite element interpolation functions, are of the cubic polynomials of the form† (see Fig. 2-7)

$$\Phi_1 = H_1^0 = 1 - 3\xi^2 + 2\xi^3 \qquad Q_1 = u_1$$

$$\Phi_2 = H_2^0 = 3\xi^2 - 2\xi^3 \qquad Q_2 = u_2$$

$$\Phi_3 = H_1^1 = \xi - 2\xi^2 + \xi^3 \qquad Q_3 = \left(\frac{\partial u}{\partial \xi}\right)_1 \qquad (2\text{-}13)$$

$$\Phi_4 = H_2^1 = \xi^3 - \xi^2 \qquad Q_4 = \left(\frac{\partial u}{\partial \xi}\right)_2$$

with $\xi = x/h$, h being the length of the element.

† Note that Φ_3, Φ_4 and Φ_5, Φ_6 must be multiplied by h and h^2, respectively, if the nodal values of derivatives are given by $\partial u/\partial x$, $\partial^2 u/\partial x^2$.

If the second as well as the first derivative is to be specified at the end nodes, we require a fifth degree Hermite polynomial such that[†]

$$u(\xi) = H_N^0(\xi)u_N + H_N^1(\xi)\left(\frac{\partial u}{\partial \xi}\right)_N + H_N^2(\xi)\left(\frac{\partial^2 u}{\partial \xi^2}\right)_N \qquad (2\text{-}14)$$

with

$$\Phi_1 = H_1^0 = 1 - 10\xi^3 + 15\xi^4 - 6\xi^5$$

$$\Phi_2 = H_2^0 = 10\xi^3 - 15\xi^4 + 6\xi^5$$

$$\Phi_3 = H_1^1 = \xi - 6\xi^3 + 8\xi^4 - 3\xi^5$$

$$\Phi_4 = H_2^1 = -4\xi^3 + 7\xi^4 - 3\xi^5$$

$$\Phi_5 = H_1^2 = \tfrac{1}{2}(\xi^2 - 3\xi^3 + 3\xi^4 - \xi^5)$$

$$\Phi_6 = H_2^2 = \tfrac{1}{2}(\xi^3 - 2\xi^4 + \xi^5)$$

The generalized coordinates Q_r consists of $Q_1 = u_1$, $Q_2 = u_2$, $Q_3 = (\partial u/\partial \xi)_1$, $Q_4 = (\partial u/\partial \xi)_2$, $Q_5 = (\partial^2 u/\partial \xi^2)_1$, and $Q_6 = (\partial^2 u/\partial \xi^2)_2$.

Additional discussions of Hermite polynomials can be found in Varga [1967] and Birkhoff, et al. [1968].

2-3 TWO-DIMENSIONAL ELEMENTS

Two-dimensional finite elements have been the most widely used because a two-dimensional idealization of domain of study is justified in many engineering and mathematical physics problems. Among the two-dimensional elements, the triangular element was the first investigated in the early days of development, although in recent years the four-sided isoparametric element has become more popular. Various features of these elements are described below.

2-3-1 Triangular Element

As noted in the one-dimensional element, we may use the standard rectangular cartesian coordinates or the natural coordinates (nondimensionalized) to obtain interpolation functions. It will be seen that the choice of a particular coordinate system influences the amount of algebra required in the formulation of finite element equations. For higher order approximations (with higher order polynomials), an evaluation of constants is particularly easy if natural coordinates are used.

Cartesian coordinate triangular element In this element, the properties of the element are determined in terms of the local rectangular cartesian coordinates

---

[†] Note that Φ_3, Φ_4 and Φ_5, Φ_6 must be multiplied by h and h^2, respectively, if the nodal values of derivatives are given by $\partial u/\partial x$, $\partial^2 u/\partial x^2$.

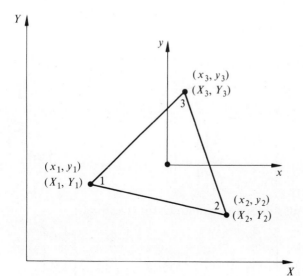

Figure 2-8 Cartesian coordinate triangular element.

(x_i) with their origin at the centroid of the triangle. If this triangle is identified from the global rectangular cartesian coordinates (X_i) with their origin outside the triangle, we note that the following relationships hold (see Fig. 2-8):

$$x_1 + x_2 + x_3 = 0 \quad \text{and} \quad y_1 + y_2 + y_3 = 0$$

or

$$\sum_{N=1}^{3} x_{Ni} = 0 \quad (N = 1, 2, 3, \quad i = 1, 2)$$

with $x_{N1} = x_N$ and $x_{N2} = y_N$. Furthermore, we have

$$x_1 = X_1 - \tfrac{1}{3}(X_1 + X_2 + X_3)$$
$$x_2 = X_2 - \tfrac{1}{3}(X_1 + X_2 + X_3)$$
$$\vdots$$
$$y_3 = Y_3 - \tfrac{1}{3}(Y_1 + Y_2 + Y_3)$$

Or simply, we write

$$x_{Ni} = X_{Ni} - \frac{1}{3} \sum_{N=1}^{3} X_{Ni} \quad (N = 1, 2, 3, \quad i = 1, 2) \tag{2-15}$$

Now consider the polynomial expansion of a variable u in the form

$$u = a_0 + a_1 x + a_2 y$$

This represents a linear variation of u in both x and y directions within the triangular element. To evaluate the three constants a_0, a_1, and a_2, we must provide three equations in terms of the known values of u, x, and y at each of the three nodes,

$$u_1 = a_0 + a_1 x_1 + a_2 y_1$$

$$u_2 = a_0 + a_1 x_2 + a_2 y_2$$

$$u_3 = a_0 + a_1 x_3 + a_2 y_3$$

Writing in a matrix form, we obtain

$$\begin{bmatrix} u_1 \\ u_2 \\ u_3 \end{bmatrix} = \begin{bmatrix} 1 & x_1 & y_1 \\ 1 & x_2 & y_2 \\ 1 & x_3 & y_3 \end{bmatrix} \begin{bmatrix} a_0 \\ a_1 \\ a_2 \end{bmatrix} \tag{2-16}$$

Solving for the constants and substituting them into (2-16) gives

$$u = \begin{bmatrix} 1 & x & y \end{bmatrix} \begin{bmatrix} 1 & x_1 & y_1 \\ 1 & x_2 & y_2 \\ 1 & x_3 & y_3 \end{bmatrix}^{-1} \begin{bmatrix} u_1 \\ u_2 \\ u_3 \end{bmatrix} = \begin{bmatrix} 1 & x & y \end{bmatrix} \frac{1}{|D|} \begin{bmatrix} \bar{a}_1 & \bar{a}_2 & \bar{a}_3 \\ \bar{b}_1 & \bar{b}_2 & \bar{b}_3 \\ \bar{c}_1 & \bar{c}_2 & \bar{c}_3 \end{bmatrix} \begin{bmatrix} u_1 \\ u_2 \\ u_3 \end{bmatrix}$$

$$= (a_1 + b_1 x + c_1 y)u_1 + (a_2 + b_2 x + c_2 y)u_2 + (a_3 + b_3 x + c_3 y)u_3$$

$$= \Phi_1 u_1 + \Phi_2 u_2 + \Phi_3 u_3$$

or

$$u = \Phi_N u_N \qquad (N = 1, 2, 3)$$

where the interpolation function Φ_N is given by

$$\Phi_N = a_N + b_N x + c_N y \tag{2-17}$$

$$a_1 = \frac{1}{|D|}(x_2 y_3 - x_3 y_2) \qquad a_2 = \frac{1}{|D|}(x_3 y_1 - x_1 y_3) \qquad a_3 = \frac{1}{|D|}(x_1 y_2 - x_2 y_1)$$

$$\tag{2-18a}$$

$$b_1 = \frac{1}{|D|}(y_2 - y_3) \qquad b_2 = \frac{1}{|D|}(y_3 - y_1) \qquad b_3 = \frac{1}{|D|}(y_1 - y_2)$$

$$\tag{2-18b}$$

$$c_1 = \frac{1}{|D|}(x_3 - x_2) \qquad c_2 = \frac{1}{|D|}(x_1 - x_3) \qquad c_3 = \frac{1}{|D|}(x_2 - x_1)$$

$$\tag{2-18c}$$

with

$$|D| = \det \begin{bmatrix} 1 & x_1 & y_1 \\ 1 & x_2 & y_2 \\ 1 & x_3 & y_3 \end{bmatrix} = 2A \qquad A = \text{area of triangle}$$

Note that the node numbers 1, 2, 3 are assigned counterclockwise in Fig. 2-8. If assigned clockwise, however, it is seen that the determinant $|D|$ yields $-2A$, twice the negative area. Observe that the fundamental requirements of the interpolation function

$$\sum_{N=1}^{3} \Phi_N = 1 \quad \text{and} \quad 0 \le \Phi_N \le 1$$

are also established in this case. Using the index notations, the foregoing derivations may be written as follows:

$$u(x_i) = a_0 + a_i x_i \qquad (i = 1, 2)$$

$$u_N = a_0 + a_i x_{Ni} \qquad (N = 1, 2, 3)$$

$$u(x_i) = \Phi_N(x_i) u_N$$

$$\Phi_N(x_i) = a_N + b_{Ni} x_i$$

where a_N and b_{Ni} are given by (2-18) with $b_{N1} = b_N$ and $b_{N2} = c_N$. Note that a_N and b_{Ni} may be written in the form

$$a_N = \frac{1}{|D|} \varepsilon_{NPQ} \varepsilon_{ij} x_{Pi} x_{Qj} \qquad b_{Ni} = \frac{1}{|D|} \varepsilon_{NPQ} \varepsilon_{kj} e_{Pk}^{(i)} e_{Qj}^{(i)}$$

with $N, P, Q = 1, 2, 3$ (node numbers); $i, j, k = 1, 2$; $e_{Pk}^{(i)} = x_{Pk}$ if $i = k$; $e_{Pk}^{(i)} = 1$ if $i \ne k$; $x_{P1} = x_P$ and $x_{P2} = y_P$.

In view of (2-15) and (2-18a), we note that

$$a_1 = \frac{1}{2A}(x_2 y_3 - x_3 y_2) = \frac{1}{2A} \left\{ \left(X_2 - \frac{1}{3} \sum_{N=1}^{3} X_N \right) \left(Y_3 - \frac{1}{3} \sum_{N=1}^{3} Y_N \right) \right.$$

$$\left. - \left(X_3 - \frac{1}{3} \sum_{N=1}^{3} X_N \right) \left(Y_2 - \frac{1}{3} \sum_{N=1}^{3} Y_N \right) \right\}$$

$$= \left(\frac{1}{2A} \right) \frac{1}{3} \begin{vmatrix} 1 & X_1 & Y_1 \\ 1 & X_2 & Y_2 \\ 1 & X_3 & Y_3 \end{vmatrix} = \left(\frac{1}{2A} \right) \frac{1}{3} \begin{vmatrix} 1 & x_1 & y_1 \\ 1 & x_2 & y_2 \\ 1 & x_3 & y_3 \end{vmatrix} = \frac{1}{2A} \frac{2A}{3} = \frac{1}{3}$$

Similarly we may prove that $a_2 = a_3 = a_1 = \frac{1}{3}$.

If the variable u is assumed to vary quadratically or cubically, then we require additional nodes along the sides and possibly at the interior. The evaluation of constants would require an inversion of the matrix of the size corresponding to the total number of nodes. An explicit inversion of a large size matrix in terms of nodal coordinate values is difficult but such complications are avoided if natural coordinates are used.

With the interpolation functions constructed for various degrees of approximations, one generally encounters integration over the spatial domain of the form

$$\int \int f(\Phi_N) \, dx \, dy = \int \int f(x, y) \, dx \, dy$$

If the functions $f(x, y)$ are of higher order, the explicit integration becomes extremely cumbersome. Let us consider an integral

$$P_{rs} = \int \int x^r y^s \, dx \, dy$$

The limits of this integral must be calculated from the slope of each side of the triangle oriented from the reference cartesian coordinates. The final form of the integral consists of the sum of the integrals performed along all three sides of the triangle. With the origin of the cartesian coordinates at the centroid (Fig. 2-9a), the following results are obtained:

$$n = r + s$$

$$n = 1 \qquad P_{rs} = \int \int x \, dx \, dy = \int \int y \, dx \, dy = 0$$

$$n = 2 \qquad P_{rs} = \frac{A}{12}(x_1^r y_1^s + x_2^r y_2^s + x_3^r y_3^s)$$

$$n = 3 \qquad P_{rs} = \frac{A}{30}(x_1^r y_1^s + x_2^r y_2^s + x_3^r y_3^s) \qquad\qquad (2\text{-}19)$$

$$n = 4 \qquad P_{rs} = \frac{A}{60}(x_1^r y_1^s + x_2^r y_2^s + x_3^r y_3^s)$$

$$n = 5 \qquad P_{rs} = \frac{2A}{105}(x_1^r y_1^s + x_2^r y_2^s + x_3^r y_3^s)$$

If the $x - y$ coordinates are oriented as shown in Fig. 2-9b, then we have

$$\int \int x^r y^s \, dx \, dy = \frac{c^{s+1} r! s! [a^{r+1} - (-b)^{r+1}]}{(r + s + 2)!} \qquad\qquad (2\text{-}20)$$

Numerical integration for a triangular element by means of a Gaussian quadrature is also possible, as described in Sec. 2-3-3.

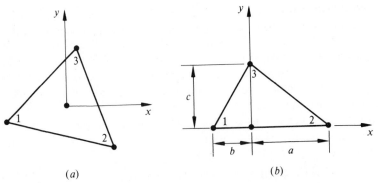

$$(a) \qquad\qquad\qquad\qquad (b)$$

Figure 2-9 Integration over the triangular element using cartesian coordinates. (a) Integration with origin at centroid; (b) integration with x-axis along one side of triangle.

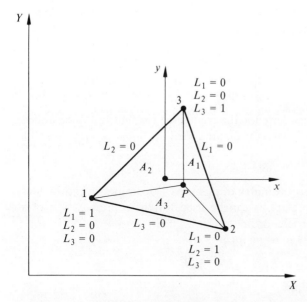

Figure 2-10 Natural coordinate triangular element (linear variation).

Natural coordinate triangular element Consider a triangle with the natural coordinates L_N whose values are zero along the sides and one on the vertices with a linear variation in between as shown in Fig. 2-10. These coordinates are defined as

$$L_i = \frac{A_i}{A}$$

Here $L_1 = A_1/A$, $L_2 = A_2/A$, and $L_3 = A_3/A$ with A_1, A_2, and A_3 being the areas obtained by connecting the three vertices from any point within the triangle such that the total area A is

$$A = A_1 + A_2 + A_3 \tag{2-21}$$

$$1 = L_1 + L_2 + L_3 \tag{2-22}$$

with $A_1 = \text{area } (P23)$, $A_2 = \text{area } (P31)$, and $A_3 = \text{area } (P12)$. It is now possible to establish a relationship between the cartesian coordinates x_i and the natural coordinates L_i in the form

$$x = L_1 x_1 + L_2 x_2 + L_3 x_3 \tag{2-23a}$$

$$y = L_1 y_1 + L_2 y_2 + L_3 y_3 \tag{2-23b}$$

Writing (2-22) and (2-23) in matrix form, we obtain

$$\begin{bmatrix} 1 \\ x \\ y \end{bmatrix} = \begin{bmatrix} 1 & 1 & 1 \\ x_1 & x_2 & x_3 \\ y_1 & y_2 & y_3 \end{bmatrix} \begin{bmatrix} L_1 \\ L_2 \\ L_3 \end{bmatrix} \tag{2-24}$$

We note that the 3×3 matrix on the right-hand side of (2-24) is the transpose of the matrix appearing on the right-hand side of (2-16). Solving for the natural coordinates, we obtain the interesting result

$$L_1 = \Phi_1 \qquad L_2 = \Phi_2 \qquad L_3 = \Phi_3$$

or
$$L_N = \Phi_N \tag{2-25}$$

The natural coordinates as used here for the triangular element are often called the area coordinates or triangular coordinates. Any variable u may now be written as

$$u = L_N u_N$$

Here we may call the natural coordinates of this kind isoparametric coordinates because the same parametric relations can be used for the representation of a variable as well as the geometric coordinates.

It is possible to write (2-23a) in the form

$$x = a_1 L_1 + a_2 L_2 + a_3 L_3 \tag{2-26}$$

Writing for each node, we obtain

$$x_1 = a_1 \qquad x_2 = a_2 \qquad x_3 = a_3$$

Substituting these into (2-26) yields the same expression as (2-23a).

For quadratic approximations, we may write

$$x = a_1 L_1 + a_2 L_2 + a_3 L_3 + a_4 L_1 L_2 + a_5 L_2 L_3 + a_6 L_3 L_1 \tag{2-27}$$

Referring to Fig. 2-11 with three additional nodes installed at midsides of the triangle, we may write (2-27) at each corner and midside node,

$$x_1 = a_1 \qquad x_2 = a_2 \qquad x_3 = a_3$$

$$x_4 = \tfrac{1}{2}a_1 + \tfrac{1}{2}a_2 + \tfrac{1}{4}a_4 \qquad x_5 = \tfrac{1}{2}a_2 + \tfrac{1}{2}a_3 + \tfrac{1}{4}a_5 \qquad x_6 = \tfrac{1}{2}a_1 + \tfrac{1}{2}a_3 + \tfrac{1}{4}a_6$$

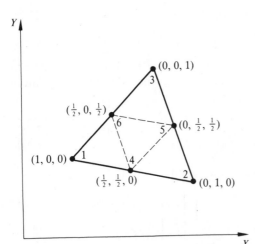

Figure 2-11 Natural coordinate triangular element (quadratic variation).

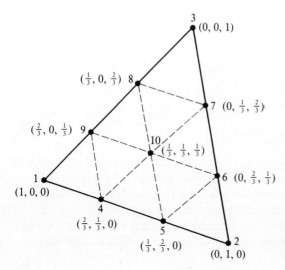

Figure 2-12 Natural coordinate triangular element (cubic variation).

Solving for the constants and substituting them into (2-27) yields

$$x = \Phi_r x_r \qquad (r = 1, 2, \ldots, 6) \tag{2-28}$$

with

$$\Phi_1 = (2L_1 - 1)L_1 \qquad \Phi_2 = (2L_2 - 1)L_2 \qquad \Phi_3 = (2L_3 - 1)L_3$$

$$\Phi_4 = 4L_1 L_2 \qquad \Phi_5 = 4L_2 L_3 \qquad \Phi_6 = 4L_3 L_1 \tag{2-29}$$

Similarly we write

$$y = \Phi_r y_r$$

and consequently, for any variable u

$$u = \Phi_r u_r$$

Using the index notations for a cubic variation, we may proceed as follows (see Fig. 2-12):

$$x = a_N L_N + b_{NM} L_N L_M + c_{NMQ} L_N L_M L_Q \tag{2-30}$$

with N, M, $Q = 1$, 2, 3 and $b_{NM} = 0$ for $N = M$ and $c_{NMQ} = 0$ for $N = M = Q$. Writing (2-30) for the three corner nodes, six side nodes (equally spaced), and the interior node, we evaluate the ten constants. Returning to (2-30) with these constants, we can now write

$$x = \Phi_r x_r \qquad (r = 1, 2, \ldots, 10) \tag{2-31}$$

Here, *for corner nodes*:

$$\Phi_N = \tfrac{1}{2}(3L_N - 1)(3L_N - 2)L_N \qquad (N = 1, 2, 3)$$

for side nodes:

$$\Phi_4 = \tfrac{9}{4}L_1L_2(3L_1 - 1) \qquad \Phi_7 = \tfrac{9}{4}L_2L_3(L_3 - 1)$$

$$\Phi_5 = \tfrac{9}{4}L_1L_2(3L_2 - 1) \qquad \Phi_8 = \tfrac{9}{4}L_3L_1(L_3 - 1)$$

$$\Phi_6 = \tfrac{9}{4}L_2L_3(L_2 - 1) \qquad \Phi_9 = \tfrac{9}{4}L_3L_1(L_1 - 1)$$

for interior node:

$$\Phi_{10} = 27L_1L_2L_3 \qquad\qquad\qquad (2\text{-}32)$$

Although the inversions required for evaluating the polynomial constants in (2-27) and (2-30) are simpler than in corresponding cases of the cartesian co-ordinate triangular element, it is interesting to note that the determination of the interpolation functions for the natural coordinate triangular element can be accomplished quite easily by noting the special geometrical features which make it possible to avoid the inversion altogether. Consider the higher order elements

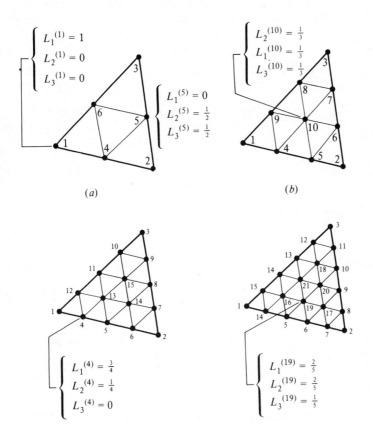

(a)   (b)

Figure 2-13 Higher order natural coordinate elements. (a) Quadratic ($m = 2$); (b) cubic ($m = 3$); (c) quartric ($m = 4$); (d) quintic ($m = 5$).

as depicted in Fig. 2-13. The Lagrangian interpolation formula may be transformed to natural coordinates by

$$B^{(r)}(L_N) = \prod_{s=1}^{s=d} \frac{1}{s}(mL_N - s + 1) \qquad \text{for } d \geq 1$$

$$= 1 \qquad \text{for } d = 0 \qquad (2\text{-}33)$$

with $d = mL_N^{(r)}$. Here m denotes the degree of approximations and $L_N^{(r)}(N = 1, 2, 3, r = 1, 2, \ldots, n, n = $ total number of nodes) represents the values of area coordinates at each node. The interpolation functions are given by

$$\Phi_r = B^{(r)}(L_1)B^{(r)}(L_2)B^{(r)}(L_3) \qquad (2\text{-}34)$$

To determine Φ_1, we write (for $m = 2$)

$$\Phi_1 = B^{(1)}(L_1)B^{(1)}(L_2)B^{(1)}(L_3)$$

$$B^{(1)}(L_1) = (2L_1 - 1 + 1)\tfrac{1}{2}(2L_1 - 2 + 1)$$

$$B^{(1)}(L_2) = 1 \qquad B^{(1)}(L_3) = 1$$

Thus,

$$\Phi_1 = L_1(2L_1 - 1)$$

The interpolation functions corresponding to other nodes may be obtained similarly and we note that the results are identical to those derived from inversion of the coefficient matrices for the polynomial expansions.

The finite element application of the triangular natural coordinates involves integration of a typical form

$$I = \int_A f(L_1, L_2, L_3)\, dA \qquad (2\text{-}35)$$

Referring to Fig. 2-14, the differential area dA is given by

$$dA = \frac{(dh)(dH)}{\sin \alpha} = \frac{(h\, dL_2)(H\, dL_1)}{\sin \alpha} = 2A\, dL_1\, dL_2$$

The limits of the integration for L_1 and L_2 are 0 to 1 and 0 to $1 - L_1$, respectively. Thus,

$$I = 2A \int_0^1 \int_0^{1-L_1} f(L_1, L_2, L_3)\, dL_1\, dL_2 \qquad (2\text{-}36)$$

where the function f may occur in the form

$$f(L_1, L_2, L_3) = L_1^m L_2^n L_3^p \qquad (2\text{-}37)$$

with m, n, p being the arbitrary powers. In view of (2-29) and (2-30), we have

$$I = 2A \int_0^1 \int_0^{1-L_1} L_1^m L_2^n L_3^p\, dL_1\, dL_2$$

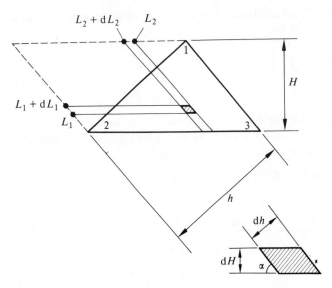

Figure 2-14 Geometry for area coordinate integration.

or
$$I = 2A \int_0^1 J L_1^m \, dL_1 \qquad (2\text{-}38)$$

where
$$J = \int_0^{1-L_1} L_2^n L_3^p \, dL_2 = \int_0^{1-L_1} L_2^n (1 - L_1 - L_2)^p \, dL_2 \qquad (2\text{-}39)$$

Integrating (2-39) by parts gives

$$J = \left[(1 - L_1 - L_2)^p \frac{L_2^{n+1}}{n+1} \right]_0^{1-L_1} + \int_0^{1-L_1} \frac{p}{n+1} L_2^{n+1} (1 - L_1 - L_2)^{p-1} \, dL_2$$

$$= \frac{p}{n+1} \int_0^{1-L_1} L_2^{n+1} (1 - L_1 - L_2)^{p-1} \, dL_2$$

$$= \frac{p(p-1)}{(n+1)(n+2)} \int_0^{1-L_1} L_2^{n+2} (1 - L_1 - L_2)^{p-2} \, dL_2$$

or
$$J = \frac{p! \, n!}{(n+p)!} \int_0^{1-L_1} L_2^{n+p} \, dL_2 = \frac{p! \, n! (1 - L_1)^{n+p+1}}{(n+p+1)!} \qquad (2\text{-}40)$$

Substituting (2-40) into (2-38) and integrating by parts again, we obtain

$$I = \frac{2A n! \, m! \, p!}{(n+m+p+2)!} \qquad (2\text{-}41)$$

For example, if $m = 2$, $n = 0$, and $p = 3$, we obtain

$$\int_A L_1^2 L_3^3 \, dx \, dy = \frac{2A(2!)(0!)(3!)}{(2+0+3+2)!} = \frac{A}{21}$$

2-3-2 Rectangular Element

If the entire domain of study is rectangular, it is more efficient to use rectangular elements rather than triangular elements. Consider a domain with a rectangular mesh. The mesh can also be generated using triangular elements with sides forming diagonals passed through each rectangle. This, of course, results in twice as many elements. That such a system of refined meshes with triangles does not necessarily provide more accurate results is well known. A simple explanation is that the additional node in the rectangular element leads to additional degrees of freedom or constants that may be specified at all nodes of an element, which contributes to more precise or adequate representation of a variable across the element than in the triangular element having an area equal to the rectangular element.

Cartesian coordinate element To construct interpolation functions for a rectangular element, one might be tempted to use a polynomial expansion in terms of the standard cartesian coordinates,

$$u = a_1 + a_2 x + a_3 y + a_4 xy + \cdots \tag{2-42}$$

The necessary terms of polynomials corresponding to the side and interior nodes, as well as the corner nodes as related to the degrees of approximations of a variable, must be chosen wisely. Polynomials are often incomplete for the desired inclusion of side and interior nodes. Furthermore, the inverses of coefficient matrices may not exist in some cases. The natural coordinates, on the other hand, usually provide an efficient means of obtaining acceptable forms of interpolation functions. Lagrange and Hermite polynomials as discussed in the one-dimensional case are also frequently used for the rectangular elements. A special element popularly known as isoparametric element is perhaps most widely adopted. Among the many desirable features of the isoparametric element is the fact that it may be used not only for the rectangular geometry but also for irregular quadrilateral geometries.

Lagrange and Hermite elements The advantage of using the Lagrange or Hermite elements for a rectangular element is that desired interpolation functions are constructed simply by a product of one-dimensional counterparts for the x and y directions, respectively.

Consider the Lagrange interpolations in two dimensions as shown in Fig. 2-15. For a linear variation of u (Fig. 2-15a), we write

$$u = \Phi_N u_N \qquad (N = 1, 2, 3, 4) \tag{2-43a}$$

with

$$\Phi_1 = L_1^{(x)} L_1^{(y)} \qquad \Phi_2 = L_2^{(x)} L_1^{(y)} \qquad \Phi_3 = L_2^{(x)} L_2^{(y)} \quad \text{and} \quad \Phi_4 = L_1^{(x)} L_2^{(y)} \tag{2-43b}$$

where
$$L_1^{(x)} = 1 - \xi, \quad L_2^{(x)} = \xi, \quad L_1^{(y)} = 1 - \eta, \quad L_2^{(y)} = \eta \tag{2-43c}$$

and $\xi = x/2a$ and $\eta = y/2b$. Similarly, for a quadratic variation of u (Fig. 2-15b), we have

$$u = \Phi_N u_N \qquad (N = 1, 2, \ldots, 9) \qquad (2\text{-}44a)$$

with

$$\Phi_1 = L_1^{(x)} L_1^{(y)} \qquad \Phi_2 = L_2^{(x)} L_1^{(y)} \qquad \Phi_3 = L_3^{(x)} L_1^{(y)}, \ldots, \Phi_9 = L_3^{(x)} L_3^{(y)} \qquad (2\text{-}44b)$$

where

$$L_1^{(x)} = \tfrac{1}{2}\xi(\xi - 1) \qquad L_2^{(x)} = 1 - \xi^2 \qquad L_3^{(x)} = \tfrac{1}{2}\xi(\xi + 1)$$
$$L_1^{(y)} = \tfrac{1}{2}\eta(\eta - 1) \qquad L_2^{(y)} = 1 - \eta^2 \qquad L_3^{(y)} = \tfrac{1}{2}\eta(\eta + 1) \qquad (2\text{-}44c)$$

and $\xi = x/a$, and $\eta = y/b$. Interpolations of cubic variations can be constructed in the same way (see Fig. 2-15c).

The Hermite polynomials may be applied similarly to the rectangular element as the Lagrange polynomials. For bicubic Hermite polynomials, we have (Fig. 2-16):

$$u = \Phi_N Q_N \qquad (N = 1, 2, \ldots, 16) \qquad (2\text{-}45a)$$

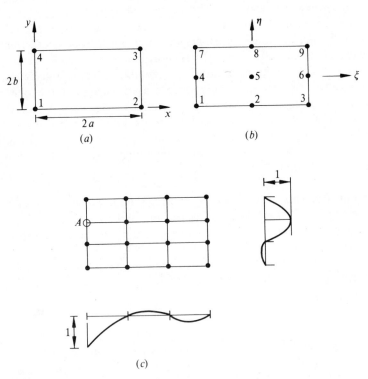

(a)

(b)

(c)

Figure 2-15 Lagrange interpolation functions. (a) Linear; (b) quadratic; (c) cubic (variations of functions along the line through A).

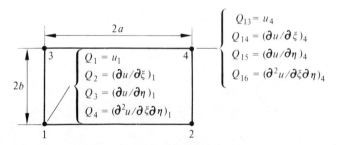

Figure 2-16 Hermite bicubic rectangular element.

with

$$\Phi_1 = H^0_{1(x)}H^0_{1(y)} \qquad \Phi_5 = H^0_{2(x)}H^0_{1(y)} \qquad \Phi_9 = H^0_{1(x)}H^0_{2(y)} \qquad \Phi_{13} = H^0_{2(x)}H^0_{2(y)}$$
$$\Phi_2 = H^1_{1(x)}H^0_{1(y)} \qquad \Phi_6 = H^1_{2(x)}H^0_{1(y)} \qquad \Phi_{10} = H^1_{1(x)}H^0_{2(y)} \qquad \Phi_{14} = H^1_{2(x)}H^0_{2(y)}$$
$$\Phi_3 = H^0_{1(x)}H^1_{1(y)} \qquad \Phi_7 = H^0_{2(x)}H^1_{1(y)} \qquad \Phi_{11} = H^0_{1(x)}H^1_{2(y)} \qquad \Phi_{15} = H^0_{2(x)}H^1_{2(y)}$$
$$\Phi_4 = H^1_{1(x)}H^1_{1(y)} \qquad \Phi_8 = H^1_{2(x)}H^1_{1(y)} \qquad \Phi_{12} = H^1_{1(x)}H^1_{2(y)} \qquad \Phi_{16} = H^1_{2(x)}H^1_{2(y)}$$

$$(2\text{-}45b)$$

and

$$Q_1 = u_1 \qquad Q_5 = u_2 \qquad Q_9 = u_3 \qquad Q_{13} = u_4$$
$$Q_2 = (\partial u/\partial \xi)_1 \qquad Q_6 = (\partial u/\partial \xi)_2 \qquad Q_{10} = (\partial u/\partial \xi)_3 \qquad Q_{14} = (\partial u/\partial \xi)_4$$
$$Q_3 = (\partial u/\partial \eta)_1 \qquad Q_7 = (\partial u/\partial \eta)_2 \qquad Q_{11} = (\partial u/\partial \eta)_3 \qquad Q_{15} = (\partial u/\partial \eta)_4$$
$$Q_4 = (\partial^2 u/\partial \xi\, \partial\eta)_1 \quad Q_8 = (\partial^2 u/\partial \xi\, \partial\eta)_2 \quad Q_{12} = (\partial^2 u/\partial \xi\, \partial\eta)_3 \quad Q_{16} = (\partial^2 u/\partial \xi\, \partial\eta)_4$$

$$(2\text{-}45c)$$

$$H^0_{1(x)} = 1 - 3\xi^2 + 2\xi^3 \qquad H^0_{1(y)} = 1 - 3\eta^2 + 2\eta^3$$
$$H^0_{2(x)} = 3\xi^2 - 2\xi^3 \qquad H^0_{2(y)} = 3\eta^2 - 2\eta^3$$
$$H^1_{1(x)} = \xi - 2\xi^2 + \xi^3 \qquad H^1_{1(y)} = \eta - 2\eta^2 + \eta^3 \qquad (2\text{-}45d)$$
$$H^1_{2(x)} = \xi^3 - \xi^2 \qquad H^1_{2(y)} = \eta^3 - \eta^2$$

Note that because of the combinations of the Hermite polynomials for both x and y directions, the mixed second derivatives must be included as nodal generalized coordinates. The reader may follow a similar procedure for higher order Hermite polynomials.

2-3-3 Isoparametric Element

The isoparametric element was first studied by Zienkiewicz and his associates [see Zienkiewicz, 1971]. The name "isoparametric" derives from the fact that the "same" parametric function which describes the geometry may be used for inter-

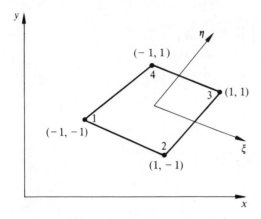

Figure 2-17 Isoparametric element (linear variation).

polating spatial variations of a variable within an element. The isoparametric element utilizes a nondimensionalized coordinate and therefore is one of the natural coordinate elements.

Consider an arbitrarily shaped quadrilateral element as shown in Fig. 2-17. The isoparametric coordinates ξ_i whose values range from 0 to ± 1 are established at the centroid of the element. The reference cartesian coordinates x_i are related to ξ_i as

$$x_i = a_i + a_{ij}\xi_j + a_{ijk}\xi_j\xi_k \tag{2-46}$$

for the two-dimensional linear element in Fig. 2-17. Here, $i, j, k = 1, 2$ and $a_{ijk} = 0$ for $j = k$. A linear variation of a variable u_i may also be written as

$$u_i = a_i + a_{ij}\xi_j + a_{ijk}\xi_j\xi_k \tag{2-47}$$

Writing (2-46) in terms of nodal values yields

$$x_{(N)i} = a_i + a_{ij}\xi_{(N)j} + a_{ijk}\xi_{(N)j}\xi_{(N)k} \tag{2-48a}$$

Here the subscript (N) denotes the node number and is not to be summed. In a matrix form, we may rewrite (2-48a) as

$$\mathbf{x}_i = \mathbf{C}\mathbf{a}_i \tag{2-48b}$$

Here the coefficient matrix $\mathbf{C}$ is given by

$$\mathbf{C} = \begin{bmatrix} 1 & -1 & -1 & 1 \\ 1 & 1 & -1 & -1 \\ 1 & 1 & 1 & 1 \\ 1 & -1 & 1 & -1 \end{bmatrix}$$

Thus,

$$\mathbf{a}_i = \mathbf{C}^{-1}\mathbf{x}_i \tag{2-49}$$

with

$$C^{-1} = \tfrac{1}{4} \begin{bmatrix} 1 & 1 & 1 & 1 \\ -1 & 1 & 1 & -1 \\ -1 & -1 & 1 & 1 \\ 1 & -1 & 1 & -1 \end{bmatrix}$$

Substituting (2-49) into (2-48b) yields

$$x_i = \Phi_N(\xi_i) x_{Ni} \tag{2-50}$$

Here $\Phi_N(\xi_i)$ is called the isoparametric function and has the form

$$\Phi_N(\xi_i) = \tfrac{1}{4}(1 + \xi_{N1}\xi_1)(1 + \xi_{N2}\xi_2) \tag{2-51}$$

Substituting the nodal values of ξ_{N1} and ξ_{N2} into (2-51) gives

$$\begin{aligned} \Phi_1(\xi_i) &= \tfrac{1}{4}(1 - \xi_1)(1 - \xi_2) \\ \Phi_2(\xi_i) &= \tfrac{1}{4}(1 + \xi_1)(1 - \xi_2) \\ \Phi_3(\xi_i) &= \tfrac{1}{4}(1 + \xi_1)(1 + \xi_2) \\ \Phi_4(\xi_i) &= \tfrac{1}{4}(1 - \xi_1)(1 + \xi_2) \end{aligned} \tag{2-52}$$

The quadratic element requires midside nodes as shown in Fig. 2-18 and x_i is given by

$$x_i = a_i + a_{ij}\xi_j + a_{ijk}\xi_j\xi_k + a_{ijkl}\xi_j\xi_k\xi_l \tag{2-53}$$

where $a_{ijkl} = 0$ for $j = k = l$.

A similar procedure as in the linear element may be used to determine **C** and $\mathbf{C}^{-1}$, and we obtain

at corner nodes:

$$\Phi_N(\xi_i) = \tfrac{1}{4}(1 + \xi_{N1}\xi_1)(1 + \xi_{N2}\xi_2)(\xi_{N1}\xi_1 + \xi_{N2}\xi_2 - 1) \tag{2-54a}$$

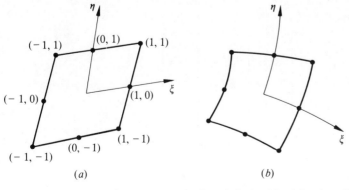

Figure 2-18 Isoparametric element (quadratic variation). (*a*) Straight edges; (*b*) curved edges.

at midside nodes:

$$\Phi_N(\xi_i) = \tfrac{1}{2}(1 - \xi_1^2)(1 + \xi_{N2}\xi_2) \qquad \text{for } \xi_{N1} = 0$$

$$\Phi_N(\xi_i) = \tfrac{1}{2}(1 + \xi_{N1}\xi_1)(1 - \xi_2^2) \qquad \text{for } \xi_{N2} = 0$$

(2-54*b*)

For a cubic element as shown in Fig. 2-19, we have

$$x_i = a_i + a_{ij}\xi_j + a_{ijk}\xi_j\xi_k + a_{ijkl}\xi_j\xi_k\xi_l + a_{ijklm}\xi_j\xi_k\xi_l\xi_m \qquad (2\text{-}55)$$

where $a_{ijklm} = 0$ for $j = k = l = m$. This provides

at corner nodes:

$$\Phi_N(\xi_i) = \tfrac{1}{32}(1 + \xi_{N1}\xi_1)(1 + \xi_{N2}\xi_2)[9(\xi_1^2 + \xi_2^2) - 10] \qquad (2\text{-}56a)$$

at side nodes:

$$\Phi_N(\xi_i) = \tfrac{9}{32}(1 + \xi_{N1}\xi_1)(1 - \xi_2^2)(1 + 9\xi_{N2}\xi_2) \qquad \text{for } \xi_{N1} = \pm 1 \quad \text{and} \quad \xi_{N2} = \pm\tfrac{1}{3}$$

$$\Phi_N(\xi_i) = \tfrac{9}{32}(1 + \xi_{N2}\xi_2)(1 - \xi_1^2)(1 + 9\xi_{N1}\xi_1) \qquad \text{for } \xi_{N2} = \pm 1 \quad \text{and} \quad \xi_{N1} = \pm\tfrac{1}{3}$$

(2-56*b*)

In engineering applications, we are concerned with a derivative and the integration of quantity associated with a variable with respect to the cartesian reference coordinates. Since the variable is represented in terms of the non-dimensionalized isoparametric coordinates, we require a transformation between the two coordinate systems. Consider a quantity given by

$$\iint \frac{\partial}{\partial x} f(\xi, \eta) \, dx \, dy \qquad (2\text{-}57)$$

with $\xi = \xi_1$, $\eta = \xi_2$, $x = x_1$, and $y = x_2$. From the chain rule, we write

$$\frac{\partial f}{\partial \xi} = \frac{\partial f}{\partial x}\frac{\partial x}{\partial \xi} + \frac{\partial f}{\partial y}\frac{\partial y}{\partial \xi}$$

$$\frac{\partial f}{\partial \eta} = \frac{\partial f}{\partial x}\frac{\partial x}{\partial \eta} + \frac{\partial f}{\partial y}\frac{\partial y}{\partial \eta}$$

(2-58)

or in a matrix form

$$\begin{bmatrix} \dfrac{\partial f}{\partial \xi} \\[2mm] \dfrac{\partial f}{\partial \eta} \end{bmatrix} = \begin{bmatrix} \dfrac{\partial x}{\partial \xi} & \dfrac{\partial y}{\partial \xi} \\[2mm] \dfrac{\partial x}{\partial \eta} & \dfrac{\partial y}{\partial \eta} \end{bmatrix} \begin{bmatrix} \dfrac{\partial f}{\partial x} \\[2mm] \dfrac{\partial f}{\partial y} \end{bmatrix}$$

Thus,

$$\begin{bmatrix} \dfrac{\partial f}{\partial x} \\[2mm] \dfrac{\partial f}{\partial y} \end{bmatrix} = \mathbf{J}^{-1} \begin{bmatrix} \dfrac{\partial f}{\partial \xi} \\[2mm] \dfrac{\partial f}{\partial \eta} \end{bmatrix} \qquad (2\text{-}59)$$

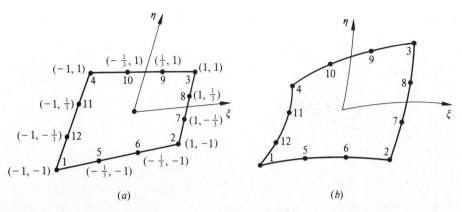

Figure 2-19 Isoparametric element (cubic variation). (*a*) Straight edges; (*b*) curved edges.

where **J** is called the Jacobian given by

$$\mathbf{J} = \begin{bmatrix} \dfrac{\partial x}{\partial \xi} & \dfrac{\partial y}{\partial \xi} \\[2mm] \dfrac{\partial x}{\partial \eta} & \dfrac{\partial y}{\partial \eta} \end{bmatrix} \tag{2-60}$$

Here the derivatives $\partial f / \partial x$ or $\partial f / \partial y$ are determined from the inverse of the Jacobian and the derivatives $\partial f / \partial \xi$ and $\partial f / \partial \eta$. The integration over the domain referenced to the cartesian coordinates must be changed to the domain now referenced to the isoparametric coordinates

$$\iint dx\, dy = \int_{-1}^{1} \int_{-1}^{1} |\mathbf{J}|\, d\xi\, d\eta \tag{2-61}$$

To prove (2-61), we consider the two coordinate systems as shown in Fig. 2-20. The directions of the cartesian coordinates and the arbitrary nonorthogonal (possibly curvilinear) isoparametric coordinates are given by the unit vectors $\mathbf{i}_1$, $\mathbf{i}_2$ and the tangent vectors $\mathbf{g}_1$, $\mathbf{g}_2$, respectively, related by

$$\mathbf{g}_1 = \frac{\partial x}{\partial \xi} \mathbf{i}_1 + \frac{\partial y}{\partial \xi} \mathbf{i}_2$$

$$\mathbf{g}_2 = \frac{\partial x}{\partial \eta} \mathbf{i}_1 + \frac{\partial y}{\partial \eta} \mathbf{i}_2$$

The differential area (shaded) is

$$d x \mathbf{i}_1 \times d y \mathbf{i}_2 = dx\, dy \mathbf{i}_3 = \mathbf{g}_1\, d\xi \times \mathbf{g}_2\, d\eta = \begin{vmatrix} \mathbf{i}_1 & \mathbf{i}_2 & \mathbf{i}_3 \\[1mm] \dfrac{\partial x}{\partial \xi} & \dfrac{\partial y}{\partial \xi} & 0 \\[2mm] \dfrac{\partial x}{\partial \eta} & \dfrac{\partial y}{\partial \eta} & 0 \end{vmatrix} d\xi\, d\eta$$

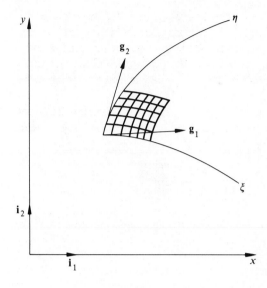

Figure 2-20 Coordinate transformation.

or
$$\mathrm{d}x\ \mathrm{d}y\mathbf{i}_3 = |\mathbf{J}|\ \mathrm{d}\xi\ \mathrm{d}\eta\mathbf{i}_3$$

with

$$|\mathbf{J}| = \begin{vmatrix} \dfrac{\partial x}{\partial \xi} & \dfrac{\partial y}{\partial \xi} \\[2mm] \dfrac{\partial x}{\partial \eta} & \dfrac{\partial y}{\partial \eta} \end{vmatrix}$$

Thus, we obtain the relations

$$\mathrm{d}x\ \mathrm{d}y = |J|\ \mathrm{d}\xi\ \mathrm{d}\eta \tag{2-62}$$

and

$$\iint \frac{\partial f}{\partial x}\,\mathrm{d}x\,\mathrm{d}y = \int_{-1}^{1}\int_{-1}^{1}\left(\overline{J}_{11}\frac{\partial f}{\partial \xi} + \overline{J}_{12}\frac{\partial f}{\partial \eta}\right)|\mathbf{J}|\ \mathrm{d}\xi\ \mathrm{d}\eta \tag{2-63}$$

where $\overline{J}_{11}$ and $\overline{J}_{12}$ are the components of the first row of the inverted Jacobian matrix (2-60).

The integration (2-63) may be performed most efficiently by means of the Gaussian quadrature [see Hildebrand, 1956]. For a one-dimensional case, we may write

$$\int_{-1}^{1} f(\xi)\,\mathrm{d}\xi = \sum_{k=1}^{n} w_k f(\xi_k) \tag{2-64}$$

or when extended to a two-dimensional case, we write

$$\int_{-1}^{1}\int_{-1}^{1} f(\xi,\eta)\,\mathrm{d}\xi\,\mathrm{d}\eta = \int_{-1}^{1}\sum_{k=1}^{n} w_k f(\xi_k,\eta)\,\mathrm{d}\eta = \sum_{j=1}^{n}\sum_{k=1}^{n} w_j w_k f(\xi_j,\eta_k) \tag{2-65}$$

where w_j and w_k are the weight coefficients and $f(\xi_k)$ and $f(\xi_j, \eta_k)$ denote the values of the functions $f(\xi_k)$ and $f(\xi_j, \eta_k)$ corresponding to the n Gaussian points. The above formulas are approximate numerical integrations based on Legendre polynomials. The weight coefficients and abscissae for the first ten Gaussian points are shown in Appendix A. The number of Gaussian points required for acceptable accuracy depends on the order of polynomials being integrated.

The Gaussian quadrature numerical integration may be easily extended to the three-dimensional element. Extension to the triangular or tetrahedral elements are also possible with some modification of the procedure.

2-4 THREE-DIMENSIONAL ELEMENTS

Three-dimensional elements are required when one- or two-dimensional idealization is not possible. Basic ingredients for three-dimensional elements have already been presented in earlier sections and no special conceptual developments are required. The three-dimensional elements may be constructed quite easily by direct extension of the ideas used for two-dimensional elements.

2-4-1 Tetrahedral Element

Consider the tetrahedral elements as shown in Fig. 2-21. For linear variation of a variable (Fig. 2-21a), we write

$$u = a_0 + a_1 x + a_2 y + a_3 z \tag{2-66}$$

It is a simple matter to write the above equation at each node, which yields a total of four equations. Evaluating the constants from these equations, we obtain

$$u = \Phi_N u_N \qquad (N = 1, 2, 3, 4) \tag{2-67}$$

where

$$\Phi_N = a_N + b_{Ni} x_i \qquad (i = 1, 2, 3) \tag{2-68a}$$

$$a_N = \frac{1}{|D|} \varepsilon_{NPQR} \varepsilon_{ijk} x_{Pi} x_{Qj} x_{Rk} \tag{2-68b}$$

$$b_{Ni} = \frac{1}{|D|} \varepsilon_{NPQR} \varepsilon_{ljk} e_{Pl}^{(i)} e_{Qj}^{(i)} e_{Rk}^{(i)} \tag{2-68c}$$

with $N, P, Q, R = 1, 2, 3, 4$ (node numbers); $i, j, k, l = 1, 2, 3$; $e_{Pl}^{(i)} = x_{Pl}$ if $i = l$ and $e_{Pl}^{(i)} = 1$ if $i \neq l$; $x_{P1} = x_P$, $x_{P2} = y_P$, $x_{P3} = z_P$;

$$|\mathbf{D}| = \begin{vmatrix} 1 & x_{11} & x_{12} & x_{13} \\ 1 & x_{21} & x_{22} & x_{23} \\ 1 & x_{31} & x_{32} & x_{33} \\ 1 & x_{41} & x_{42} & x_{43} \end{vmatrix} = \begin{vmatrix} 1 & x_1 & y_1 & z_1 \\ 1 & x_2 & y_2 & z_2 \\ 1 & x_3 & y_3 & z_3 \\ 1 & x_4 & y_4 & z_4 \end{vmatrix} = 6V \tag{2-68d}$$

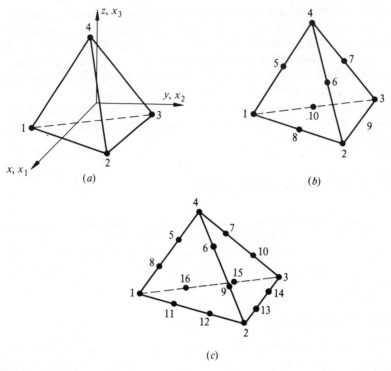

Figure 2-21 Tetrahedral element (cartesian coordinate). (*a*) Linear variation; (*b*) quadratic variation; (*c*) cubic variation.

where V is the volume of the tetrahedron. The operations involved in (2-68) lead to

$$\Phi_N = a_N + b_N x + c_N y + d_N z \qquad (2\text{-}69)$$

with $b_{N1} = b_N$, $b_{N2} = c_N$, $b_{N3} = d_N$, and

$$a_1 = \begin{bmatrix} x_2 & y_2 & z_2 \\ x_3 & y_3 & z_3 \\ x_4 & y_4 & z_4 \end{bmatrix} \frac{1}{6V} \qquad b_1 = -\begin{bmatrix} 1 & y_2 & z_2 \\ 1 & y_3 & z_3 \\ 1 & y_4 & z_4 \end{bmatrix} \frac{1}{6V}$$

$$c_1 = \begin{bmatrix} x_2 & 1 & z_2 \\ x_3 & 1 & z_3 \\ x_4 & 1 & z_4 \end{bmatrix} \frac{1}{6V} \qquad d_1 = -\begin{bmatrix} x_2 & y_2 & 1 \\ x_3 & y_3 & 1 \\ x_4 & y_4 & 1 \end{bmatrix} \frac{1}{6V} \qquad (2\text{-}70)$$

$$a_2 = \begin{bmatrix} x_3 & y_3 & z_3 \\ x_4 & y_4 & z_4 \\ x_1 & y_1 & z_1 \end{bmatrix} \frac{1}{6V}, \text{ etc.}$$

For higher order approximations, the coefficient matrix becomes very large in size and a resort to natural coordinates is inevitable. The most suitable choice is the volume coordinate system extended from the area coordinates for a two-dimensional triangular element.

If the three-dimensional natural coordinates (tetrahedral or volume co-ordinates) are used, a node having the coordinate of one decreases to zero as it moves to the opposite triangular surface formed by the rest of the nodes (Fig. 2-22). For the linear element (Fig. 2-22a), the interpolation functions are

$$\Phi_N = L_N \qquad (N = 1, 2, 3, 4) \tag{2-71}$$

For higher order interpolations (Fig. 2-22b,c), we invoke the formula similar to (2-34),

$$\Phi_r = B^{(r)}(L_1)B^{(r)}(L_2)B^{(r)}(L_3)B^{(r)}(L_4) \tag{2-72}$$

where $B^{(r)}(L_N)$ is given by (2-33). This provides the following results:

For quadratic variation (Fig. 2-22b)
at corner nodes:

$$\Phi_N = (2L_N - 1)L_N$$

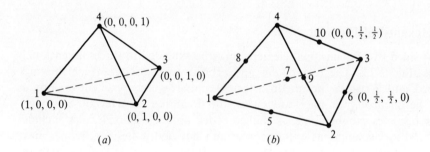

(a)   (b)

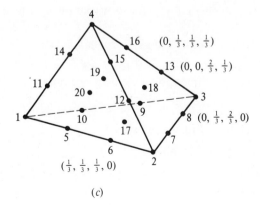

(c)

Figure 2-22 Tetrahedral element (natural, volume, or tetrahedral coordinates). (a) Linear variation; (b) quadratic variation; (c) cubic variation.

at midside nodes:

$$\Phi_6 = 4L_2L_3, \quad \Phi_{10} = 4L_2L_4, \text{ etc.}$$

For cubic variation (Fig. 2-22c)
at corner nodes:

$$\Phi_N = \tfrac{1}{2}(3L_N - 1)(3L_N - 2)L_N$$

at side nodes:

$$\Phi_8 = \tfrac{9}{2}L_3L_2(3L_3 - 1), \quad \Phi_{13} = \tfrac{9}{2}L_3L_4(L_3 - 1), \text{ etc.}$$

at midside nodes:

$$\Phi_{17} = 27L_1L_2L_3, \quad \Phi_{18} = 27L_2L_3L_4, \text{ etc.}$$

The spatial integration of the tetrahedral coordinates may be derived similarly as in the triangular coordinates. This results in

$$I = \int_v L_1^m L_2^n L_3^p L_4^q \, dv$$

or

$$I = \frac{6Vm!n!p!q!}{(m + n + p + q + 3)!} \tag{2-73}$$

2-4-2 Hexahedral Element

The four-sided two-dimensional elements may be extended to three-dimensional elements (Fig. 2-23). The rectangular and arbitrary quadrilateral elements are developed into a regular hexahedron (brick) and irregular hexahedron. For a regular hexahedron, we may use either the Lagrange or Hermite element, but this becomes cumbersome as higher order approximations must include interior and surface nodes as well as corner and side nodes. Besides, neither may be applicable for irregular hexahedrons. An element which is free from these disadvantages is the isoparametric element.

In the isoparametric element for a linear variation of the geometry and variable, we write

$$x_i = a_i + a_{ij}\xi_j + a_{ijk}\xi_j\xi_k + a_{ijkl}\xi_j\xi_k\xi_l \tag{2-74}$$

where $i, j, k, l = 1, 2, 3$ and $a_{ijk} = a_{ijkl} = 0$ for $j = k$, $k = l$, or $l = j$. Using the same procedure as in the two-dimensional element, we obtain

$$\Phi_N = \tfrac{1}{8}(1 + \xi_{N1}\xi_1)(1 + \xi_{N2}\xi_2)(1 + \xi_{N3}\xi_3) \tag{2-75}$$

For a quadratic variation (Fig. 2-23b), we have

$$x_i = a_i + a_{ij}\xi_j + a_{ijk}\xi_j\xi_k + a_{ijkl}\xi_j\xi_k\xi_l + a_{ijklm}\xi_j\xi_k\xi_l\xi_m \tag{2-76}$$

where $i, j, k, l, m = 1, 2, 3$ and $a_{ijkl} = a_{ijklm} = 0$ for $j = k$, $k = l$, $l = m$, $m = k$, $l = j$. The interpolation functions are

at corner nodes:

$$\Phi_N = \tfrac{1}{8}(1 + \xi_{N1}\xi_1)(1 + \xi_{N2}\xi_2)(1 + \xi_{N3}\xi_3)(\xi_{N1}\xi_1 + \xi_{N2}\xi_2 + \xi_{N3}\xi_3 - 2) \tag{2-77a}$$

at midside nodes:

$$\Phi_N = \tfrac{1}{4}(1 - \xi_1^2)(1 + \xi_{N2}\xi_2)(1 + \xi_{N3}\xi_3)$$

for
$$\xi_{N1} = 0, \quad \xi_{N2} = \pm 1, \quad \xi_{N3} = \pm 1; \text{ etc.}$$

(2-77*b*)

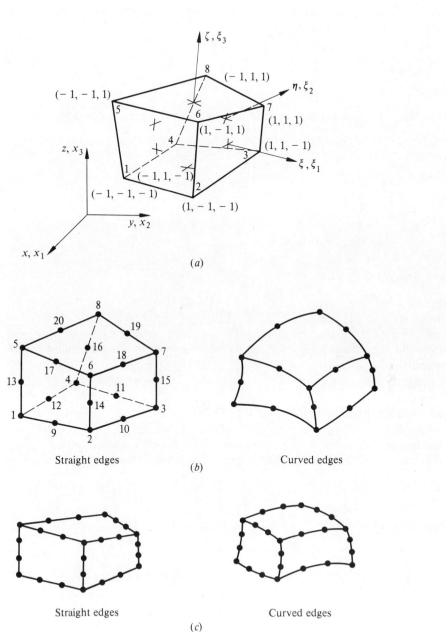

Figure 2-23 Hexahedral isoparametric elements. (*a*) Linear variation (8 nodes); (*b*) quadratic variation (20 nodes); (*c*) cubic variation (32 nodes).

We now require integration of the form

$$\iiint \frac{\partial}{\partial x} f(\xi, \eta, \zeta) \, dx \, dy \, dz \qquad (2\text{-}78)$$

with $\xi = \xi_1$, $\eta = \xi_2$, $\zeta = \xi_3$, $x = x_1$, $y = x_2$, and $z = x_3$. Proceeding similarly as in the two-dimensional case, we obtain

$$\iiint \frac{\partial f}{\partial x} \, dx \, dy \, dz = \int_{-1}^{1} \int_{-1}^{1} \int_{-1}^{1} \left(\bar{J}_{11} \frac{\partial f}{\partial \xi} + \bar{J}_{12} \frac{\partial f}{\partial \eta} + \bar{J}_{13} \frac{\partial f}{\partial \zeta} \right) |\mathbf{J}| \, d\xi \, d\eta \, d\zeta$$
$$(2\text{-}79)$$

where $\bar{J}_{11}$, $\bar{J}_{12}$, and $\bar{J}_{13}$ are the first row of the 3×3 inverted Jacobian matrix

$$\mathbf{J} = \begin{bmatrix} \dfrac{\partial x}{\partial \xi} & \dfrac{\partial y}{\partial \xi} & \dfrac{\partial z}{\partial \xi} \\[2ex] \dfrac{\partial x}{\partial \eta} & \dfrac{\partial y}{\partial \eta} & \dfrac{\partial z}{\partial \eta} \\[2ex] \dfrac{\partial x}{\partial \zeta} & \dfrac{\partial y}{\partial \zeta} & \dfrac{\partial z}{\partial \zeta} \end{bmatrix} \qquad (2\text{-}80)$$

and

$$\int_{-1}^{1} \int_{-1}^{1} \int_{-1}^{1} f(\xi, \eta, \zeta) \, d\xi \, d\eta \, d\zeta = \sum_{i=1}^{n} \sum_{j=1}^{n} \sum_{k=1}^{n} w_i w_j w_k f(\xi_i, \eta_j, \zeta_k) \qquad (2\text{-}81)$$

The weight coefficients w_i, w_j, w_k, and the abscissae $f(\xi_i, \eta_j, \zeta_k)$ are obtained from Appendix A.

2-5 AXISYMMETRIC RING ELEMENT

If the three-dimensional domain of study is axisymmetric, then any two-dimensional element may be used with the spatial integral replaced by

$$\iiint f(x, y, z) \, dx \, dy \, dz = \int_0^{2\pi} \iint f(r, z) r \, d\theta \, dr \, dz \qquad (2\text{-}82)$$

where $dx = dr$, $dy = r \, d\theta$, and $dz = dz$ (see Fig. 2-24). For isoparametric elements, we have

$$\int_0^{2\pi} \int_{-1}^{1} \int_{-1}^{1} g(\xi, \eta) r \, d\theta \, |\mathbf{J}| \, d\xi \, d\eta = 2\pi \int_{-1}^{1} \int_{-1}^{1} g(\xi, \eta) r(\xi, \eta) |\mathbf{J}| \, d\xi \, d\eta \qquad (2\text{-}83)$$

This represents a three-dimensional ring element generated by a two-dimensional element.

Note that the applications arise in the flow fields of missiles and rockets for

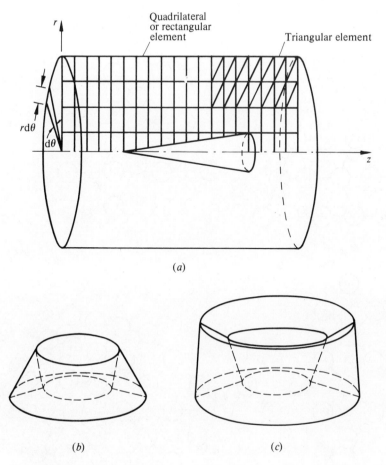

Figure 2-24 Axisymmetric ring elements. (*a*) Discretized geometry of a cylinder; (*b*) triangular ring; (*c*) quadrilateral ring.

the zero angle of attack. For a nonzero angle of attack, the flow fields become asymmetric. In this case, the axisymmetric ring element can no longer be used and three-dimensional elements must be invoked instead.

2-6 LAGRANGE AND HERMITE FAMILIES

All finite elements, regardless of their geometrical shapes, may be grouped into two categories: Lagrange and Hermite families. The Lagrange family consists of finite elements in which the values of a variable are specified at nodes whereas the Hermite family includes derivatives of the variable as well as its values defined at nodes.

Both Lagrange and Hermite families may be represented by the polynomials

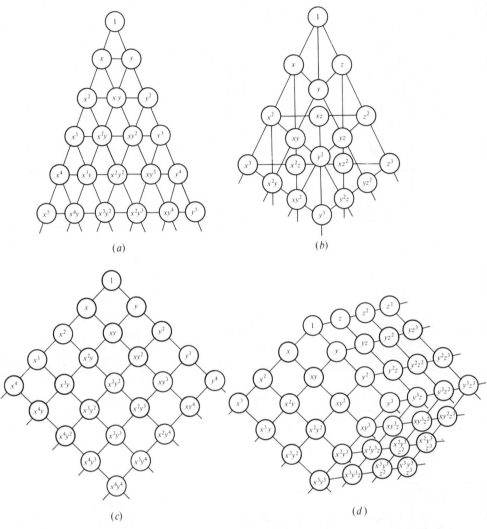

Figure 2-25 Polynomial expansions in finite elements. (*a*) Pascal triangle simplex; (*b*) Pascal tetrahedron simplex; (*c*) two-dimensional hypercube; (*d*) three-dimensional hypercube.

derived from the Pascal triangle simplex (Fig. 2-25*a*) and Pascal tetrahedron simplex (Fig. 2-25*b*), or from two-dimensional hypercube (Fig. 2-25*c*) and three-dimensional hypercube (Fig. 2-25*d*).

In the Lagrange family, the polynomial terms contained in the circles represent the corresponding number of nodes required. However, in general, interior nodes lead to cumbersome bookkeeping and subsequent removal of some of the polynomial terms, resulting in an incomplete polynomial.

In the case of a Hermite family, the number of nodes and polynomial terms required increases since derivatives in addition to the variable itself are to be

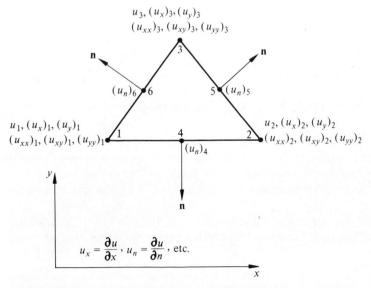

Figure 2-26 Hermite triangle, 21 constants to be determined.

specified. However, a reasonable compromise can be met by eliminating the values of a variable and specifying only the normal derivatives at side nodes. Let us consider 21 terms of quintic polynomial for a triangular element given by

$$u = a_1$$
$$+ a_2 x + a_3 y$$
$$+ a_4 x^2 + a_5 xy + a_6 y^2$$
$$+ a_7 x^3 + a_8 x^2 y + a_9 xy^2 + a_{10} y^3$$
$$+ a_{11} x^4 + a_{12} x^3 y + a_{13} x^2 y^2 + a_{14} xy^3 + a_{15} y^4$$
$$+ a_{16} x^5 + a_{17} x^4 y + a_{18} x^3 y^2 + a_{19} x^2 y^3 + a_{20} xy^4 + a_{21} y^5$$

(2-84)

The nodal values of u and derivatives of u are shown in Fig. 2-26. The polynomials for the Hermite hypercube elements can be generated similarly. Let the degree of complete polynomial be m. Then the number of corresponding polynomial terms is $(k + 1)^2$ and the derivatives to be specified are those correspond-

Table 2-1

k	$(k + 1)^2$	$m = k - 1$	Derivatives to be specified	C^m
1	4	0	None	C^0
2	9	1	$u_x u_{xy} u_y$	C^1
3	16	2	$u_x u_{xy} u_y$ $u_{xx} u_{xxy} u_{xxyy} u_{xyy} u_{yy}$	C^2

ing to the complete polynomial of degree $m = k - 1$. The finite element models thus generated are referred to as C^m-conforming element. For example, let $k = 1$, $k = 2$, and $k = 3$. As shown in Table 2-1, the C^0-element requires no derivatives, identical to the Lagrange family discussed earlier.

2-7 GLOBAL INTERPOLATION FUNCTIONS

Our discussions here so far have been concerned only with local interpolation functions. As demonstrated in Sec. 1-5-2 for one-dimensional problems, the local interpolation functions can be mapped into global forms. The global interpolation functions for two- and three-dimensions are generated the same way as in one-dimension.

First consider one-dimensional domain ($0 \le x \le 1$) again but with quadratic functions of (2-8). The global and local functions u are as given by (1-115) through (1-120),

$$u = \Phi_i u_i \tag{2-85}$$

$$u_N^{(e)} = \Phi_N^{(e)} u_N^{(e)} \tag{2-86}$$

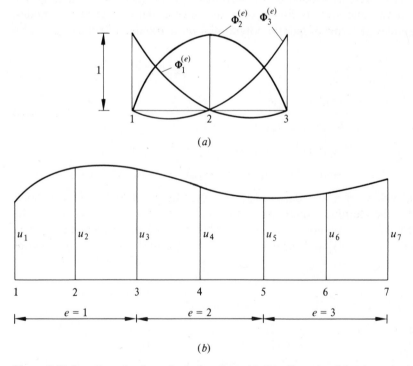

Figure 2-27 One-dimensional quadratic functions. (a) One-dimensional local quadratic functions $\Phi_N^{(e)}$; (b) one-dimensional global quadratic functions Φ_i.

$$\Phi_i = \sum_{e=1}^{E} \Phi_N^{(e)} \Delta_{Ni}^{(e)} \tag{2-87}$$

$$\Phi_1^{(e)} = \tfrac{1}{2}\xi(\xi - 1) \tag{2-88a}$$

$$\Phi_2^{(e)} = 1 - \xi^2 \tag{2-88b}$$

$$\Phi_3^{(e)} = \tfrac{1}{2}\xi(\xi + 1) \tag{2-88c}$$

with the local nodes $z_N^{(e)}$ and global nodes Z_i related by the Boolean matrix

$$z_N^{(e)} = \Delta_{Ni}^{(e)} Z_i \tag{2-89}$$

The local interpolation functions $\Phi_N^{(e)}$ are plotted in Fig. 2-27a. If the global domain consists of three local elements ($E = 3$, $N = 7$), then the global functions u_i are determined from (2-85) and (2-87) as

$$u = \Phi_1^{(1)}u_1 + \Phi_2^{(1)}u_2 + (\Phi_3^{(1)} + \Phi_1^{(2)})u_3 + \Phi_2^{(2)}u_4 + (\Phi_3^{(2)} + \Phi_1^{(3)})u_5$$

$$+ \Phi_2^{(3)}u_6 + \Phi_3^{(3)}u_7$$

with $\qquad \Phi_1 = \Phi_1^{(1)}, \quad \Phi_2 = \Phi_2^{(2)}, \quad \Phi_3 = \Phi_3^{(1)} + \Phi_1^{(2)}, \dots, \text{etc.}$

Note that $\Phi_N^{(e)}$ is zero if the point x does not lie in the element e. For example, let us seek u at $x = \tfrac{5}{12}$. This point lies in element 2 ($e = 2$) and halfway between nodes 3 and 4 . Thus,

$$u = (\Phi_1^{(2)}u_3 + \Phi_2^{(2)}u_4 + \Phi_3^{(2)}u_5)_{\xi = -1/2}$$

$$= 0.375u_3 + 0.75u_4 - 0.125u_5$$

At node $3(x = \tfrac{1}{3})$, $\Phi_3^{(1)}$ or $\Phi_1^{(2)}$ assumes a value of 1 at the disjoined local nodes. But from the property $\Phi_i(Z_j) = \delta_{ij}$ as given by (1-120b), we have $\Phi_3(Z_3) = 1$.

Let us now consider a two-dimensional domain Ω and its boundary Γ (Fig. 2-28a). Let this domain consist of two triangular elements (Fig. 2-28b). Let a sub-domain or a typical element e be denoted as Ω_e with its boundary Γ_e. The

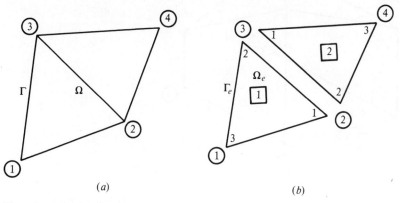

Figure 2-28 Global and local nodes. (a) Global nodes; (b) local nodes.

local and global finite element models are, respectively,

$$\overline{\Omega}_e = \Omega_e \cup \Gamma_e$$

$$\overline{\Omega} = \Omega \cup \Gamma = \bigcup_{e=1}^{e=2} \overline{\Omega}_e$$

The relation between the local and the global nodes is still given by (2-89). Here the Boolean matrices $\Delta_{Ni}^{(e)}$ take the forms

$$\Delta_{Ni}^{(1)} = \begin{bmatrix} 0 & 1 & 0 & 0 \\ 0 & 0 & 1 & 0 \\ 1 & 0 & 0 & 0 \end{bmatrix}$$

$$\Delta_{Ni}^{(2)} = \begin{bmatrix} 0 & 0 & 1 & 0 \\ 0 & 1 & 0 & 0 \\ 0 & 0 & 0 & 1 \end{bmatrix}$$

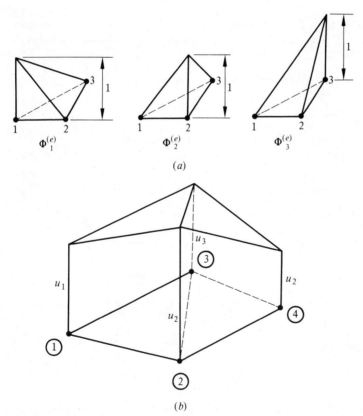

(a)

(b)

Figure 2-29 Two-dimensional linear functions (triangular elements). (a) Local linear triangular element interpolation functions $\Phi_N^{(e)}$; (b) two-dimensional global linear functions Φ_i—two-element system (Fig. 2-28a).

All other properties governing the local element and global interpolation functions are the same as those discussed in Sec. 1-5-1. Thus, a function $u(x, y)$ is written as

$$u(x, y) = \sum_{e=1}^{E} u^{(e)}(x, y) = \Phi_i(x, y)u_i \qquad (2\text{-}90)$$

with

$$u^{(e)}(x, y) = \Phi_N^{(e)}(x, y)u_N^{(e)} \qquad (2\text{-}91)$$

$$u_N^{(e)} = \Delta_{Ni}^{(e)}u_i \qquad (2\text{-}92)$$

$$\Phi_i(x, y) = \sum_{e=1}^{E} \Phi_N^{(e)}(x, y)\Delta_{Ni}^{(e)} \qquad (2\text{-}93)$$

Here the local and global interpolation functions $\Phi_N^{(e)}$ and Φ_j have the properties (1-117, 1-120)

$$\Phi_N^{(e)}(z_M) = \delta_{NM} \qquad \Phi_i(Z_j) = \delta_{ij}$$

From these observations for the present example of the two-element system, we obtain

$$u(x, y) = \Phi_3^{(1)}u_1 + (\Phi_1^{(1)} + \Phi_2^{(2)})u_2 + (\Phi_2^{(1)} + \Phi_1^{(2)})u_3 + \Phi_3^{(2)}u_4 \qquad (2\text{-}94)$$

where the explicit forms of $\Phi_N^{(e)}$ are given by (2-17). The global linear function $u(x, y)$ as determined from the global interpolation functions (2-93) and (2-94) is shown in Fig. 2-29. Similarly, the global functions for other types of elements, linear or nonlinear, can be obtained (see Fig. 2-30).

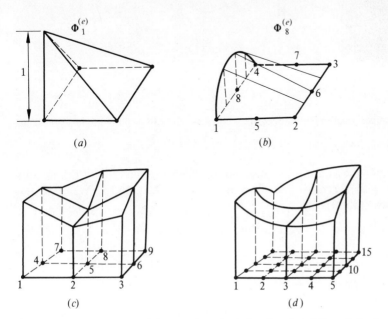

Figure 2-30 Two-dimensional linear and quadratic functions (quadrilateral elements). (*a*) Local four-node element; (*b*) local eight-node element; (*c*) global functions for (*a*); (*d*) global functions for (*b*).

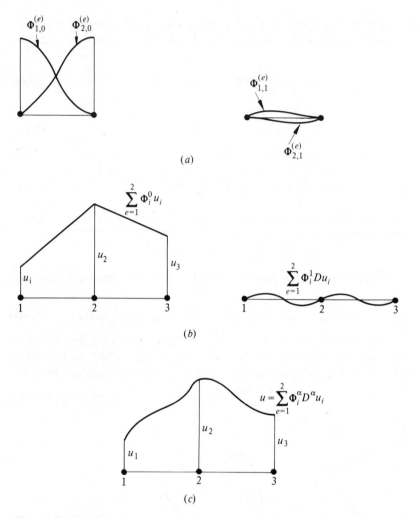

Figure 2-31 One-dimensional Hermite interpolation functions. (*a*) Local functions; (*b*) global functions; (*c*) combined global functions.

Now consider the Hermite polynomial in which the interpolation functions are associated with derivatives of a variable specified at nodes. We have the following representation:

$$u^{(e)} = \Phi_{N0}^{(e)} u_N + \Phi_{N1}^{(e)} (u_N^{(e)})' = \Phi_{N\alpha}^{(e)} D^\alpha u_N^{(e)} \tag{2-95}$$

with

$$(u_N^{(e)})' = \left(\frac{du}{dx}\right)_N \quad \text{and} \quad D^\alpha \Phi_{N\beta}^{(e)}(z_M) = \delta_{NM}\,\delta_{\alpha\beta}$$

where $|\alpha| = \alpha_1 + \alpha_2 + \cdots \leq q$, called the local interpolation functions of order q. Here $q = 1$ since $|\alpha| = 0 + 1$. Thus the global function takes the form

$$u = \sum_{e=1}^{E} u^{(e)} = \sum_{e=1}^{E} \Phi_{N\alpha}^{(e)} D^{\alpha} u_N^{(e)} = \sum_{e=1}^{E} \Phi_{N\alpha}^{(e)} D^{\alpha} \Delta_{Ni}^{(e)} u_i$$

or

$$u = \Phi_i^{\alpha} D^{\alpha} u_i \qquad (2\text{-}96)$$

with

$$\Phi_i^{\alpha} = \sum_{e=1}^{E} \Phi_{N\alpha}^{(e)} \Delta_{Ni}^{(e)} \qquad (2\text{-}97)$$

The local Hermite functions shown in Fig. 2-7 are mapped into a global form in Fig. 2-31.

2-8 PROJECTIONS AND DUAL FUNCTIONS

Let the function $u(\mathbf{x})$ be an element of Hilbert space, $u(\mathbf{x}) \in H^m(\Omega)$, and

$$\tilde{u}(\mathbf{x}) = c^i \Phi_i(\mathbf{x}) \qquad (2\text{-}98)$$

where c^i represents the values of derivatives of $u(\mathbf{x})$ and $\Phi_i(\mathbf{x})$, being linearly independent, forms a basis for a subspace $S_h(\Omega) \in H^m(\Omega)$. Let us construct an inner product in $H^s(\Omega)$ with $s \le m$

$$(\Phi_i, \Phi_j) = G_{ij} \qquad (2\text{-}99)$$

which is called the *Gram matrix*, invertible, symmetric, and positive definite. Now we introduce a *dual basis function (conjugate function)* $\Phi^i(\mathbf{x})$ defined as

$$\Phi^i = G^{ij} \Phi_j \qquad (2\text{-}100)$$

where $G^{ij} = (G_{ij})^{-1}$. Note that Φ^i are linear combinations of the functions Φ^i and Gram matrix and possess an important property

$$(\Phi^i, \Phi_j) = \delta_j^i \qquad (2\text{-}101)$$

Constructing an inner product

$$(\tilde{u}, \Phi^i) = c^j(\Phi_j, \Phi^i) = c^j \delta_j^i = c^i$$

and define Π as the projection operator given by

$$\Pi = \Pi_p \Pi_r \qquad (2\text{-}102)$$

where Π_r and Π_p are restriction and projection operators such that

$$\Pi_r u = [(u, \Phi^1), \quad (u, \Phi^2), \ldots] = (u^1, u^2, \ldots)$$

with $u^i = (u, \Phi^i)$ and

$$\Pi_p(u^1, u^2, \ldots) = u^i \Phi_i$$

Thus we have

$$\Pi^2 u = \Pi \Pi u = \Pi_p \Pi_r u^i \Phi_i$$

$$= \Pi_p[u^i(\Phi_i, \Phi^1), u^i(\Phi_i, \Phi^2), \ldots]$$

$$= \Pi_p(u^1, u^2, \ldots) = u^i \Phi_i = \Pi u$$

which indicates that $\Pi^2 = \Pi$. Furthermore

$$\| u - \tilde{u} \|^2_{H^s(\Omega)} = \| u \|^2_{H^s(\Omega)} - 2c^i(u, \Phi_i) + c^i c^j(\Phi_i, \Phi_j)$$

A minimum of $\| u - \tilde{u} \|^2$ with respect to c^i

$$\frac{\partial}{\partial c^i} \left(\| u - \tilde{u} \|^2_{H^s(\Omega)} \right) = 0$$

provides

$$(u, \Phi_i) - c^j G_{ij} = 0$$

or

$$c^j = G^{ij}(u, \Phi_i) = (u, G^{ij}\Phi_i)$$

$$= (u, \Phi^j) = u^j$$

In virtue of these results, we obtain

$$\| u - \Pi u \|_{H^s(\Omega)} = \inf_{\tilde{u} \in S_h} \| u - \tilde{u} \|_{H^s(\Omega)} \tag{2-103}$$

Note that the $\tilde{u}(\mathbf{x})$ that provides the infimum in (2-103) is $u^i \Phi_i = \Pi u$.

Finally, denote E to be the interpolation error in $H^s(\Omega)$ with $s < m$

$$E = u - \Pi u$$

Then

$$(E, \tilde{u}) = (u - u^i \Phi_i, c^j \Phi_j) = c^j [(u, \Phi_j) - u^i G_{ij}]$$

$$= c^j [(u, G_{jk}\Phi^k) - u^i G_{ij}] = c^j G_{jk}[u^k - u^k] = 0 \tag{2-104}$$

Here the projection operator Π is an orthogonal projection of $H^m(\Omega)$ onto $S_h(\Omega)$. The interpolant $\Pi u = \tilde{u}$ is the best approximation to u in $H^s(\Omega)$ norm. This makes the norm of error E as small as possible.

THREE

SOLUTION AND ACCURACY OF
FINITE ELEMENT EQUATIONS

In this chapter, we discuss procedures for obtaining finite element equations in two- and three-dimensional boundary-initial value problems and their solutions. We further explore interpolation and approximation errors—accuracy of the analysis attainable.

3-1 LOCAL AND GLOBAL FORMS OF FINITE ELEMENT EQUATIONS

We have illustrated procedures of constructing finite element equations for one-dimensional problems in Sec. 1-5. Extension into two- and three-dimensional cases follows the same general guidelines. The only difference is to take care of additional terms and area or volume integrals involved. For multivariable problems, simultaneous partial differential equations will be dealt with. In the sequel, we shall use the Galerkin method to demonstrate the construction of finite element equations.

Consider the second order partial differential equations of the form

$$\nabla^2 u + f = 0 \qquad \text{in } \Omega \tag{3-1}$$

$$u = 0 \qquad \text{on } \Gamma_1 \tag{3-2a}$$

$$u = u_0 \qquad \text{on } \Gamma_2 \tag{3-2b}$$

$$\frac{\partial u}{\partial n} = 0 \qquad \text{on } \Gamma_3 \tag{3-3a}$$

$$\frac{\partial u}{\partial n} = g \qquad \text{on } \Gamma_4 \tag{3-3b}$$

Recall that (3-2) and (3-3) are Dirichlet (essential) and Neumann (natural) boundary conditions, respectively, as discussed in Sec. 1-2-4. Using index notations, we rewrite (3-1) as

$$u_{,ii} + f = 0 \qquad (3\text{-}4a)$$

and the normal derivative takes the form

$$\frac{\partial u}{\partial n} = u_{,i} n_i \qquad (3\text{-}4b)$$

where $i = 1, 2$ for two-dimensional problems and $i = 1, 2, 3$ for three-dimensional problems, and n_i denotes the components of a vector normal to the boundary surface.

The local and global interpolation functions $\Phi_N^{(e)}$ and Φ_α for the variable u are related by

$$u^{(e)} = \Phi_N^{(e)} u_N^{(e)} \qquad (3\text{-}5a)$$

$$u = \sum_{e=1}^{E} u^{(e)} = \sum_{e=1}^{E} \Phi_N^{(e)} u_N^{(e)} = \sum_{e=1}^{E} \Phi_N^{(e)} \Delta_{N\alpha}^{(e)} u_\alpha$$

or

$$u = \Phi_\alpha u_\alpha \qquad (3\text{-}5b)$$

$$\Phi_\alpha = \sum_{e=1}^{E} \Delta_{N\alpha}^{(e)} \Phi_N^{(e)} \qquad (3\text{-}5c)$$

where $N = 1, 2, \ldots, F$ and $\alpha = 1, 2, \ldots, G$ with $F =$ the total number of nodes in a local element Ω_e and $G =$ the total number of nodes in the global domain of study (Ω); E is the total number of elements; and $\Delta_{N\alpha}^{(e)}$ is the Boolean matrix.

If the approximations of (3-5) are inserted into (3-4a), then (3-4a) may not be satisfied. Thus we introduce a residual ε such that

$$u_{,ii} + f = \varepsilon$$

The Galerkin finite element equation takes the form

$$(\varepsilon, \Phi_\alpha) = \int_\Omega (u_{,ii} + f) \Phi_\alpha \, d\Omega = 0 \qquad (3\text{-}6)$$

Integrating by parts yields

$$(\varepsilon, \Phi_\alpha) = \int_\Gamma u_{,i} n_i \overset{*}{\Phi}_\alpha \, d\Gamma - \int_\Omega (u_{,i} \Phi_{\alpha,i} - f\Phi_\alpha) \, d\Omega$$

$$= \int_\Gamma g \overset{*}{\Phi}_\alpha(s) \, d\Gamma - \left(\int_\Omega \Phi_{\alpha,i} \Phi_{\beta,i} \, d\Omega \right) u_\beta + \int_\Omega f\Phi_\alpha \, d\Omega = 0$$

Here the interpolation function $\Phi_\alpha \in \Omega$ which interpolates u in Ω changes its role on the boundaries. Namely, by virtue of the Green–Gauss theorem, the variation of $g = u_{,i} n_i$ is to be interpolated along the boundary surface Γ, not within the domain Ω. Thus $\Phi_\alpha(\mathbf{x})$ is changed to $\overset{*}{\Phi}_\alpha(s)$, which is regarded as an interpolation function for $u_{,i}n_i$. The local boundary interpolation function $\overset{*}{\phi}{}_N^{(e)}$ is related by

$$\overset{*}{\phi}_\alpha(s) = \sum_{e=1}^{E} \Delta_{N\alpha}^{(e)} \overset{*}{\phi}{}_N^{(e)}(s)$$

It should be noted that the indices α and N refer only to the boundary nodes and the summation $\Sigma_{e=1}^{E}$ also involves only boundary surface elements. An example of $\overset{*}{\phi}{}_N^{(e)}(s)$ for a linear variation of $g = u_{,i}n_i$ between the two boundary nodes in two-dimensional problems may be taken as

$$\overset{*}{\phi}{}_N^{(e)} = \left[1 - \frac{s}{l}, \frac{s}{l} \right] \qquad N = 1, 2$$

where l is the length of the boundary line within an element. With these in mind and in view of (3-5c), we obtain the global finite element equations

$$A_{\alpha\beta} u_\beta = F_\alpha \tag{3-7}$$

where

$$A_{\alpha\beta} = \sum_{e=1}^{E} A_{NM}^{(e)} \Delta_{N\alpha}^{(e)} \Delta_{M\beta}^{(e)} \tag{3-8a}$$

$$F_\alpha = \sum_{e=1}^{E} F_N^{(e)} \Delta_{N\alpha}^{(e)} \tag{3-8b}$$

$$A_{NM}^{(e)} = \int_{\Omega_e} \Phi_{N,i}^{(e)} \Phi_{M,i}^{(e)} \, d\Omega_e \tag{3-9a}$$

$$F_N^{(e)} = \int_{\Omega_e} f\Phi_N^{(e)} \, d\Omega_e + \int_{\Gamma_e} g\overset{*}{\phi}{}_N^{(e)} \, d\Gamma_e \tag{3-9b}$$

It is important to note that the limits of integration in (3-6) encompass the entire domain of study (Ω, Γ) but they have now been transformed into individual subdomains (Ω_e, Γ_e) which will then be connected and summed together throughout the entire domain.

Traditionally, derivations of finite element equations took a local approach instead of global. In the local approach, the differential equations are initially confined to a subdomain element. Thus (3-6) is replaced by

$$(\varepsilon^{(e)}, \Phi_N^{(e)}) = \int_{\Omega_e} (u_{,ii}^{(e)} + f^{(e)})\Phi_N^{(e)} \, d\Omega_e = 0$$

Proceeding similarly as before, we obtain

$$\int_{\Gamma_e} u_{,i}^{(e)} n_i \overset{*}{\phi}{}_N^{(e)} \, d\Gamma_e - \int_{\Omega_e} (u_{,i}^{(e)} \Phi_{N,i} - f\Phi_N) \, d\Omega_e = 0$$

In view of (3-5a,b,c), we arrive at the local finite element equations

$$A_{NM}^{(e)} u_M^{(e)} = F_N^{(e)} \tag{3-10}$$

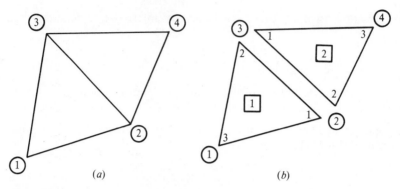

Figure 3-1 Two-element system. (a) Global domain; (b) isolated elements.

Since the derivations of local finite element equations involve less steps than those of global equations, we shall employ the local approach in later chapters. To obtain the global matrices $A_{\alpha\beta}$ and F_β, let us return to the two-element system of Fig. 2-28 reproduced in Fig. 3-1. Although the expansion of (3-8) can be performed by summing of repeated indices, we may show such operations by matrix multiplications as follows:

$$A_{\alpha\beta} = \sum_{e=1}^{E} (\boldsymbol{\Delta}^{(e)})^T \mathbf{A}^{(e)} \boldsymbol{\Delta}^{(e)} = (\boldsymbol{\Delta}^{(1)})^T \mathbf{A}^{(1)} \boldsymbol{\Delta}^{(1)} + (\boldsymbol{\Delta}^{(2)})^T \mathbf{A}^{(2)} \boldsymbol{\Delta}^{(2)}$$

$$= \begin{bmatrix} 0 & 0 & 1 \\ 1 & 0 & 0 \\ 0 & 1 & 0 \\ 0 & 0 & 0 \end{bmatrix} \begin{bmatrix} A_{11}^{(1)} & A_{12}^{(1)} & A_{13}^{(1)} \\ A_{21}^{(1)} & A_{22}^{(1)} & A_{23}^{(1)} \\ A_{31}^{(1)} & A_{32}^{(1)} & A_{33}^{(1)} \end{bmatrix} \begin{bmatrix} 0 & 1 & 0 & 0 \\ 0 & 0 & 1 & 0 \\ 1 & 0 & 0 & 0 \end{bmatrix}$$

$$+ \begin{bmatrix} 0 & 0 & 0 \\ 0 & 1 & 0 \\ 1 & 0 & 0 \\ 0 & 0 & 1 \end{bmatrix} \begin{bmatrix} A_{11}^{(2)} & A_{12}^{(2)} & A_{13}^{(2)} \\ A_{21}^{(2)} & A_{22}^{(2)} & A_{23}^{(2)} \\ A_{31}^{(2)} & A_{32}^{(2)} & A_{33}^{(2)} \end{bmatrix} \begin{bmatrix} 0 & 0 & 1 & 0 \\ 0 & 1 & 0 & 0 \\ 0 & 0 & 0 & 1 \end{bmatrix}$$

or

$$A_{\alpha\beta} = \begin{bmatrix} A_{11} & A_{12} & A_{13} & A_{14} \\ A_{21} & A_{22} & A_{23} & A_{24} \\ A_{31} & A_{32} & A_{33} & A_{34} \\ A_{41} & A_{42} & A_{43} & A_{44} \end{bmatrix} = \begin{bmatrix} A_{33}^{(1)} & A_{31}^{(1)} & A_{32}^{(1)} & 0 \\ A_{13}^{(1)} & A_{11}^{(1)} + A_{22}^{(2)} & A_{12}^{(1)} + A_{21}^{(2)} & A_{23}^{(2)} \\ A_{23}^{(1)} & A_{21}^{(1)} + A_{12}^{(2)} & A_{22}^{(1)} + A_{11}^{(2)} & A_{13}^{(2)} \\ 0 & A_{32}^{(2)} & A_{31}^{(2)} & A_{33}^{(2)} \end{bmatrix}$$

$$F_\beta = \sum_{e=1}^{E} (\boldsymbol{\Delta}^{(e)})^T \mathbf{F}^{(e)} = (\boldsymbol{\Delta}^{(1)})^T \mathbf{F}^{(1)} + (\boldsymbol{\Delta}^{(2)})^T \mathbf{F}^{(2)}$$

or

$$
F_\beta = \begin{bmatrix} F_1 \\ F_2 \\ F_3 \\ F_4 \end{bmatrix} = \begin{bmatrix} F_3^{(1)} \\ F_1^{(1)} + F_2^{(2)} \\ F_2^{(1)} + F_1^{(2)} \\ F_3^{(2)} \end{bmatrix}
$$

The procedure of assembly implied in (3-8a,b) requiring determination of Boolean matrices for all elements appears to be quite cumbersome. An intuitive and more convenient approach is schematically shown below.

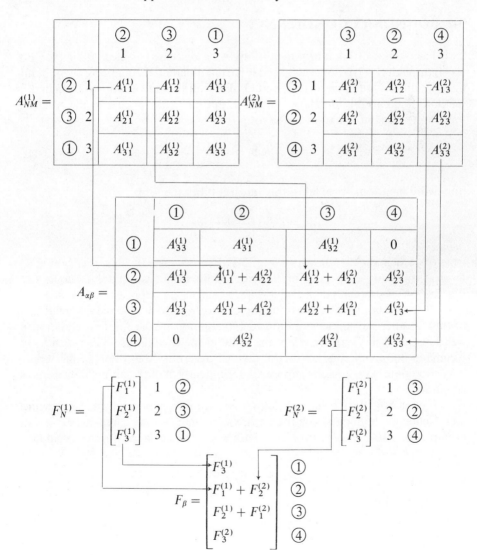

It is seen that the assembled global matrix is obtained by finding the appropriate entries from the local matrices with the local node numbers replaced by the corresponding incident global node numbers. For example, $A_{11}^{(1)}$ of the first element goes to the second row and second column in the global matrix because the local node 1 is incident with the global node 2. Similarly, $A_{12}^{(1)}$ enters in the second row and third column of the global matrix since the global node number 2 is incident with the global node 3. All entries in the same rows and columns are algebraically added together as we move to the second element. The same procedure applies in order to obtain F_β.

3-2 BOUNDARY CONDITIONS

Two basic types of boundary conditions were discussed in Sec. 1-2-4 [see Eq. (1-44) through (1-47)]. For second order partial differential equations, these boundary conditions are given by (3-2) and (3-3). Recall that in variational or Galerkin methods, the natural boundary conditions automatically appear in the resulting local finite element equation in the form

$$F_{N(b)}^{(e)} = \int_\Gamma u_{,i} n_i \overset{*}{\Phi}_N^{(e)} \, d\Gamma = \int_\Gamma g \overset{*}{\Phi}_N^{(e)} \, d\Gamma \tag{3-11a}$$

For two-dimensional problems, this integral takes the form

$$F_{N(b)}^{(e)} = \int_0^l g \overset{*}{\Phi}_N^{(e)}(s) \, ds \tag{3-11b}$$

which is schematically shown in Fig. 3-2. If $g = u_{,i} n_i = 0$, then this boundary integral is simply dropped. For the case of an ideal flow (Laplace equation), u is the velocity potential ϕ, and we have $\phi_{,i} n_i = V_i n_i$ representing the velocity normal to the boundary surface or prescribed free stream velocity. Calculation of this integral and implementation to the solution of the Laplace equation is well illustrated in Chap. 5. While the finite element equation (3-7) satisfies the Neumann boundary condition naturally, it is necessary that the Dirichlet boundary condition, say $u = u_0$, be imposed independently of the resulting finite element equations.

Often in boundary value problems, there are instances in which the Dirichlet and Neumann boundary conditions are combined at the same location. For example, in heat transfer problems with a resistance layer on the boundary, we may specify

$$u_{,i} n_i + \alpha u = -q \tag{3-12}$$

This is referred to as Cauchy boundary condition and can be handled by substituting $u_{,i} n_i = -\alpha u - q$ into (3-11a).

Since the Neumann boundary conditions are automatically satisfied by

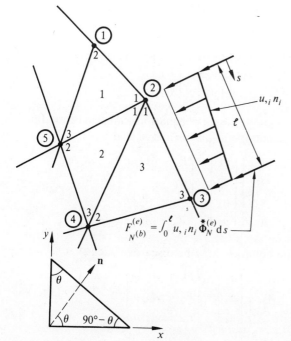

Figure 3-2 Neumann boundary conditions.

calculating (3-11) as a part of the finite element equations, we describe below how to impose Dirichlet boundary conditions by modifying the global equations. Toward this end, let us consider global equations of the type

$$A_{ij}u_j = F_i \qquad i,j = 1, 2, \ldots, n \qquad (3\text{-}13a)$$

with n being the total number of nodes in the entire domain of study Ω. Let us take the 4×4 equations given by

$$
\begin{bmatrix}
A_{11} & A_{12} & A_{13} & A_{14} \\
A_{21} & A_{22} & A_{23} & A_{24} \\
A_{31} & A_{32} & A_{33} & A_{34} \\
A_{41} & A_{42} & A_{43} & A_{44}
\end{bmatrix}
\begin{bmatrix}
u_1 \\ u_2 \\ u_3 \\ u_4
\end{bmatrix}
=
\begin{bmatrix}
F_1 \\ F_2 \\ F_3 \\ F_4
\end{bmatrix}
\qquad (3\text{-}13b)
$$

Assume that the given boundary condition requires $u_2 = 0$. Substituting $u_2 = 0$ into (3-13b) and replacing the second equation by $u_2 = 0$, we obtain

$$
\begin{bmatrix}
A_{11} & 0 & A_{13} & A_{14} \\
0 & 1 & 0 & 0 \\
A_{31} & 0 & A_{33} & A_{34} \\
A_{41} & 0 & A_{43} & A_{44}
\end{bmatrix}
\begin{bmatrix}
u_1 \\ u_2 \\ u_3 \\ u_4
\end{bmatrix}
=
\begin{bmatrix}
F_1 \\ 0 \\ F_3 \\ F_4
\end{bmatrix}
\qquad (3\text{-}14a)
$$

It is quite obvious that the values of u_1, u_3, and u_4 are not affected if we strike out the second row and second column and write

$$\begin{bmatrix} A_{11} & A_{13} & A_{14} \\ A_{31} & A_{33} & A_{34} \\ A_{41} & A_{43} & A_{44} \end{bmatrix} \begin{bmatrix} u_1 \\ u_3 \\ u_4 \end{bmatrix} = \begin{bmatrix} F_1 \\ F_3 \\ F_4 \end{bmatrix} \tag{3-14b}$$

We may now solve either (3-14a) or (3-14b). A set of equations can also be modified if u_2 is not zero but $u_2 = a$. The same operations as above lead to

$$\begin{bmatrix} A_{11} & 0 & A_{13} & A_{14} \\ 0 & 1 & 0 & 0 \\ A_{31} & 0 & A_{33} & A_{34} \\ A_{41} & 0 & A_{43} & A_{44} \end{bmatrix} \begin{bmatrix} u_1 \\ u_2 \\ u_3 \\ u_4 \end{bmatrix} = \begin{bmatrix} F_1 - A_{12}a \\ a \\ F_3 - A_{32}a \\ F_4 - A_{42}a \end{bmatrix} \tag{3-15a}$$

Once again, if desired, the second equation may be discarded such that

$$\begin{bmatrix} A_{11} & A_{13} & A_{14} \\ A_{31} & A_{33} & A_{34} \\ A_{41} & A_{43} & A_{44} \end{bmatrix} \begin{bmatrix} u_1 \\ u_3 \\ u_4 \end{bmatrix} = \begin{bmatrix} F_1 - A_{12}a \\ F_3 - A_{32}a \\ F_4 - A_{42}a \end{bmatrix} \tag{3-15b}$$

Applications of boundary conditions of this type in this manner become cumbersome if so many prescribed values are to be imposed. In this case, we write

$$\left[\begin{array}{c|c} \mathbf{A}_{(aa)} & \mathbf{A}_{(ab)} \\ \hline \mathbf{A}_{(ba)} & \mathbf{A}_{(bb)} \end{array} \right] \begin{bmatrix} \mathbf{u}_{(a)} \\ \hline \mathbf{u}_{(b)} \end{bmatrix} = \begin{bmatrix} \mathbf{F}_a \\ \hline \mathbf{F}_b \end{bmatrix} \tag{3-16a}$$

where the subscript (a) refers to the n unknowns to be determined whereas the subscript (b) denotes m prescribed values of $\mathbf{u}$. Solving for $\mathbf{u}_{(a)}$ from the first set of equations yields

$$\underset{n \times 1}{\mathbf{u}_{(a)}} = \underset{n \times n}{\mathbf{A}_{(aa)}^{-1}} \left[\underset{n \times 1}{\mathbf{F}_{(a)}} - \underset{n \times m}{\mathbf{A}_{(ab)}} \underset{m \times 1}{\mathbf{u}_{(b)}} \right] \tag{3-16b}$$

Note that (3-16b) provides the identical results as in (3-15b).

All Dirichlet conditions may be imposed through Lagrange multipliers. To illustrate, let us consider a hypothetical case:

$$A_{ij}u_j = F_i \qquad (i = 1, 2, \ldots, n) \tag{3-17}$$

with the following boundary conditions

$$u_1 = 0 \tag{3-18a}$$

$$u_2 = a \tag{3-18b}$$

$$u_4 = u_6 \tag{3-18c}$$

The boundary conditions (3-18) may be written in the form

$$q_{ri}u_i = b_r \tag{3-19}$$

with $r = 1, 2, \ldots, m$, m is the total number of boundary conditions ($m = 3$ in this case). Expanding (3-19) corresponding to (3-18) yields

$$\begin{bmatrix} 1 & 0 & 0 & 0 & 0 & 0 & \cdots \\ 0 & 1 & 0 & 0 & 0 & 0 & \cdots \\ 0 & 0 & 0 & 1 & 0 & -1 & \cdots \end{bmatrix} \begin{bmatrix} u_1 \\ u_2 \\ u_3 \\ u_4 \\ u_5 \\ u_6 \\ \vdots \\ u_n \end{bmatrix} = \begin{bmatrix} 0 \\ a \\ 0 \\ 0 \\ 0 \\ 0 \\ \vdots \\ 0 \end{bmatrix} \tag{3-20}$$

Here q_{ri} is called the boundary condition matrix. Let us now introduce quantities λ_r referred to as the Lagrange multipliers and regarded as restraints or forces required to maintain the boundary conditions such as (3-18). Then the product

$$\lambda_r(q_{ri}u_i - b_r) = 0 \tag{3-21}$$

can be considered as an invariant or energy required to maintain the boundary conditions. The global finite element equation (3-17) takes an invariant functional

$$\delta I = (A_{ij}u_j - f_i)\delta u_i = 0$$

or

$$\delta I = \delta(\tfrac{1}{2}A_{ij}u_iu_j - f_iu_i) = 0$$

for which the stationary condition is given by

$$I = \tfrac{1}{2}A_{ij}u_iu_j - f_iu_i \tag{3-22}$$

Adding (3-21) to (3-22) yields

$$J = \tfrac{1}{2}A_{ij}u_iu_j - f_iu_i + \lambda_r(q_{ri}u_i - b_r) \tag{3-23}$$

The expression (3-23) refers to the correct energy in which the total energy in the domain is modified for the boundary conditions. The minimum of (3-23) with respect to every u_i and λ_r represents the solution of (3-17) subject to the imposed boundary conditions. Therefore,

$$\delta J = \frac{\partial J}{\partial u_i}\delta u_i + \frac{\partial J}{\partial \lambda_r}\delta\lambda_r = 0$$

Since δu_i and $\delta\lambda_r$ are arbitrary, we must have $\partial J/\partial u_i = 0$ and $\partial J/\partial\lambda_r = 0$. These conditions yield

$$\begin{bmatrix} A_{ij} & q_{ri} \\ q_{rj} & 0 \end{bmatrix} \begin{bmatrix} u_j \\ \lambda_r \end{bmatrix} = \begin{bmatrix} F_i \\ b_r \end{bmatrix} \tag{3-24a}$$

or

$$
\begin{bmatrix}
A_{11} & A_{12} & \cdots & A_{1n} & q_{11} & q_{21} & \cdots & q_{m1} \\
A_{21} & A_{22} & \cdots & A_{2n} & q_{12} & q_{22} & \cdots & q_{m2} \\
\vdots & \vdots & & \vdots & \vdots & \vdots & & \vdots \\
A_{n1} & A_{n2} & \cdots & A_{nn} & q_{1n} & q_{2n} & \cdots & q_{mn} \\
\hline
q_{11} & q_{12} & \cdots & q_{1n} & 0 & 0 & \cdots & 0 \\
q_{21} & q_{22} & \cdots & q_{2n} & 0 & 0 & \cdots & 0 \\
\vdots & \vdots & & \vdots & \vdots & \vdots & & \vdots \\
q_{m1} & q_{m2} & \cdots & q_{mn} & 0 & 0 & \cdots & 0
\end{bmatrix}
\begin{bmatrix}
u_1 \\ u_2 \\ \vdots \\ u_n \\ \hline \lambda_1 \\ \lambda_2 \\ \vdots \\ \lambda_m
\end{bmatrix}
=
\begin{bmatrix}
F_1 \\ F_2 \\ \vdots \\ F_n \\ \hline b_1 \\ b_2 \\ \vdots \\ b_m
\end{bmatrix}
\qquad (3\text{-}24b)
$$

For example, let $i, j = 1, 2, \ldots, n$ ($n = 7$). Then the boundary conditions given by (3-18) or (3-19) are imposed upon the finite element equations (3-17) as

$$
\begin{bmatrix}
A_{11} & A_{12} & A_{13} & A_{14} & A_{15} & A_{16} & A_{17} & 1 & 0 & 0 \\
A_{21} & & & \cdots & & & A_{27} & 0 & 1 & 0 \\
A_{31} & & & \cdots & & & A_{37} & 0 & 0 & 0 \\
A_{41} & & & \cdots & & & A_{47} & 0 & 0 & 1 \\
A_{51} & & & \cdots & & & A_{57} & 0 & 0 & 0 \\
A_{61} & & & \cdots & & & A_{67} & 0 & 0 & -1 \\
A_{71} & A_{72} & A_{73} & A_{74} & A_{75} & A_{76} & A_{77} & 0 & 0 & 0 \\
\hline
1 & 0 & 0 & 0 & 0 & 0 & 0 & 0 & 0 & 0 \\
0 & 1 & 0 & 0 & 0 & 0 & 0 & 0 & 0 & 0 \\
0 & 0 & 0 & 1 & 0 & -1 & 0 & 0 & 0 & 0
\end{bmatrix}
\begin{bmatrix}
u_1 \\ u_2 \\ u_3 \\ u_4 \\ u_5 \\ u_6 \\ u_7 \\ \hline \lambda_1 \\ \lambda_2 \\ \lambda_3
\end{bmatrix}
=
\begin{bmatrix}
F_1 \\ F_2 \\ F_3 \\ F_4 \\ F_5 \\ F_6 \\ F_7 \\ \hline 0 \\ a \\ 0
\end{bmatrix}
$$

The solution to these equations provides the values of Lagrange multipliers λ_r as well as the unknowns u_j. Here λ_r merely assisted in imposing the boundary conditions and they may be discarded upon solution. Note that the transpose of the boundary matrix q_{rj} (3 × 7) is q_{ri} (7 × 3) on the upper right side of the coefficient matrix. It should be further noted that interchange of rows and columns is necessary to avoid zeros on the diagonal before a standard equation solver is applied.

In conclusion, our illustrations have shown that there are many different ways of imposing boundary conditions of the Dirichlet type. Imposition of these boundary conditions results simply in modification of the global finite element equations. Details of applications are shown in Chap. 5 for example problems of an ideal flow.

3-3 SOLUTION OF FINITE ELEMENT EQUATIONS

3-3-1 General

The initial and/or boundary value problems of mathematical physics are represented in general by the partial differential equations of elliptic, parabolic, or hyperbolic type. A linear second order partial differential equation may be expressed in the form

$$a\frac{\partial^2 u}{\partial x^2} + b\frac{\partial^2 u}{\partial x\,\partial y} + c\frac{\partial^2 u}{\partial y^2} + d\frac{\partial u}{\partial x} + e\frac{\partial u}{\partial y} + fu + g = 0$$

where the coefficients may be functions of the variables x and y. A partial differential equation is elliptic if $b^2 - 4ac < 0$, parabolic if $b^2 - 4ac = 0$, hyperbolic if $b^2 - 4ac > 0$. It can easily be verified that

Laplace equation $\qquad \nabla^2 u = 0$

Diffusion equation $\qquad \nabla^2 u = \dfrac{1}{\alpha}\dfrac{\partial u}{\partial t}$

Wave equation $\qquad \nabla^2 u = \dfrac{1}{\beta^2}\dfrac{\partial^2 u}{\partial t^2}$

are the elliptic, parabolic, and hyperbolic type partial differential equations, respectively. Here α and β are constants and t is the time coordinate.

The finite element equations obtained from these partial differential equations lend themselves to a numerical solution by various existing techniques. The resulting finite element equations may be of steady state or transient state, either linear or nonlinear. The linear equations are solved by Gaussian elimination, Cholesky method, etc., and nonlinear equations by iterative methods such as Newton–Raphson method or continuation method and by minimization methods of various forms. For the case of a transient state, various temporal operators of either explicit or implicit type are utilized (see Sec. 3-3-3). Furthermore, we are sometimes concerned with finding the eigenvalues and eigenvectors for wave motions (see Sec. 5-7-3). Standard texts on numerical analysis [Hildebrand, 1956; Ortega and Rheinboldt, 1970; Richtmyer and Morton, 1967; Wilkinson, 1965] provide an excellent treatise on these subjects.

3-3-2 Steady State Nonlinear Problems

Let us consider an equation of the type

$$\alpha\frac{\partial^2 u}{\partial x^2} - u\frac{\partial u}{\partial x} = -f$$

with the finite element interpolation of the form

$$u = \Phi_N u_N$$

Here, for simplicity, we drop the symbol (e) denoting a local element. The Galerkin-finite element equation in local form is written as

$$\int_\Omega \left[\alpha\frac{\partial^2 u}{\partial x^2} - \frac{\partial}{\partial x}\left(\frac{u^2}{2}\right) + f\right]\Phi_N \, d\Omega = 0$$

Proceeding as in Sec. 3-1, we obtain

$$\int_\Gamma \alpha \frac{\partial u}{\partial x} \overset{*}{\Phi}_N \, d\Gamma - \int_\Omega \alpha \frac{\partial u}{\partial x} \frac{\partial \Phi_N}{\partial x} \, d\Omega - \int_\Gamma \frac{u^2}{2} \overset{*}{\Phi}_N \, d\Gamma$$

$$+ \int_\Omega \frac{u^2}{2} \frac{\partial \Phi_N}{\partial x} \, d\Omega + \int_\Omega f \Phi_N \, d\Omega = 0$$

Writing in a compact notation yields

$$B_{NM} u_M + A_{NMQ} u_M u_Q = F_N$$

where

$$B_{NM} = \int_\Omega \frac{\partial \Phi_N}{\partial x} \frac{\partial \Phi_M}{\partial x} \, d\Omega$$

$$A_{NMQ} = -\frac{1}{2} \int_\Omega \frac{\partial \Phi_N}{\partial x} \Phi_M \Phi_Q \, d\Omega$$

$$F_N = \int_\Omega f \Phi_N \, d\Omega + \int_\Gamma \left(\alpha \frac{\partial u}{\partial x} - \frac{u^2}{2} \right) \overset{*}{\Phi}_N \, d\Gamma$$

The assembled global equation takes the form

$$B_{ij} u_j + A_{ijk} u_j u_k = F_i \tag{3-25}$$

It is evident that the second term on the left-hand side of (3-25) is nonlinear. There are numerous methods of solution of nonlinear equations of the type (3-25). In the sequel, we discuss some of the more commonly used techniques.

Newton–Raphson methods[†] The nonlinear finite element equations of the type (3-25) may be written in the form

$$R_i(u) = B_{ij} u_j + A_{ijk} u_j u_k - F_i$$
$$= A_{ijk} u_j u_k - \hat{F}_i = 0 \tag{3-26}$$

with

$$\hat{F}_i = B_{ij} u_j - F_i$$

Expanding (3-26) in the Taylor series and retaining only the first order terms yields

$$R_i(u) = R_i(u^{(0)} + \Delta u) = R_i(u^{(0)}) + \frac{\partial R_i(u^{(0)})}{\partial u_j} (u_j - u_j^{(0)}) = 0 \tag{3-27}$$

Solving for u_j in (3-27) gives

$$u_j = u_j^{(0)} - (J_{ij}^{(0)})^{-1} R_i(u^{(0)}) \tag{3-28}$$

---

† See Ortega and Rheinboldt [1970] for additional details.

where $J_{ij}^{(0)}$ is the Jacobian defined as

$$J_{ij}^{(0)} = \frac{\partial R_i(u^{(0)})}{\partial u_j} \qquad (3\text{-}29)$$

Here we must assume the initial values $u_j^{(0)}$ and the successive solutions of u_j at $r + 1$th step are obtained by

$$u_j^{(r+1)} = u_j^{(r)} - (J_{ij}^{(r)})^{-1} R_i(u^{(r)}) \qquad (3\text{-}30)$$

The Jacobian $J_{ij}^{(r)}$ needs to be updated at every iterative cycle. However, at the expense of slower convergence, the initial Jacobian $J_{ij}^{(0)}$ may be kept and used at all cycles. These operations are graphically depicted in Figs. 3-3a and 3-3b.

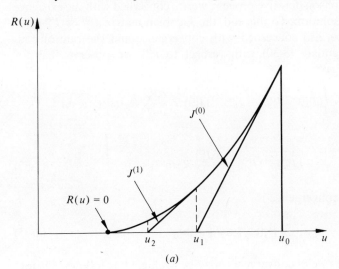

(a)

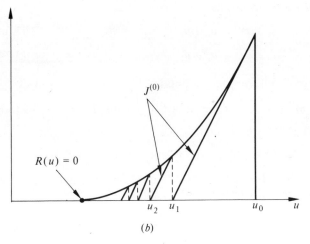

(b)

Figure 3-3 Newton–Raphson method one-dimensional representation. (a) Variable Jacobian; (b) initial constant Jacobian.

The iterative method (3-30) is called the Newton–Raphson method. If $J_{ij}^{(0)}$ instead of $J_{ij}^{(r)}$ is used in (3-30), we refer to this case as the modified Newton–Raphson method.

Often in practice, the calculation of the Jacobian matrix is complicated. The partial derivatives may therefore be approximated by difference quotients of the form

$$J_{ij}^{(r)} \approx \frac{\partial R_i(u^{(r)})}{\partial u_j} = \frac{R_i(u^{(r)}) - R_i(u^{(r-1)})}{\Delta u_j} \tag{3-31}$$

with $\Delta u_j = u_j^{(r)} - u_j^{(r-1)}$. The iterative solution (3-30) via approximate Jacobian (3-31) is known as a discretized Newton–Raphson method.

In the Newton–Raphson iterative process, we are concerned with the existence and uniqueness of the solution. To this end, the Jacobian matrix $J_{ij}^{(0)}$ or $J_{ij}^{(r)}$ must be nonsingular. We are also concerned with convergence and the rate of convergence. If we differentiate (3-30) with respect to $u_j^{(r)}$, we observe that for convergence we must have

$$\frac{\partial u_i^{(r+1)}}{\partial u_j^{(r)}} = \delta_{ij} - \left\{ (J_{ik}^{(r)})^{-1} J_{kj}^{(r)} - (J_{ik}^{(r)})^{-1} (J_{kl}^{(r)})^{-1} \frac{\partial J_{lm}^{(r)}}{\partial u_j^{(r)}} R_m(u^{(r)}) \right\} \le \delta_{ij}$$

or

$$\frac{\partial u_i^{(r+1)}}{\partial u_j^{(r)}} = (J_{ik}^{(r)})^{-1} (J_{kl}^{(r)})^{-1} \frac{\partial J_{lm}^{(r)}}{\partial u_j^{(r)}} R_m(u^{(r)}) \le \delta_{ij} \tag{3-32}$$

Thus the criterion for convergence is

$$\frac{\partial J_{lm}^{(r)}}{\partial u_m^{(r)}} R_j(u^{(r)}) \le J_{ik}^{(r)} J_{kl}^{(r)} \delta_{ij} \tag{3-33}$$

On the other hand, the rate of convergence may be evaluated as follows: Denote the error at the rth and $r + 1$th steps as

$$\varepsilon_i^{(r)} = u_i - u_i^{(r)} \tag{3-34a}$$

$$\varepsilon_i^{(r+1)} = u_i - u_i^{(r+1)} \tag{3-34b}$$

Subtracting both sides of (3-30) from u_i, we obtain

$$u_i - u_i^{(r+1)} = u_i - u_i^{(r)} + \{J_{ij}^{(r)}(u^{(r)})\}^{-1} R_j(u^{(r)}) \tag{3-35}$$

In view of (3-34a,b) and (3-35), we obtain

$$\varepsilon_i^{(r+1)} = \varepsilon_i^{(r)} + \{J_{ij}^{(r)}(u - \varepsilon^{(r)})\}^{-1} R_j(u - \varepsilon^{(r)}) \tag{3-36}$$

Expanding the last terms on the right-hand side of (3-36) in Taylor series and neglecting higher order terms, it can be shown that

$$\varepsilon_i^{(r+1)} = \frac{1}{2} \varepsilon_l^{(r)} \varepsilon_k^{(r)} \{J_{ij}^{(r)}\}^{-1} \frac{\partial J_{jk}}{\partial u_l} + \cdots$$

or

$$\varepsilon^{(r+1)} = 0\{(\varepsilon^{(r)})^2\} \tag{3-37}$$

This implies that the error decreases with the square of the error at the previous step. In other words, the rate of convergence is second order.

Continuation methods In the Newton–Raphson method, if the initial approximations are not sufficiently close to a solution to the nonlinear equation, the convergence to the correct solution may not be achieved. To remedy such a situation, the continuation methods are designed to widen the domain of convergence or to obtain sufficiently close starting points. The well-known incremental loading method in solid mechanics is a continuation method. We now return to the nonlinear equation (3-26) and set

$$R_i(u, s) = 0 \tag{3-38}$$

where s is a real parameter corresponding to a prescribed load. Consider incremental changes Δu_j and Δs such that

$$R_i(u + \Delta u, s + \Delta s) = 0 \tag{3-39}$$

Expanding (3-39) in Taylor series

$$R_i(u + \Delta u, s + \Delta s) = R_i(u, s) + \frac{\partial R_i(u, s)}{\partial u_j} \Delta u_j + \frac{\partial R_i(u, s)}{\partial s} \Delta s + \cdots = 0 \tag{3-40}$$

Neglecting the higher order terms and using (3-38), we obtain

$$\Delta u_j = - \left\{ \frac{\partial R_i(u, s)}{\partial u_j} \right\}^{-1} \frac{\partial R_i(u, s)}{\partial s} \Delta s \tag{3-41}$$

For the $r + 1$th step $\Delta u_j^{(r+1)} = u_j^{(r+1)} - u_j^{(r)}$, it follows that

$$u_j^{(r+1)} = u_j^{(r)} - \left\{ \frac{\partial R_i(u^{(r)}, s^{(r)})}{\partial u_j} \right\}^{-1} \frac{\partial R_i(u^{(r)}, s^{(r)})}{\partial s} \Delta s^{(r+1)} \tag{3-42}$$

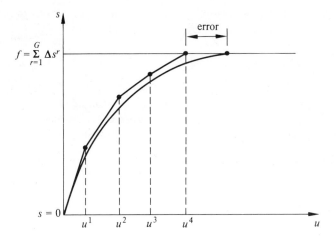

Figure 3-4 Continuation method (incremental loading).

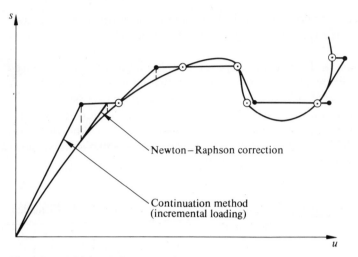

Figure 3-5 Multiple solutions — combination of incremental loading and Newton–Raphson.

with $\Delta s^{(r+1)} = s^{(r+1)} - s^{(r)}$, and the total load f_i is $f_i = \Sigma_{r=1}^{G} \Delta s^{(r)} \delta_{ri} = s^G > s^{G-1} \cdots > s^2 > s^1 > s^0 = 0$ and $r = 0, 1, \ldots, G - 1$. It should be noted that as the incremental loads are increased, errors are possibly amplified (Fig. 3-4). These errors may be corrected by combining the Newton–Raphson method as suggested in Fig. 3-5. Such combinations are particularly useful when there is a possibility for multiple solutions (see Fig. 3-4).

Minimization methods The problem of solving a system of nonlinear equations may be replaced by an equivalent problem of minimizing a function $Q(u) = q_i(u) q_i(u)$. The iterates are expected to decrease the function value at each stage for which

$$Q(u^{(r+1)}) \leq Q(u^{(r)}) \tag{3-43}$$

Consider a system

$$u_j^{(r+1)} = u_j^{(r)} + \alpha^{(r)} p_j^{(r)} \tag{3-44}$$

where $\alpha^{(r)}$ are scalars. The vector $p_j^{(r)}$ may be thought of as defining a direction along which a new iterate $u_j^{(r+1)}$ will be chosen, and the scalars α^r as determining the step length from $u_j^{(r)}$ to $u_j^{(r+1)}$. If $\| p^r \| = 1$, then α^r represents the distance from $u_j^{(r)}$ to $u_j^{(r+1)}$. If we consider the function $Q(u^{(r)} + \alpha^{(r)} p^{(r)})$ and expand it in Taylor series, we have

$$Q(u^{(r)} + \alpha^{(r)} p^{(r)}) = Q(u^{(r)}) + \alpha^{(r)} \frac{\partial Q(u^{(r)})}{\partial u_j} p_j^{(r)} + \frac{1}{2} (\alpha^{(r)})^2 \frac{\partial^2 Q(u^{(r)})}{\partial u_j \partial u_k} p_j^{(r)} p_k^{(r)} + \cdots \tag{3-45}$$

Neglecting the higher order terms, we see that

$$Q(u^{(r)} + \alpha^{(r)} p^{(r)}) = Q(u^{(r)}) + \alpha^{(r)} G_{,j}^{(r)} p_j^{(r)} + \tfrac{1}{2} (\alpha^{(r)})^2 H_{,jk}^{(r)} p_j^{(r)} p_k^{(r)} \tag{3-46}$$

To determine the minimum condition of (3-46) with respect to the parameter α^r, we must have

$$\frac{\partial}{\partial \alpha^r} \{ Q(u^{(r)} + \alpha^{(r)} p^{(r)}) \} = G_j^{(r)} p_j^{(r)} + \alpha^{(r)} H_{jk}^{(r)} p_j^{(r)} p_k^{(r)} = 0 \tag{3-47}$$

Therefore, we obtain $\alpha^{(r)}$ as

$$\alpha^{(r)} = -(H_{jk}^{(r)} p_j^{(r)} p_k^{(r)})^{-1} G_m^{(r)} p_m^{(r)} \tag{3-48}$$

Here the simplest choice of the directional vector $p_m^{(r)}$ is that $p_m^{(r)} = \delta_m^{(r)}$. This is known as the *univariant relaxation method* in which

$$\alpha^{(r)} = -(H^{(r-1)}) G^{(r)}$$

If the directional vector $p_m^{(r)}$ is set equal to the gradient of Q, then

$$p_m^{(r)} = -G_m^{(r)}$$

and $-G_m^{(r)}$ is the direction of maximum descent at u_j. This leads to the well-known *method of steepest descent*.

There are still various other minimization methods that have been reported in the literature. For additional information, see Ortega and Rheinboldt [1970].

Perturbation method The perturbation method [Morse and Feshbach, 1953] is often considered quite practical in many of the nonlinear problems. Here we introduce a *perturbation parameter* which assists in obtaining solutions to the perturbed system. The basic idea of the perturbation method is to seek a solution in the form of a series in the powers of ε such that

$$u_j = u_j^{(0)} + \varepsilon u_j^{(1)} + \varepsilon^2 u_j^{(2)} + \cdots \varepsilon^{(n)} u_j^{(n)} \tag{3-49}$$

where $u_j^{(\cdot)}$ are to be determined. We evaluate them by requiring that u_j must satisfy the equation such as (3-26). Inserting (3-49) into (3-26) yields

$$A_{ijk} u_j^{(0)} u_k^{(0)} + \varepsilon [A_{ijk} u_j^{(0)} u_k^{(1)} + A_{ijk} u_j^{(1)} u_k^{(0)}] + \cdots$$
$$+ \varepsilon^{(n)} \left[A_{ijk} u_j^{(0)} u_k^{(n)} + A_{ijk} u_j^{(n)} u_k^{(0)} + \sum_{\alpha=1}^{n} A_{ijk} u_j^{(\alpha)} u_k^{(n-\alpha)} \right] = \hat{F}_i^{(0)} + \varepsilon \hat{F}_i^{(1)} + \cdots + \varepsilon^{(n)} \hat{F}_i^{(n)} \tag{3-50a}$$

If this is to be satisfied identically in ε, then the coefficient of each power of ε must independently vanish. Thus we obtain a set of equations for u_j

$$A_{ijk} u_j^{(0)} u_k^{(0)} = \hat{F}_i^{(0)} \tag{3-50b}$$

$$A_{ijk} u_j^{(0)} u_k^{(1)} + A_{ijk} u_j^{(1)} u_k^{(0)} = \hat{F}_i^{(1)} \tag{3-50c}$$

$$\vdots$$

$$A_{ijk} u_j^{(0)} u_k^{(n)} + A_{ijk} u_j^{(n)} u_k^{(0)} = \hat{F}_i^{(n)} - \sum_{\alpha=1}^{n-1} A_{ijk} u_j^{(\alpha)} u_k^{(n-\alpha)} \tag{3-50d}$$

Note that (3-50b) represents the unperturbed system. Substituting the initial values $u_j^{(0)}$ and $\hat{F}_i^{(0)}$ into the equations (3-50c)–(3-50d) and solving these linear equations recursively, we obtain the increments $u_j^{(r)}$ $(r = 1, 2, \ldots, n)$. Finally, substituting these increments into (3-50a) leads to our desired solution. See Stoker [1950, p. 223] for convergence of perturbation series.

3-3-3 Time-Dependent Problems

In general, there are two approaches to the time-dependent problems. The first approach is to regard the interpolation function as being dependent on time as well as the spatial domain such that[†]

$$\frac{\partial u(\mathbf{x}, t)}{\partial t} = \frac{\partial \Phi_i(\mathbf{x}, t)}{\partial t} u_i$$

A disadvantage of this method is the increase in computational dimension requiring the finite element in time. The second approach is the so-called *semidiscrete* method in which the time derivative of a variable at nodes is replaced by a *temporal operator* (finite difference operator) from the relation

$$\frac{\partial u(\mathbf{x}, t)}{\partial t} = \dot{u}(\mathbf{x}, t) = \Phi_i(\mathbf{x})\dot{u}_i(t)$$

where $\dot{u}_i(t)$ is the time derivative of u prescribed at node i, which may then be written in a finite difference form. Because of limited use for the finite element in time we discuss below only the semidiscrete method.

Parabolic equations The typical parabolic equation is of the form

$$\frac{\partial u}{\partial t} - \alpha \frac{\partial^2 u}{\partial x^2} = 0 \tag{3-51}$$

with its finite element analog in local form

$$\int_\Omega \left(\frac{\partial u}{\partial t} - \alpha \frac{\partial^2 u}{\partial x^2} \right) \Phi_N \, d\Omega = 0 \tag{3-52a}$$

Integrating by parts yields

$$A_{NM}\dot{u}_M + B_{NM}u_M = F_N \tag{3-52b}$$

where

$$A_{NM} = \int_\Omega \Phi_N \Phi_M \, d\Omega \qquad B_{NM} = \int_\Omega \alpha \frac{\partial \Phi_N}{\partial x} \frac{\partial \Phi_M}{\partial x} \, d\Omega$$

$$F_N = \int_\Gamma \alpha \frac{\partial u}{\partial x} \overset{*}{\Phi}_N \, d\Gamma \qquad \dot{u}_M = \frac{\partial u_M}{\partial t}$$

Assembling (3-52b) into a global form gives

$$A_{ij}\dot{u}_j + B_{ij}u_j = F_i \tag{3-53}$$

Note that this represents a system of simultaneous ordinary differential equations. We now introduce a discretized time domain consisting of n number of small time increments Δt

---

[†] See Oden [1972, p. 148] for details.

$$\Delta t = \frac{t}{n}$$

The time derivative $\dot{u}_j$ will be replaced by finite differences of various forms called *temporal operators*, such as

$$
\dot{u}_j =
\begin{cases}
\dfrac{u_j^{(n+1)} - u_j^{(n)}}{\Delta t} & \text{(forward difference)} & (3\text{-}54a) \\[2ex]
\dfrac{u_j^{(n)} - u_j^{(n-1)}}{\Delta t} & \text{(backward difference)} & (3\text{-}54b) \\[2ex]
\dfrac{u_j^{(n+1)} - u_j^{(n-1)}}{2\Delta t} & \text{(central difference)} & (3\text{-}54c)
\end{cases}
$$

Here these schemes correspond to $u_j = u_j^{(n)}$. Thus, if the scheme (3-54a) is used, we get

$$A_{ij} \frac{u_j^{(n+1)} - u_j^{(n)}}{\Delta t} + B_{ij}u_j^{(n)} = F_i^{(n)} \qquad (3\text{-}55)$$

With the forward difference scheme (3-54a), we may propose various forms for u_j as

$$
u_j =
\begin{cases}
u_j^{(n)} & (3\text{-}56a) \\[1.5ex]
u_j^{(n+1)} & (3\text{-}56b) \\[1.5ex]
[u_j^{(n+1)} + u_j^{(n)}]/2 & (3\text{-}56c)
\end{cases}
$$

In view of (3-54a) and (3-56), we write

$$[A_{ij} + (1 - \theta)\Delta t B_{ij}]u_j^{(n+1)} = \Delta t F_i^{(n)} + (A_{ij} - \theta \Delta t B_{ij})u_j^{(n)} \qquad (3\text{-}57)$$

where $\theta = 1$, $\theta = 0$, and $\theta = \frac{1}{2}$ for the schemes (3-56a), (3-56b), and (3-56c), respectively. Now the boundary and initial conditions can be applied to (3-57) in a manner similar to steady state problems. Recursive solutions of $u_j^{(n+1)}$ can be carried out in terms of the previous history $u_j^{(n)}$ by inverting the quantity in the bracket on the right-hand side of (3-57). Note that for $\theta = 1$, the second term in the bracket vanishes and $u_j^{(n+1)}$ can be obtained by inversion of A_{ij} explicitly. Thus we refer to this case as *explicit scheme*. If $\theta = 0$ and $\theta = \frac{1}{2}$, the inversion involves both terms, the process called *implicit scheme*.

Let us first examine the explicit scheme. Solving for $u_j^{(n+1)}$ yields

$$u_j^{(n+1)} = A_{jl}^{-1}[\Delta t F_l + (A_{lk} - \Delta t B_{lk})u_k^{(n)}] \qquad (3\text{-}58)$$

Let the errors at the $n + 1$th step and nth step be given by $\varepsilon_j^{(n+1)}$ and $\varepsilon_j^{(n)}$, and if these errors are added to (3-58), then

$$u_j^{(n+1)} + \varepsilon_j^{(n+1)} = A_{jl}^{-1}[\Delta t F_l + (A_{lk} - \Delta t B_{lk})(u_k^{(n)} + \varepsilon_k^{(n)})] \qquad (3\text{-}59)$$

Subtracting (3-58) from (3-59) yields

$$\varepsilon_j^{(n+1)} = (\delta_{jk} - A_{jl}^{-1}B_{lk}\Delta t)\varepsilon_k^{(n)} \qquad (3\text{-}60)$$

For stable solutions, we must assure that errors at the nth step must not grow at the $n + 1$th step; that is

$$\left| \varepsilon_j^{(n+1)} \right| \leq \left| \varepsilon_j^{(n)} \right|$$

This requirement can be met when

$$\left| g_{jk} \right| = \left| \delta_{jk} - A_{jl}^{-1} B_{lk} \Delta t \right| \leq \left| \delta_{jk} \right| \tag{3-61}$$

where g_{jk} is called the *amplification matrix* [Lax and Richtmyer, 1956; Richtmyer and Morton, 1967]. Thus, in view of (3-60) and (3-61) and setting $\varepsilon_j^{(n+1)} = \lambda \varepsilon_j^{(n)}$, we write

$$(g_{jk} - \lambda \delta_{jk}) \varepsilon_k^{(n)} = 0 \tag{3-62}$$

The stability of the solution (3-58) can be assured if each and every eigenvalue λ_i of the amplification matrix (3-61) is made smaller than unity

$$\left| \lambda_i \right| \leq 1 \tag{3-63}$$

Obviously, the largest eigenvalue, called the *spectral radius*, governs the stability. Since there exists a bound for Δt outside of which stability can no longer be maintained, the explicit scheme here is said to be *conditionally stable*.†

If a one-dimensional problem is considered together with linear interpolation functions $\Phi_1 = 1 - x/h$ and $\Phi_2 = x/h$, the local finite element coefficient matrices in (3-52b) become

$$A_{NM} = \int_0^h \Phi_N \Phi_M \, dx = \frac{h}{6} \begin{bmatrix} 2 & 1 \\ 1 & 2 \end{bmatrix}$$

$$B_{NM} = \int_0^L \alpha \frac{\partial \Phi_N}{\partial x} \frac{\partial \Phi_M}{\partial x} \, dx = \frac{\alpha}{h} \begin{bmatrix} 1 & -1 \\ -1 & 1 \end{bmatrix}$$

$$F_N = \alpha \frac{\partial u}{\partial x} \overset{*}{\Phi}_N \Big|_0^h$$

Let the domain of study be $0 \leq x \leq 3h$ and the boundary condition be given as

$$u(0) = u(3h) = 0$$

and $F_N = 0$ at all boundary nodes. Assuming that the domain consists of three equal elements with four nodes ($h = \Delta x$), we obtain the global equations of the form

$$\frac{\Delta x}{6} \begin{bmatrix} 2 & 1 & & \\ 1 & 4 & 1 & \\ & 1 & 4 & 1 \\ & & 1 & 2 \end{bmatrix} \begin{bmatrix} \dot{u}_1 \\ \dot{u}_2 \\ \dot{u}_3 \\ \dot{u}_4 \end{bmatrix} + \frac{\alpha}{\Delta x} \begin{bmatrix} 1 & -1 & & \\ -1 & 2 & -1 & \\ & -1 & 2 & -1 \\ & & -1 & 1 \end{bmatrix} \begin{bmatrix} u_1 \\ u_2 \\ u_3 \\ u_4 \end{bmatrix} = \begin{bmatrix} F_1 \\ F_2 \\ F_3 \\ F_4 \end{bmatrix}$$

$$\tag{3-64a}$$

---

† The solution stability described here is unrelated to the error estimates or rates of convergence to be discussed in Sec. 3-4-5.

Upon modification of (3-64a) for the given Dirichlet boundary conditions, we obtain

$$\frac{\Delta x}{6} \begin{bmatrix} 4 & 1 \\ 1 & 4 \end{bmatrix} \begin{bmatrix} \dot{u}_2 \\ \dot{u}_3 \end{bmatrix} + \frac{\alpha}{\Delta x} \begin{bmatrix} 2 & -1 \\ -1 & 2 \end{bmatrix} \begin{bmatrix} u_2 \\ u_3 \end{bmatrix} = \begin{bmatrix} 0 \\ 0 \end{bmatrix} \tag{3-64b}$$

with

$$A_{jl} = \frac{\Delta x}{6} \begin{bmatrix} 4 & 1 \\ 1 & 4 \end{bmatrix} \qquad B_{lk} = \frac{\alpha}{\Delta x} \begin{bmatrix} 2 & -1 \\ -1 & 2 \end{bmatrix}$$

Thus to meet the stability criterion (3-63), we calculate the eigenvalues of the matrix

$$\begin{bmatrix} g_{11} - \lambda & g_{12} \\ g_{21} & g_{22} - \lambda \end{bmatrix}$$

where

$$g_{11} = g_{22} = \frac{18\alpha\Delta t}{5(\Delta x)^2} - 1 \qquad g_{12} = g_{21} = -\frac{12\alpha\Delta t}{5(\Delta x)^2}$$

and obtain

$$\lambda_1 = \frac{6\alpha\Delta t}{(\Delta x)^2} - 1 \qquad \lambda_2 = \frac{6\alpha\Delta t}{5(\Delta x)^2} - 1$$

with λ_1 being the spectral radius. This leads to

$$\Delta t \le \frac{(\Delta x)^2}{3\alpha}$$

Now the limiting Δt is seen to be governed by various factors such as the constant α, mesh size Δx, and the interpolation functions which constitute the matrices A_{ij} and B_{ij}. For two- or three-dimensional problems, however, an evaluation of stability explicitly in terms of the mesh size Δx is not possible. The eigenvalues are calculated with overall mesh configurations implicitly built into the matrices A_{ij} and B_{ij}.

Similar derivations for the implicit schemes (3-56b) and (3-56c) reveal that

$$\varepsilon_j^{(n+1)} = g_{jk}\varepsilon_k^{(n)}$$

where the amplification matrices for (3-56b) and (3-56c) are, respectively,

$$g_{jk} = C_{jl}^{-1}\delta_{lk} = C_{jk}^{-1} \tag{3-65}$$

$$g_{jk} = D_{jl}^{-1}E_{lk} \tag{3-66}$$

with

$$C_{jk} = \delta_{jk} + A_{jm}^{-1}B_{mk}\Delta t \tag{3-67a}$$

$$D_{jl} = \delta_{jl} + A_{jm}^{-1}B_{ml}\frac{\Delta t}{2} \tag{3-67b}$$

$$E_{lk} = \delta_{lk} - A_{lm}^{-1}B_{mk}\frac{\Delta t}{2} \tag{3-67c}$$

It is seen that for all values of Δt, we have $g_{jk} \leq \delta_{jk}$. Thus we say that the implicit scheme is *unconditionally stable*.†

As a last example, consider the Newton backward difference (finite differenced Taylor series) [Chung, 1975],

$$u_j^{(n+s)} = u_j^{(n)} + \Delta u_j^{(n)}s + \frac{1}{2!}\Delta^2 u_j^{(n)}s(s+1) + \frac{1}{3!}\Delta^3 u_j^{(n)}s(s+1)(s+2) + \cdots$$

$$+ \frac{1}{m!}\Delta^m u_j^{(n)}s(s+1)(s+2)(s+3)\cdots(s+m-1) \quad (3\text{-}68)$$

where n denotes the time step and $s = (t - t_n)/\Delta t$. The time derivative of u at the time step n takes the form

$$\frac{\partial u_j}{\partial t} = \dot{u}_j = \frac{\partial u_j^{(n)}}{\partial s}\frac{\partial s}{\partial t} = \left(\frac{\partial u_j^{(n+s)}}{\partial s}\frac{\partial s}{\partial t}\right)_{s=0} = \frac{1}{\Delta t}\left(\Delta u_j^{(n)} + \frac{1}{2}\Delta^2 u_j^{(n)}\right.$$

$$\left. + \frac{1}{3}\Delta^3 u_j^{(n)} + \cdots + \frac{1}{m}\Delta^m u_j^{(n)}\right) \quad (3\text{-}69)$$

and

$$\frac{\partial^2 u_j}{\partial t^2} = \ddot{u}_j = \left(\frac{\partial \dot{u}_j^{(n+s)}}{\partial t}\right)_{s=0} = \frac{1}{\Delta t^2}\left\{\Delta^2 u_j^{(n)} + \Delta^3 u_j^{(n)} + \cdots\right.$$

$$\left. + \frac{2}{m}\Delta^m u_j^{(n)}\left(1 + \frac{1}{2} + \cdots + \frac{1}{m-1}\right)\right\} \quad (3\text{-}70)$$

where

$$\Delta u_j^{(n)} = u_j^{(n)} - u_j^{(n-1)} \quad (3\text{-}71)$$

$$\Delta^2 u_j^{(n)} = \Delta u_j^{(n)} - \Delta u_j^{(n-1)} = u_j^{(n)} - 2u_j^{(n-1)} + u_j^{(n-2)} \quad (3\text{-}72)$$

$$\Delta^3 u_j^{(n)} = \Delta^2 u_j^{(n)} - \Delta^2 u_j^{(n-1)} = u_j^{(n)} - 3u_j^{(n-1)} + 3u_j^{(n-2)} - u_j^{(n-3)} \quad (3\text{-}73)$$

Substituting the above into (3-69) and (3-70) results in, respectively,

$$\dot{u}_j = \frac{1}{6\Delta t}(11u_j^{(n)} - 18u_j^{(n-1)} + 9u_j^{(n-2)} - 2u_j^{(n-3)}) \quad (3\text{-}74)$$

$$\ddot{u}_j = \frac{1}{\Delta t^2}(2u_j^{(n)} - 5u_j^{(n-1)} + 4u_j^{(n-2)} - u_j^{(n-3)}) \quad (3\text{-}75)$$

where only those terms of third order and lower are retained. In view of (3-74) and (3-53) and moving up one step, we obtain

$$(11A_{ij} + 6\Delta t(1-\theta)B_{ij})u_j^{(n+1)} = 6\Delta t F_i^{(n)} - (-18A_{ij} + 6\Delta t\theta B_{ij})u_j^{(n)}$$

$$- A_{ij}(9u_j^{(n-1)} - 2u_j^{(n-2)}) \quad (3\text{-}76)$$

---

† Oscillations of solutions due to inadequate choices of trial and test functions can occur in implicit schemes, as well as in explicit schemes.

The errors associated with various time steps are

$$(11A_{ij} + 6\Delta t(1 - \theta)B_{ij})\varepsilon_j^{(n+1)} = (18A_{ij} - 6\Delta t\theta B_{ij})\varepsilon_j^{(n)} - A_{ij}(9\varepsilon_j^{(n-1)} - 2\varepsilon_j^{(n-2)})$$

$$(3\text{-}77)$$

To investigate stability, we assume that the error in the numerical solution at time $t_{(n)} = n\Delta t$ is given by

$$\varepsilon_j^{(n)} = \exp(\gamma n\Delta t)C_j \tag{3-78a}$$

and similarly at $t_{(n-1)} = (n-1)\Delta t$ by

$$\varepsilon_j^{(n-1)} = \exp(\gamma(n-1)\Delta t)C_j \tag{3-78b}$$

etc. Here γ is the constant and C_j denotes arbitrary nodal point errors. Defining the characteristic value

$$\omega = \exp(\gamma\Delta t) \tag{3-79}$$

and substituting (3-78) into (3-77), we obtain the eigenvalue problem

$$(g_{ij} - \lambda\delta_{ij})\varepsilon_j^{(n)} = 0 \tag{3-80}$$

where for $\theta = 1$ (explicit scheme), the amplification matrix g_{ij} is

$$g_{ij} = \tfrac{1}{11}(18\delta_{ij} - 6\Delta t A_{ik}^{-1}B_{kj}) \tag{3-81}$$

and λ denotes the eigenvalue

$$\lambda = \frac{\omega^3 + 9\omega - 2}{\omega^2} \tag{3-82}$$

For the example problem (3-64b), we obtain

$$\lambda_1 = \frac{36\alpha\Delta t}{11(\Delta x)^2} - \frac{18}{11} \qquad \lambda_2 = \frac{36\alpha\Delta t}{55(\Delta x)^2} - \frac{18}{11}$$

This gives the limiting Δt as

$$\Delta t \le \frac{29(\Delta x)^2}{36\alpha}$$

which is 2.4 times the limiting Δt for the forward differences (3-54a and 3-56a).

Stability analysis for the implicit scheme can be performed similarly as shown for the forward differences, and we find that the scheme is *unconditionally stable*.

Hyperbolic equations We consider the second order hyperbolic equation

$$\frac{\partial^2 u}{\partial t^2} - \beta^2 \frac{\partial^2 u}{\partial x^2} = 0 \tag{3-83}$$

The finite element analogue takes the form

$$A_{ij}\ddot{u}_j + B_{ij}u_j = F_i \tag{3-84}$$

The temporal operators in (3-54) can be extended to the second derivative, and we write

$$\ddot{u}_j = \frac{u_j^{(n+1)} - 2u_j^{(n)} + u_j^{(n-1)}}{\Delta t^2} \tag{3-85}$$

In view of (3-84), (3-85), and (3-56), we obtain

$$(A_{ij} + \Delta t^2(1 - \theta)B_{ij})u_j^{(n+1)} = \Delta t^2 F_i - (-2A_{ij} + \Delta t^2 \theta B_{ij})u_j^{(n)} - A_{ij}u_j^{(n-1)} \tag{3-86}$$

This leads to

$$(A_{ij} + \Delta t^2(1 - \theta)B_{ij})\varepsilon_j^{(n+1)} = (2A_{ij} - \Delta t^2 \theta B_{ij})\varepsilon_j^{(n)} - A_{ij}\varepsilon_j^{(n-1)} \tag{3-87}$$

Again, by virtue of (3-78) and (3-79), we can show that the eigenvalue problem for stability analysis takes a form identical to (3-80). Here, for $\theta = 1$, the amplification matrix becomes

$$g_{ij} = 2\delta_{ij} - (\Delta t)^2 A_{ik}^{-1} B_{kj} \tag{3-88}$$

and the eigenvalue is defined as

$$\lambda = \frac{\omega^2 + 1}{\omega} \tag{3-89}$$

The eigenvalues for an example problem similar to (3-64b) are

$$\lambda_1 = \frac{6\beta^2(\Delta t)^2}{(\Delta x)^2} - 2 \qquad \lambda_2 = \frac{6\beta^2(\Delta t)^2}{5(\Delta x)^2} - 2$$

with the limiting Δt being

$$\Delta t \leq \frac{\Delta x}{\sqrt{2}\,\beta}$$

This indicates that Δt is now proportional to Δx, not $(\Delta x)^2$ as required for the parabolic equation.

Let us now examine the third order Newton backward difference (3-75). The finite element equation is of the form

$$(2A_{ij} + (\Delta t)^2(1 - \theta)B_{ij})u_j^{(n+1)} = (\Delta t)^2 F_i^{(n)} - (-5A_{ij} + (\Delta t)^2 \theta B_{ij})u_j^{(n)}$$
$$- A_{ij}(4u_j^{(n-1)} - u_j^{(n-2)}) \tag{3-90}$$

Determinations of the amplification matrix and subsequent analysis for stability follow similarly as in earlier cases.

To investigate the performance of various temporal operators, consider a one-dimensional problem given by

$$m\ddot{u} + ku = 0$$

From (3-87) for $\theta = 1$, we have

$$u^{(n+1)} = \left(2 - \frac{k}{m}(\Delta t)^2\right)u^{(n)} - u^{(n-1)} \tag{3-91}$$

Let us assume that the response u at the nth time step is of the form

$$u^{(n)} = a\phi^n \tag{3-92}$$

where a is the arbitrary constant to be determined from initial conditions and ϕ is a value to be chosen to satisfy (3-91). Substituting (3-92) into (3-91) yields

$$\phi^2 + \left(\frac{k}{m}(\Delta t)^2 - 2\right)\phi + 1 = 0$$

Suppose that Δt is set equal to $\sqrt{2m/k}$. Then

$$\phi^2 + 1 = 0$$

from which

$$\phi = \pm\sqrt{-1} = \pm i = \exp\left(\pm\frac{i\pi}{2}\right) \tag{3-93}$$

In view of (3-92) and (3-93), we get

$$u^{(n)} = a_1 \exp\left(\frac{i\pi n}{2}\right) + a_2 \exp\left(-\frac{i\pi n}{2}\right) = b_1 \sin\frac{\pi n}{2} + b_2 \cos\frac{\pi n}{2} \tag{3-94}$$

Noting that $n = t/\Delta t = t\sqrt{k/2m}$, we obtain the general solution corresponding to (3-91) as

$$u = b_1 \sin\left(1.11\sqrt{\frac{k}{m}}\,t\right) + b_2 \cos\left(1.11\sqrt{\frac{k}{m}}\,t\right) \tag{3-95}$$

whereas the exact solution is

$$u = b_1 \sin\sqrt{\frac{k}{m}}\,t + b_2 \cos\sqrt{\frac{k}{m}}\,t \tag{3-96}$$

Comparing, we note that the difference amounts to increasing the effective natural frequency from $\sqrt{k/m}$ to $1.11\sqrt{k/m}$. With increasing time t, however, the solution u is stable and does not diverge.

At this point, let us examine what happens if $\Delta t > \sqrt{2m/k}$ is used. For example, let $\Delta t = 3\sqrt{m/k}$. We then have

$$\phi^2 + 7\phi + 1 = 0 \tag{3-97}$$

which gives

$$\phi = -\exp(-1.925)$$
$$\phi = -\exp(1.925)$$

and

$$u = \cos\left(1.047\sqrt{\frac{k}{m}}\,t\right)\left[b_1 \sinh\left(0.642\sqrt{\frac{k}{m}}\,t\right) + b_2 \cosh\left(0.642\sqrt{\frac{k}{m}}\,t\right)\right] \tag{3-98}$$

It is evident that the solution is divergent as time t approaches infinity.

It is interesting to note that a similar study conducted on the temporal operators of the third order Newton backward difference shows somewhat different results. Let us consider an implicit scheme of (3-76), $\theta = 0$. For $\Delta t = \sqrt{2m/k}$ and $98\sqrt{m/k}$ we have, respectively,

$$u = a_1 \exp\left(-0.701\sqrt{\frac{k}{m}}\,t\right) + \exp\left(-0.139\sqrt{\frac{k}{m}}\,t\right)\left[b_1 \sin\left(0.712\sqrt{\frac{k}{m}}\,t\right)\right.$$

$$\left. + b_2 \cos\left(0.712\sqrt{\frac{k}{m}}\,t\right)\right] \quad (3\text{-}99)$$

and

$$u = a_1 \exp\left(-0.1805\sqrt{\frac{k}{m}}\,t\right) + \exp\left(-0.1424\sqrt{\frac{k}{m}}\,t\right)\left[b_1 \sin\left(0.183\sqrt{\frac{k}{m}}\,t\right)\right.$$

$$\left. + b_2 \cos\left(0.183\sqrt{\frac{k}{m}}\,t\right)\right] \quad (3\text{-}100)$$

Note that the first term provides decaying action. The significant damping occurs due to the exponential terms for b_1 and b_2, and frequency decreases considerably as Δt increases.

Nonlinear equations We have so far dealt with only linear problems. The well-known Navier–Stokes equation is nonlinear and its solution has been the subject of active research for many years. For simplicity, let us consider the Burger's equation,

$$\frac{\partial u}{\partial t} + u\frac{\partial u}{\partial x} - \nu\frac{\partial^2 u}{\partial x^2} = 0 \quad (3\text{-}101)$$

The global finite element analogue takes the form [see (3-25)]

$$A_{ij}\dot{u}_j + C_{ijk}u_ju_k + B_{ij}u_j = F_i \quad (3\text{-}102)$$

From the approximations (3-54a) and (3-56a)

$$[A_{ij} + \Delta t(1 - \theta)(C_{ijk}u_k^{(n+1)} + B_{ij})]u_j^{(n+1)}$$

$$= \Delta t F_i^{(n)} + [A_{ij} - \Delta t\theta(C_{ijk}u_k^{(n)} + B_{ij})]u_j^{(n)} \quad (3\text{-}103)$$

To carry out the solution of (3-103), the Newton–Raphson or perturbation method discussed in Sec. 3-3-2 must be combined. Details will be presented in Sec. 5-6-2. A glance at (3-103), however, reveals that any attempt to perform stability analysis is a formidable task, although many successful computations have been reported in the literature (see discussions in Sec. 5-6-2).

3-3-4 Matrix Reduction

In the solution of matrix equations involving a large number of degrees of freedom, it is often advantageous to reduce the size of the matrix to be inverted, retaining only those unknowns absolutely needed and discarding those undesired.

Consider global finite element equations of the form

$$A_{ij}u_j = f_i \qquad (3\text{-}104)$$

This can be partitioned as

$$\left[\begin{array}{c|c} \mathbf{A}_{(aa)} & \mathbf{A}_{(ab)} \\ \hline \mathbf{A}_{(ba)} & \mathbf{A}_{(bb)} \end{array}\right] \left[\begin{array}{c} \mathbf{u}_{(a)} \\ \mathbf{u}_{(b)} \end{array}\right] = \left[\begin{array}{c} \mathbf{f}_{(a)} \\ \mathbf{f}_{(b)} \end{array}\right] \qquad (3\text{-}105)$$

where $\mathbf{u}_{(a)}$ and $\mathbf{u}_{(b)}$ refer to unknowns to be retained and discarded, respectively. Solving for $\mathbf{u}_{(b)}$ yields

$$\mathbf{u}_{(b)} = \mathbf{A}_{(bb)}^{-1}(\mathbf{f}_{(b)} - \mathbf{A}_{(ba)}\mathbf{u}_{(a)}) \qquad (3\text{-}106)$$

The unknowns, $\mathbf{u}_{(a)}$ and $\mathbf{u}_{(b)}$, are combined as

$$\mathbf{u} = \left[\begin{array}{c} \mathbf{u}_{(a)} \\ \mathbf{u}_{(b)} \end{array}\right] = \left[\begin{array}{c} \mathbf{I} \\ -\mathbf{A}_{(bb)}^{-1}\mathbf{A}_{(ba)} \end{array}\right]\mathbf{u}_{(a)} + \left[\begin{array}{c} 0 \\ \mathbf{A}_{(bb)}^{-1}\mathbf{f}_{(b)} \end{array}\right] \qquad (3\text{-}107)$$

or

$$\mathbf{u} = \mathbf{C}\mathbf{u}_{(a)} + \mathbf{D} \qquad (3\text{-}108)$$

where

$$\mathbf{C} = \left[\begin{array}{c} \mathbf{I} \\ -\mathbf{A}_{(bb)}^{-1}\mathbf{A}_{(ba)} \end{array}\right] \qquad \mathbf{D} = \left[\begin{array}{c} 0 \\ \mathbf{A}_{(bb)}^{-1}\mathbf{f}_{(b)} \end{array}\right]$$

It is seen that the vector $\mathbf{u}$ containing the total unknowns is now defined by the vector $\mathbf{u}_{(a)}$ representing only those unknowns to be retained. Writing (3-108) in index notations yields

$$u_j = C_{j\alpha}u_\alpha + D_j \qquad (3\text{-}109)$$

with $\alpha = 1,\ldots,m$, m is the total number of unknowns to be retained. The invariant energy functional of (3-104) is written as

$$\chi = \tfrac{1}{2}A_{ij}u_iu_j - f_iu_i \qquad (3\text{-}110)$$

The minimum condition of (3-110) with respect to every component of u_α is found as

$$\delta\chi = \frac{\partial\chi}{\partial u_\alpha}\,\delta u_\alpha = 0 \qquad (3\text{-}111)$$

Since δu_α is arbitrary, we must have

$$\frac{\partial\chi}{\partial u_\alpha} = 0 \qquad (3\text{-}112)$$

Substituting (3-109) into (3-110) and differentiating gives

$$A_{ij}C_{i\alpha}C_{j\beta}u_\beta = f_iC_{i\alpha} \qquad (3\text{-}113)$$

or

$$\bar{A}_{\alpha\beta}u_\beta = \bar{f}_\alpha \qquad (3\text{-}114)$$

If the original and reduced numbers of degrees of freedom are n and m, respectively, we may write in matrix form

$$\underset{m \times n}{\mathbf{C}^T} \; \underset{n \times n}{\mathbf{A}} \; \underset{n \times m}{\mathbf{C}} \; \underset{m \times 1}{\mathbf{u}_{(a)}} = \underset{m \times n}{\mathbf{C}^T} \; \underset{n \times 1}{\mathbf{f}} \tag{3-115}$$

or

$$\underset{m \times m}{\overline{\mathbf{A}}} \; \underset{m \times 1}{\mathbf{u}_{(a)}} = \underset{m \times 1}{\overline{\mathbf{f}}} \tag{3-116}$$

where

$$\overline{\mathbf{A}} = \mathbf{C}^T \mathbf{A} \mathbf{C} \tag{3-117}$$

$$\overline{\mathbf{f}} = \mathbf{C}^T \mathbf{f} \tag{3-118}$$

Now the modified matrix $\overline{\mathbf{A}}$ is reduced to $m \times m$ and the solution of (3-114) yields the unknowns which have been retained. If desired, the unknowns $\mathbf{u}_{(b)}$ can be calculated from (3-106).

3-4 MATHEMATICAL PROPERTIES OF FINITE ELEMENTS

3-4-1 General

As discussed in Chap. 1, the finite element method is rooted in the realm of the classical methods of Rayleigh [1877] and Ritz [1909] and the classical projection methods of Galerkin [1915]. The finite element method is also a tool for the numerical analysis of partial differential equations endowed by the modern theory of approximations.†/ The mathematical theory of finite element methods have been the subject of much research in the 1970's, and today a fairly complete theory of convergence and stability of finite element approximations of linear boundary and initial value problems has evolved.

A study of all of these mathematical properties is not the main purpose of this book. Our aim here is to develop the finite element method operationally as an elegant and very appealing method for the analysis of problems in fluid dynamics. But since we are to deal with a mathematical method, we cannot overlook its most important mathematical features, particularly those which have direct bearing on its success. Our interests are practical. We wish to know what criteria must be met if the method is to converge, in what sense is the method convergent, what criteria are available for selecting interpolation functions, and what rate of convergence can be expected for a given choice of basis functions.

In this chapter, we address these questions in an elementary way, leaving those more mathematically inclined, to see additional details in some of the excellent works that have appeared on the subject in recent years. In particular, we mention the extensive study of Babuska and Aziz [Aziz (Ed.), 1972] on the

---

† It was commented in Sec. 1-2-4 that the finite element computations can be performed without having studied the mathematical foundations in the context of functional analysis. This is because the engineer invented the method and the mathematician proved it. The reader may postpone this section until he has studied the rest of the book.

mathematical theory of finite elements. In addition, the book of Strang and Fix [1972] is devoted to a comprehensive treatment of various mathematical aspects of the finite element theory, and Oden and Reddy [1976] render in their book a well-organized structure of the finite element theory from the viewpoint of mathematics.

3-4-2 Interpolation Theories

Elementary approach Consider a quadratic function u which is defined on a one-dimensional element with three nodes located at $x = -h$, $x = 0$, and $x = h$, as shown in Fig. 2-5c. Since three data points are available, the variable u may be approximated by $\tilde{u}$,

$$\tilde{u} = a_1 + a_2 x + a_3 x^2 \tag{3-119}$$

Writing (3-119) at each node and evaluating the constants, we obtain

$$\tilde{u} = \Phi_N u_N = \Phi_1 u_1 + \Phi_2 u_2 + \Phi_3 u_3 \tag{3-120}$$

in which u_N refers to the value of $\tilde{u}$ at the Nth node and the interpolation function Φ_N is of the form

$$\Phi_1 = \xi(\xi - 1)/2 \qquad \Phi_2 = 1 - \xi^2 \qquad \Phi_3 = \xi(\xi + 1)/2$$

with $\xi = x/h$, as we have shown in Chap. 2. Recall that Φ_N is known also as basis (trial) function denoting a member of a linearly independent set of elements that span a finite dimensional linear space. Let a Taylor series expansion of any function $f(x)$ in the interval (a, x) be given by

$$f(x) = f(a) + f'(a)(x - a) + \frac{f''(a)}{2!}(x - a)^2 + \frac{f'''(a)}{3!}(x - a)^3 + \cdots$$

and consider that u_1 and u_2 are approximate solutions of u at nodes 1 and 2, respectively. Then it is possible to expand u_1 and u_3 around $x = 0$ in the form

$$u_1 = u_2 - u_2' + \frac{u_2''}{2!} - \frac{u_2'''}{3!} + \cdots \tag{3-121a}$$

and

$$u_3 = u_2 + u_2' + \frac{u_2''}{2!} + \frac{u_2'''}{3!} + \cdots \tag{3-121b}$$

where the prime indicates a derivative with respect to ξ and u_2 is infinitely differentiable. In view of (3-120) and (3-121), we arrive at an approximate value $\tilde{u}$,

$$\tilde{u} = u_2(\Phi_1 + \Phi_2 + \Phi_3) + u_2'(\Phi_3 - \Phi_1) + \frac{1}{2!}u_2''(\Phi_3 + \Phi_1)$$

$$+ \frac{1}{3!}u_2'''(\Phi_3 - \Phi_1) + \cdots \tag{3-122}$$

Since $\Phi_1 + \Phi_2 + \Phi_3 = 1$, $\Phi_3 - \Phi_1 = \xi$, and $\Phi_3 + \Phi_1 = \xi^2$, we obtain

$$\tilde{u} = u_2 + u_2'\xi + \frac{1}{2!}u_2''\xi^2 + \frac{1}{3!}u_2'''\xi + \cdots \tag{3-123}$$

On the other hand, the exact function u can be expanded around $x = 0$ in the form

$$u = u_2 + u_2'\xi + \frac{1}{2!}u_2''\xi^2 + \frac{1}{3!}u_2'''\xi^3 + \cdots \tag{3-124}$$

In view of (3-123) and (3-124), it follows that

$$u - \tilde{u} = \frac{1}{3!}u_2'''(\xi^3 - \xi) + \cdots \tag{3-125}$$

Since $x = \xi h$ and $dx = h\,d\xi$, the third derivative appearing in (3-125) is evaluated as

$$\frac{d^3u}{d\xi^3} = \frac{d^3u}{dx^3}\left(\frac{dx}{d\xi}\right)^3 = \frac{d^3u}{dx^3}h^3 \tag{3-126}$$

Therefore, from (3-125) and (3-126), we write the absolute value of the difference between $\tilde{u}$ and u as

$$|u - \tilde{u}| = \left|\frac{h^3}{6}\frac{d^3u_2}{dx^3}(\xi^3 - \xi) + \cdots\right|$$

This, then, leads to

$$|u - \tilde{u}| \le ch^3\left|d^3u/dx^3\right|_{\max}$$

where c is a positive constant independent of h and $\left|d^3u/dx^3\right|$ is bounded. For the interpolation function Φ_N with a polynomial degree k, we may write

$$|u - \tilde{u}| \le ch^{k+1}C_{k+1} \tag{3-127}$$

where C_{k+1} is a bound on the $(k + 1)$ derivatives of u with respect to x, $\left|d^{(k+1)}u/dx^{(k+1)}\right|$. The mth derivative of (3-127) takes the form

$$\left|\partial^m(u - \tilde{u})/\partial x^m\right| \le ch^{k+1-m}C_{k+1} \tag{3-128}$$

Thus, for an element of diameter h, we have

$$|u - \tilde{u}| = 0(h^{(k+1)})$$

and the mth derivatives can be approximated up to $0(h^{k+1-m})$. For the problems of the $2m$th order, the energy includes derivatives of u up to the mth order; and the error in the energy norm becomes

$$\|u - \tilde{u}\|_m \le ch^{k+1-m}C_{k+1} \tag{3-129}$$

where the double strokes $\|\ \|$ imply square root of the energy known as energy norm.

Based on this very rudimentary introduction to error estimates and an assurance of convergence to a true solution, a few remarks are in order. As shown by Mikhlin [1964], a necessary condition for assuring the convergence in the variational methods is that the trial (interpolation) function must be *complete* in energy. In the finite element method under this criterion, approximate energy in a coarse mesh approaches the exact energy with refinement of the mesh size. In problems of the $2m$th order (for harmonic problem $m = 1$, for biharmonic problem $m = 2$), the energy expression includes derivatives up to the mth order. The accuracy obtainable with the finite element interpolation depends on how closely this function can approximate the exact solution up to its mth derivative. Indeed, in proving the convergence of the finite element method, we note that any smooth function can be approximated by linear segments of size h up to $O(h^2)$. The first derivatives of this smooth function can be approximated by the linear elements up to $O(h)$. In second order problems (heat transfer, potential flow, elasticity, etc.), the energy expression includes squares of the first derivatives. Therefore, this type of element permits the energy to be approximated up to $O(h)$. Since for obtaining the finite element approximate solution, the energy is minimized, the actual error in the energy is at most $O(h^2)$. Evidently, using higher order interpolation schemes inside the element permits the attainment of a better approximation with the same number of elements and hence a higher rate of convergence. If the interpolation functions inside the element include a complete polynomial of the kth degree, then the rate of convergence in the energy norm is $O(h^{k+1-m})$. For discretization and computational errors in higher order finite elements, see Fried [1971].

Interpolation in Euclidian spaces In the preceding section, we presented an elementary approach to a derivation of criterion for error estimate and convergence in a one-dimensional problem. However, this approach cannot be used for general situations in multidimensional problems. We must then resort to more rigorous mathematical analysis. Some mathematical notations and definitions are presented in Sec. 1-2-4. In the following discussions, the plan of Ciarlet and Raviart [1972] which was further detailed by Oden and Reddy [1976] will be followed.

Let h_e be the diameter of element Ω_e given by

$$h_e = \max_{\mathbf{x},\mathbf{y} \in \Omega_e} |\mathbf{x} - \mathbf{y}| = \max_{\mathbf{x},\mathbf{y} \in \Omega_e} \left(\sum_{i=1}^{n} |x_i - y_i|^2 \right)^{1/2}$$

$$(3\text{-}130a)$$

with h defined as the maximum diameter of all of the elements (Fig. 3-6)

$$h = \max(h_1, h_2, \dots, h_E) = \max_{1 \le e \le E} (h_e) \qquad (3\text{-}130b)$$

where E denotes the total number of elements in the finite element model $\overline{\Omega}$ in the domain Ω, together with the boundaries Γ such that

$$\overline{\Omega} = \Omega \cup \Gamma = \bigcup_{e=1}^{E} \overline{\Omega}_e$$

$$\overline{\Omega}_e = \Omega_e \cup \Gamma_e$$

We further define that

$$d_e = \text{sup (diameter of all spheres contained in } \Omega_e) \qquad (3\text{-}131a)$$

and
$$d = \min_{1 \le e \le E} (d_e) \qquad (3\text{-}131b)$$

If we now consider that the elements in Fig. 3-6 represent the global positions as they exist in the finite element model $\bar{\Omega}$, they can be thought of as having been mapped from some fixed master element $\hat{\Omega}_M$, which possesses the geometry described by

$$\hat{h} = \text{diameter of } \hat{\Omega}_M = \max_{x,y \in \hat{\Omega}} |\hat{\mathbf{x}} - \hat{\mathbf{y}}| \qquad (3\text{-}132a)$$

$$\hat{d} = \text{sup (diameters of all spheres contained in } \hat{\Omega}_M) \qquad (3\text{-}132b)$$

The two sets of finite elements Ω_e and $\hat{\Omega}_M$ have the nodal points

$$\mathscr{L}_e = \{\mathbf{x}^N\}_{N=1}^{N_e} \qquad \hat{\mathscr{L}}_M = \{\hat{\mathbf{x}}^N\}_{N=1}^{N_e}$$

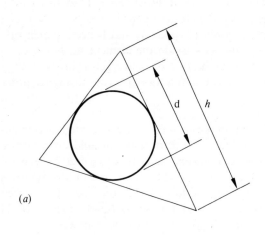

(a)

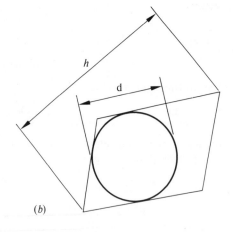

(b)

Figure 3-6 One-dimensional element.

which are related by a linear transformation

$$\mathbf{x} = T\hat{\mathbf{x}} + \mathbf{b} \tag{3-133}$$

where T is the transformation matrix (invertible square matrix) and $\mathbf{b}$ is the translation vector relating the local element coordinates $\mathbf{x}$ and the fixed master coordinates $\hat{\mathbf{x}}$. Then the norm of the transformation matrix on the Euclidean space $\mathbf{R}^n$ is given by

$$\| T \| = \sup_{|\hat{\mathbf{x}}| \neq 0} \frac{|T\hat{\mathbf{x}}|}{|\hat{\mathbf{x}}|} = \frac{1}{\hat{d}} \sup_{|\hat{\mathbf{x}}| = \hat{d}} \| T\hat{\mathbf{x}} \| \tag{3-134}$$

With two points $\tilde{\mathbf{y}}$ and $\tilde{\mathbf{z}}$ chosen in Ω_M on the sphere of diameter $\hat{d}$, we have

$$|\hat{\mathbf{y}} - \hat{\mathbf{z}}| = \hat{d} \tag{3-135}$$

In view of (3-135) and (3-134), we obtain

$$\| T \| = \frac{1}{\hat{d}} \sup |T(\hat{\mathbf{y}} - \hat{\mathbf{z}})| = \frac{1}{\hat{d}} \sup |\mathbf{y} - \mathbf{z}|$$

Thus

$$\| T \| \leq \frac{h}{\hat{d}} \tag{3-136a}$$

Similarly, it can be shown that

$$\| T^{-1} \| \leq \frac{\hat{h}}{d} \tag{3-136b}$$

Furthermore, the components of the matrix T are related by

$$dx = dx_1 \, dx_2 \cdots dx_n = \varepsilon_{ij\ldots l} T_{i1} T_{j2} \cdots T_{ln} \, d\hat{x}_1 \, d\hat{x}_2 \cdots d\hat{x}_n$$

with n not to be summed and $\varepsilon_{ij\ldots l}$ being the n-dimensional permutation symbol. Thus we may write

$$dx = \det |T| \, d\hat{x} \tag{3-137}$$

Finally, derivatives of functions $u(\hat{\mathbf{x}})$ and $u(\mathbf{x})$ on the master and local elements are

$$D_{\hat{x}_i} u(\hat{\mathbf{x}}) = T_{ij} D_{x_j} u(\mathbf{x})$$

$$D_{\hat{x}_i}^2 u(\hat{\mathbf{x}}) = T_{ij} T_{jk} D_{x_k}^2 u(\mathbf{x})$$

$$\vdots$$

$$D_{\hat{x}_i}^{\alpha i} u(\hat{\mathbf{x}}) = T_{ij} T_{jk} \cdots T_{pq} D_{x_q}^{\alpha i} u(\mathbf{x})$$

$$\vdots$$

This process results in an inequality form

$$|D^\alpha u(\hat{\mathbf{x}})| \leq \| T \|^{|\alpha|} |D^\alpha u(\mathbf{x})| \tag{3-138}$$

with $\alpha = (\alpha_1, \alpha_2, \ldots, \alpha_n)$.

With these preliminaries, we are prepared to discuss our main subject, the interpolation theory. There are basically two approaches in handling interpolation errors. The first approach is to prescribe a set of points in $\mathbf{R}^n$ and interpolate the values of a given function u by a polynomial of a given degree. The second approach is to utilize the space which consists of functions having all generalized derivatives of order m and integrable in powers of k called the Sobolev space.

We assume that the given function $u(\mathbf{x})$ to be interpolated can be represented by a Taylor series of the form

$$u(\mathbf{x}^N) = \sum_{r=0}^{k} \frac{1}{r!} \hat{D}^r u(\mathbf{x})(\mathbf{x}^N - \mathbf{x})^r + \frac{1}{(k+1)!} \hat{D}^{k+1} u(\eta_N(\mathbf{x}))(\mathbf{x}^N - \mathbf{x})^{k+1}$$

$$(3\text{-}139)$$

with

$$\eta_N(\mathbf{x}) = \theta_N \mathbf{x} + (1 - \theta_N)\mathbf{x}^N \qquad 0 < \theta_N < 1; \qquad 1 \le N \le N_e \qquad (3\text{-}140)$$

Here D^k denotes the Fréchet derivative related by the partial derivative

$$D^{\alpha} u(\mathbf{x}) = \hat{D}^{|\alpha|} u(\mathbf{x}) \cdot (\mathbf{e}_{\alpha 1}, \mathbf{e}_{\alpha 2}, \ldots, \mathbf{e}_{\alpha n}), \qquad \mathbf{e}_i \cdot \mathbf{e}_j = \delta_{ij}$$

As indicated in Sec. 2-6, all finite elements may be grouped into two categories: Lagrange family (values of function u are specified at nodes) and Hermite family (derivatives as well as the values of function u are specified at nodes).

For the Lagrange family, we note that the finite element Ω_e has the nodal points

$$\mathscr{L}_e = \{\mathbf{x}^N\}_{N=1}^{N_e}$$

Let $\tilde{u}(\mathbf{x})$ be the unique interpolating polynomial of degree $\le p$ of $u(x)$ given by

$$\tilde{u}(\mathbf{x}) = \sum_{N=1}^{N_e} \alpha_N(\mathbf{x}) u(\mathbf{x}^N)$$

where $\alpha_N(\mathbf{x})$ is the interpolation function. Then it can be shown that

$$\hat{D}^m \tilde{u}(\mathbf{x}) = \hat{D}^m u(\mathbf{x}) + \frac{1}{(k+1)!} \sum_{N=1}^{N_e} [\hat{D}^{k+1} u(\eta_N(x))(\mathbf{x}^N - x)^{k+1}] \hat{D}^m \alpha_N(\mathbf{x}) \quad (3\text{-}141)$$

In view of (3-139) and (3-141), we obtain

$$\| \hat{D}^m(u - \tilde{u}) \| \le \frac{1}{(k+1)!} \sum_{N=1}^{N_e} |\hat{D}^{k+1} u(\eta_N(\mathbf{x}))(\mathbf{x}^N - x)^{k+1}| \, \| \hat{D}^m \alpha_N(\mathbf{x}) \| \quad (3\text{-}142)$$

We denote here that

$$C_{k+1} = \sup \| \hat{D}^{k+1} u(\mathbf{x}) \| < \infty \qquad (3\text{-}143)$$

where $\| \hat{D}^{k+1}(u) \|$ is called the operator norm

$$\| D^{k+1} u \| = \sup \{ \| D^{k+1} u(\mathbf{x}) \cdot (\xi_1, \xi_2, \ldots, \xi_n) \|, \, |\xi_i| \le 1, 2, \ldots, n \} \qquad (3\text{-}144)$$

It follows from (3-143) that

$$\left| \hat{D}^{k+1} u(\eta_N(\mathbf{x}))(\mathbf{x}_N - \mathbf{x})^{k+1} \right| \le C_{k+1} h^{k+1}$$

Transforming the element from the set $\mathscr{L}_e$ to an equivalent set $\mathscr{L}_M$,

$$\alpha_N(\mathbf{x}) = \hat{\alpha}_N(\hat{\mathbf{x}}) = \hat{\alpha}_N(T^{-1}(\mathbf{x} - \mathbf{b}))$$

and

$$\hat{D}^m \alpha_N(\mathbf{x})(\xi_1, \xi_2, \ldots, \xi_m) = \hat{D}^m \hat{\alpha}_N(T^{-1}(\mathbf{x} - \mathbf{b}))(T^{-1}\xi_1, T^{-1}\xi_2, \ldots, T^{-1}\xi_m)$$

Noting that $\Omega_M = \hat{\Omega}$, we have

$$\sup_{\mathbf{x} \in \Omega_e} \| \hat{D}^m \alpha_N(\mathbf{x}) \| \le \| T^{-1} \|^m \sup_{\hat{\mathbf{x}} \in \hat{\Omega}} \| \hat{D}^m \hat{\alpha}_N \| \le \frac{\hat{h}^m}{r^m} \sup_{\hat{\mathbf{x}} \in \hat{\Omega}} \| D^m \alpha_N(\hat{\mathbf{x}}) \| \qquad (3\text{-}145)$$

At this point, we introduce a constant c defined as

$$c = \frac{\hat{h}^m}{(p+1)!} \sum_{N=1}^{N_e} \sup_{\hat{\mathbf{x}} \in \hat{\Omega}} \| \hat{D}^m \hat{\alpha}_N(\hat{\mathbf{x}}) \| \qquad (3\text{-}146)$$

Here the constant c is independent of u, h, or d. Now combining expressions (3-142), (3-143), and (3-144), we arrive at the interpolation error formula

$$\sup_{\mathbf{x} \in \Omega_e} \| \hat{D}^m u(\mathbf{x}) - \hat{D}^m \tilde{u}(\mathbf{x}) \| \le c \, \frac{h^{k+1}}{d^m} C_{k+1} \qquad (3\text{-}147)$$

For the Hermite family, on the other hand, we express the finite element Ω_e in the form

$$\mathscr{L}_e = \bigcup_{\mu=0}^{\nu} \left\{ \mathbf{x}_{(\mu)}^N \right\}_{N=1}^{N_\mu}$$

Here the set $\mathscr{L}_e = \{x^N\}_{N=1}^{N_e}$ of local nodes is divided into the $\nu + 1$ subsets which allow us to distinguish subsets of nodal points at which derivatives of a specific order are prescribed. We first note $\tilde{u}(\mathbf{x})$ is of the form

$$\tilde{u}(\mathbf{x}) = \sum_{N=1}^{N_0} u(\mathbf{x}_{(0)}^N) \alpha_N^0(\mathbf{x}) + \sum_{N=1}^{N_1} \sum_{i=1}^{\gamma} [\hat{D}u(\mathbf{x}_{(1)}^N) \cdot e_i^N] \alpha_N^{1,i}(\mathbf{x})$$

Thus,

$$\hat{D}^m \tilde{u}(\mathbf{x}) = \hat{D}^m u(\mathbf{x}) + \frac{1}{(k+1)!} \sum_{N=1}^{N_e} [\hat{D}^{k+1} u(\eta_N^0(\mathbf{x}))(\mathbf{x}_{(0)}^N - \mathbf{x})^{k+1}] \hat{D}^m \alpha_N^0(\mathbf{x})$$

$$+ \frac{1}{k!} \sum_{N=1}^{N_e} \sum_{i=1}^{\gamma} [\hat{D}_N^{k+1} u(\eta_{Ni}^1(\mathbf{x})) \cdot (e_i^N, (\mathbf{x}_{(1)}^N - \mathbf{x})^k)] \hat{D}^m \alpha_N^{1,i}(\mathbf{x}) \qquad (3\text{-}148)$$

Here $\alpha_N^0(\mathbf{x})$ and $\alpha_N^{1,q}$ are unique polynomials of degree $\leq k$ such that

$$\alpha_N^0(\mathbf{x}_{(0)}^M) = \delta_N^M \qquad M, N = 1, 2, \ldots, N_0$$

$$\hat{D}\alpha_N^0(\mathbf{x}_{(1)}^M) \cdot \mathbf{e}_i^M = 0 \qquad \begin{aligned} N &= 1, 2, \ldots, N_0 \\ M &= 1, 2, \ldots, N_1 \\ i &= 1, 2, \ldots, \gamma \end{aligned}$$

$$\alpha_N^{1,i}(\mathbf{x}_{(0)}^M) = 0 \qquad \begin{aligned} N &= 1, 2, \ldots, N_1 \\ M &= 1, 2, \ldots, N_0 \\ i &= 1, 2, \ldots, \gamma \end{aligned}$$

$$\hat{D}\alpha_N^{1,i}(\mathbf{x}_{(1)}^M) \cdot \mathbf{e}_j^M = \delta_N^M \delta_j^i \qquad \begin{aligned} M, N &= 1, 2, \ldots, N \\ i, j &= 1, 2, \ldots, \gamma \end{aligned}$$

$$\eta_N^0(\mathbf{x}) = \theta_N^0 \mathbf{x} + (1 - \theta_N^0)_{(0)} \mathbf{x}^N \qquad \theta_N^0 > 0$$

$$\eta_{Ni}^1(\mathbf{x}) = \theta_{Ni}^1 \mathbf{x} + (1 - \theta_{Ni}^1)_{(1)} \mathbf{x}^N \qquad \theta_{Ni}^1 > 1$$

From these results, proceeding similarly as in (3-77), we obtain for $v = 1$,

$$\| \hat{D}^m(u - \tilde{u}) \| \leq \frac{1}{(k+1)!} \sum_{N=1}^{N_0} | \hat{D}^{k+1} u(\eta_N^0(\mathbf{x})) \cdot (\mathbf{x}_{(0)}^N - \mathbf{x})^{k+1} | \cdot \| \hat{D}^m \alpha_N^0 \alpha_N^0 \|$$

$$+ \frac{1}{k!} \sum_{N=1}^{N_1} \sum_{i=1}^{N} | \hat{D}^{k+1} u(\eta_{Ni}^1(\mathbf{x})) \cdot \mathbf{e}_{i,}^N (\mathbf{x}_{(1)}^N - \mathbf{x})^k | \| \hat{D}^m \alpha_N^{1,i}(\mathbf{x}) \|$$

We note here that $\| e_i \| \leq h \| \hat{e}_i^N \| / \hat{d}$ and

$$\sup_{\mathbf{x} \in \Omega_e} \| \hat{D}^m \alpha_N^0(\mathbf{x}) \| \leq \frac{\hat{h}^m}{d^m} \sup_{\hat{\mathbf{x}} \in \hat{\Omega}} \| \hat{D}^m \hat{\alpha}_N^0 \|$$

$$\sup_{\mathbf{x} \in \Omega_e} \| \hat{D}^m \alpha_N^{1,i}(\mathbf{x}) \| \leq \frac{\hat{h}^m}{d^m} \sup_{\hat{\mathbf{x}} \in \hat{\Omega}} \| \hat{D}^m \alpha_N^{1,i}(\hat{\mathbf{x}}) \|$$

In view of these results and letting

$$c = \frac{\hat{h}^m}{(k+1)!} \sum_{N=1}^{N_e} \sup_{\hat{\mathbf{x}} \in \hat{\Omega}} \| \hat{D}^m \hat{\alpha}_N^0(\hat{\mathbf{x}}) \| + \frac{1}{k!} \frac{\hat{h}^m}{\hat{d}} \sum_{N=1}^{N_1} \sum_{i=1}^{\gamma} \| \hat{e}_i^N \| \sup_{\hat{\mathbf{x}} \in \hat{\Omega}} \| \hat{D}^m \alpha_N^{1,i}(\hat{\mathbf{x}}) \| \qquad (3\text{-}149)$$

we obtain (3-147).

Although the derivation (3-147) undergoes a tedious process, it is interesting to note that if the mesh parameter d is equal to h (this is true only for a circular element), then (3-147) reduces to

$$\sup_{\mathbf{x} \in \Omega_e} \| \hat{D}^m(u - \tilde{u}) \| \leq ch^{k+1-m} C_{k+1} \qquad (3\text{-}150)$$

This is identical to (3-130) which was derived for a one-dimensional element. It is clear from (3-147) that an increase of discretization error results from $h \gg r$, which represents a needle-like element as schematically shown in Fig. 3-7.

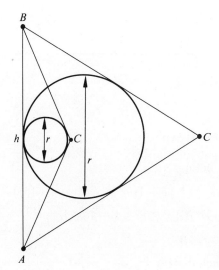

Figure 3-7 Two-dimensional element. (*a*) Triangular element; (*b*) quadrilateral element.

Interpolation in Sobolev spaces Instead of interpolating functions u in Euclidean space, we now consider the accuracy of finite element interpolations in Sobolev spaces $H^{p+1}(\Omega)$. From the definition of (3-138), we may write

$$\int_{\hat{\Omega}} |D^\alpha \hat{u}|^2 \, d\hat{x} \leq \|T\|^{2s} \int_{\Omega} |D^s u(\mathbf{x})|^2 (\det T)^{-1} \, dx$$

with $|\alpha| = s$ and $\hat{u} = u(\hat{x})$. Thus we obtain the relationships,

$$|\hat{u}|_{H^s(\hat{\Omega})} \leq \|T\|^s |\det T|^{-1/2} |u|_{H^s(\Omega)} \tag{3-151}$$

and

$$|\hat{u}|_{H^s(\Omega)} \leq \|T^{-1}\|^s |\det T|^{1/2} |\hat{u}|_{H^s(\hat{\Omega})} \tag{3-152}$$

Here $|u|$ is called the seminorm defined as

$$|u|_{H_p^s(\Omega)} = \|u\|_{L_p^s(\Omega)} = \left(\sum_{|\alpha| \leq m} \|D^\alpha u\|_{L_p(\Omega)}^p \right)^{1/p} \tag{3-153}$$

With $\|T^{-1}\| \geq 1$, the norm of u in $H^m(\Omega)$ takes the form

$$\|u\|_{H^m(\Omega)} = \left(\sum_{|\alpha| \leq m} |u|_{H^{|\alpha|}(\Omega)}^2 \right)^{1/2}$$

$$\leq |\det T|^{1/2} \left(\sum_{|\alpha| \leq m} \|T^{-1}\|^{2|\alpha|} |\hat{u}|_{H^{|\alpha|}(\hat{\Omega})}^2 \right)^{1/2}$$

$$\leq |\det T|^{1/2} \|T^{-1}\|^m \|\hat{u}\|_{H^m(\Omega)} \tag{3-154}$$

Introducing a projection operator Π that casts each u in $W_p^m(\Omega)$ into some polynomial in $\mathscr{P}_{m-1}(\Omega)$, we may decompose $W_p^m(\Omega)$ into the direct sum of two subspaces $\mathscr{P}_{m-1}(\Omega)$ and a space $M_p^m(\Omega) = W_p^m(\Omega) - \mathscr{P}_{m-1}$

$$\hat{\Pi}u = u - \Pi u = (I - \Pi)u$$

$$\hat{\Pi}^2 = \hat{\Pi}$$

$$u = \Pi u + \hat{\Pi}u$$

It then follows that

$$W_p^m(\Omega) = \mathscr{P}_{m-1}(\Omega) + M_p^m(\Omega)$$

With these definitions and from (3-154), we obtain

$$\| u - \Pi u \|_{H^m(\Omega)} \leq |\det T|^{1/2} \| T^{-1} \|^m \| \hat{u} - \hat{\Pi}\hat{u} \|_{H^m(\hat{\Omega})} \qquad (3\text{-}155)$$

Let Π be any linear operator from H^{k+1} into $H^m(\Omega)$, $0 \leq m \leq k+1$ such that

$$\Pi u = u \qquad \forall \in \mathscr{P}_k(\Omega)$$

Then, $\forall u \in H^{k+1}$, there is a constant $k \geq 0$ such that†

$$\| u - \Pi u \|_m \leq k \| I - \Pi \|_{(H^{k+1}, H^m)} |u|_{k+1}$$

where $|u|_{k+1}$ is the seminorm

$$|u|_{k+1}^2 = \int_\Omega \sum_{|\alpha|=k+1} |D^\alpha u|^2 \, dx$$

Thus

$$\| \hat{u} - \hat{\Pi}\hat{u} \|_{H^m(\hat{\Omega})} \leq c_1 \| \hat{I} - \hat{\Pi} \|_{(H^{k+1}(\hat{\Omega}), H^m(\hat{\Omega}))} |\hat{u}|_{H^{k+1}(\hat{\Omega})} \qquad (3\text{-}156)$$

In view of (3-151),

$$\| \hat{u} - \hat{\Pi}\hat{u} \|_{H^m(\hat{\Omega})} \leq c_2 \| \hat{I} - \hat{\Pi} \|_{(H^{k+1}(\hat{\Omega}), H^m(\hat{\Omega}))} \| T \|^{k+1} |\det T|^{-1/2} |u|_{H^{k+1}(\Omega)}$$

Returning to (3-155) with these inequalities, we have

$$\| u - \Pi u \|_{H^m(\Omega)}$$

$$\leq |\det T|^{1/2} \| T^{-1} \|^m c_1 \| I - \Pi \|_{(H^{k+1}(\hat{\Omega}), H^m(\hat{\Omega}))} \| T \|^{k+1} |\det T|^{-1/2} |u|_{H^{k+1}(\Omega)}$$

From (3-136)

$$\| T^{-1} \|^m \| T \|^{k+1} \leq \frac{\hat{h}^m h^{k+1}}{\hat{d}^{k+1} d^m}$$

Let the constant c, independent of u and h, be denoted by

$$c = c_1 \frac{\hat{h}^m}{\hat{d}^{p+1}} \| \hat{I} - \hat{\Pi} \|_{(H^{k+1}(\hat{\Omega}), H^m(\hat{\Omega}))} \qquad (3\text{-}157)$$

then we obtain the interpolation error formula in Sobolev spaces in the form

$$\| u - \Pi u \|_{H^m(\Omega)} \leq c \frac{h^{k+1}}{d^m} |u|_{H^{k+1}(\Omega)} \qquad (3\text{-}158)$$

---

† See Bramble and Hilbert [1970, 1971].

If we refine a mesh such that h/d is kept constant,

$$\frac{h}{d} = \frac{h}{\min_{1 \le e \le E} (h_e)} \le \sigma \qquad \sigma > 0$$

then (3-158) takes the form

$$\| u - \Pi u \|_{H^m(\Omega)} \le c h^{k+1-m} |u|_{H^{k+1}(\Omega)} \tag{3-159}$$

It is clear that if polynomials of degree k are used, we require $k > m$ for convergence in $H^m(\Omega)$. For arbitrary $u \in H^r(\Omega)$ there exists a constant $c > 0$ and a function $\tilde{u} \in S_h(\Omega)$, $S_h(\Omega) \subset H^m(\Omega)$, such that

$$\| u - \tilde{u} \|_{H^s(\Omega)} \le c h^\mu \| u \|_{H^r(\Omega)} \tag{3-160}$$

where $0 \le s \le \min(m, r)$ and $\mu = \min(k + 1 - s, r - s)$.

3-4-3 Boundary Value Problems

Consider the linear elliptic boundary value problem

$$Au = f \quad \text{in } \Omega \tag{3-161a}$$

$$u = 0 \quad \text{on } \Gamma \tag{3-161b}$$

where f is in $L_2(\Omega)$ and A is the second-order, linear partial-differential operator

$$Au = -\left(\frac{\partial^2 u}{\partial x^2} + \frac{\partial^2 u}{\partial x \, \partial y} + \frac{\partial^2 u}{\partial y^2} \right) + u \qquad (x, y) \in \Omega \subset \mathbf{R}^2$$

If $f \in H^r(\Omega)$, $r \ge 0$, then the solution u exists in $H^{r+2}(\Omega)$. Note that $u \in H^2(\Omega)$ in (3-160). Now consider a problem of finding $u \in H_0^1(\Omega)$ such that

$$\int_\Omega \left[\frac{\partial u}{\partial x} \frac{\partial v}{\partial x} + \frac{\partial u}{\partial y} \frac{\partial v}{\partial x} + \frac{\partial u}{\partial x} \frac{\partial v}{\partial y} + \frac{\partial u}{\partial y} \frac{\partial v}{\partial y} + uv \right] dx \, dy = \int_\Omega f v \, dx \, dy \quad \forall v \in H_0^1(\Omega)$$

$$\tag{3-162}$$

which is a variational boundary value problem associated with (3-161). The boundary conditions are included in (3-162) with $v \in H_0^1(\Omega)$. The expression (3-162) can be written as

$$(Au, v)_{H^0(\Omega)} = (f, v)_{H^0(\Omega)}$$

or

$$B(u, v) = (f, v) \qquad \forall v \in H_0^1(\Omega)$$

where $(\cdot, \cdot)_{H^0(\Omega)}$ is the $L_2(\Omega)$ inner product, and for each u, $B(u, v)$ is a linear functional on $H_0^1(\Omega)$.

For the general case of a linear partial-differential operator of order $2m$ with homogeneous boundary conditions,

$$Au = f \quad \text{in } \Omega$$

$$B_k u = 0 \quad \text{on } \Gamma \qquad 0 \le k \le m - 1$$

with
$$Au = \sum_{|\alpha|,|\beta| \le m} (-1)^{|\alpha|} D^{\alpha}(a_{\alpha\beta}(\mathbf{x})D^{\beta}u)$$

$$B_k u = \sum_{|\alpha| \le q_k} b_{k\alpha}(\mathbf{x})D^{\alpha}u \qquad 0 \le q_k \le 2m-1 \qquad b_{k\alpha} \in C^{\infty}(\Gamma)$$

Then by partial integration, we obtain

$$\int_{\Omega} (Au)v \, dx = \int_{\Omega} \sum_{|\alpha|,|\beta| \le m} a_{\alpha\beta}(\mathbf{x})D^{\alpha}uD^{\beta}v \, dx + \sum_{k=0}^{m-1} \int_{\Gamma} G_k u F_k v \, ds$$

where G_k is a complementary system of boundary operators of order $2m-1$ and F_k forms a Dirichlet system of order $m-1$. Now the elliptic boundary value problem is represented by

$$B(u, v) = (f, v)$$

where

$$B(u, v) = \int_{\Omega} \sum_{|\alpha|,|\beta| \le m} a_{\alpha\beta}(\mathbf{x})D^{\alpha}uD^{\beta}v \, dx$$

If the boundary conditions are nonhomogeneous, we have

$$B_k u = g_k \qquad \text{on } \Gamma$$

In this case, we introduce a function w such that $B_k w = g_k$; then the function $\hat{u} = u - w$ satisfies

$$A\hat{u} = \hat{f} \tag{3-163a}$$

$$B_k \hat{u} = 0 \tag{3-163b}$$

with the modified interior data $\hat{f} = f - Aw$ and $g_k = B_k w$.

Note that the boundary operator system $\{B_k\}_{k=0}^{m-1}$ consists of two distinct classes of boundary operators. $\{B_k\}_{k=0}^{r-1}$ with $q_k < m$ represents *essential* (Dirichlet) boundary conditions and $\{B_k\}_{k=r}^{m-1}$ with $q_k \ge m$ implies *natural* (Neumann) boundary conditions.

The boundary operators are intrinsically related to the operator A; thus, we may combine (3-163a) and (3-163b) and write

$$Pu = F$$

where $P \equiv (A; B_k)$. The operator P maps $H^s(\Omega)$, $s \ge 2m$, into the product space [see the trace theorem and (1-65)],

$$J_{s-q_k-1/2}^{s-2m}(\bar{\Omega}) \equiv H^{s-2m}(\Omega) \times \prod_{k=0}^{m-1} H^{s-q_k-1/2}(\Gamma)$$

provided with the norm

$$\| Pu \|_{J_{s-q_k-1/2(\Omega)}^{s-2m}}^2 = \| Au \|_{H^{s-2m}(\Omega)}^2 + \sum_{k=0}^{m-1} \| B_k u \|_{H^{s-q_k-1/2}(\Gamma)}^2$$

The boundary value problem (3-163) is said to be *well posed* if there exists a unique solution $u(\mathbf{x}) \in H^s(\Omega)$, $s \geq 2m$, that depends continuously on the data

$$F = (f; g_k) \in J^{s-2m}_{s-q_k-1/2}(\overline{\Omega})$$

The existence and uniqueness of a solution depends on whether or not the operator P is one-to-one (uniqueness) and onto (existence).

For the homogeneous boundary conditions and for $f \in H^r(\Omega)$, $r \geq 0$, u must be in $H^{r+2m}(\Omega)$ because u must be differentiated $2m$ times to produce a function with the same smoothness of f. For the nonhomogeneous boundary conditions

$$B_k u = g_k \qquad \text{on } \Gamma$$

and if $g_k \in H^r(\Gamma)$ and $f = 0$, it follows from the trace theorem (1-65) that $u \in H^{r-q_k-1/2}(\Omega)$. These observations concerning the "degree of smoothness" are referred to as *regularity*. Let us consider $u \in H^s(\Omega)$, $s \geq 2m$; then there exists a constant $c > 0$ such that

$$\| u \|_{H^s(\Omega)} \leq c \left[\| f \|_{H^{s-2m}(\Omega)} + \sum_{k=0}^{m-1} \| g_k \|_{H^{s-q_k-1/2}(\Gamma)} \right] \tag{3-164}$$

This is known as regularity inequality. For proof, see Lions and Magenes [1972].

We now discuss the Lax–Milgram theorem which is well known for linear variational boundary value problems. Let U and V be Hilbert spaces. We wish to find $u \in U$ such that

$$B(u, v) = l(v) \qquad \forall \, v \in V \tag{3-165}$$

where $B(\cdot, \cdot)$ represents a bilinear form from $U \times V$ into $\mathbf{R}$ and $l \in V'$. Here our concern is to find what conditions can be imposed on $B(\cdot, \cdot)$ such that a unique solution to (3-165) will exist and depend continuously on the data l. We consider that

1. There exists a constant $M > 0$ such that

$$B(u, v) \leq M \| u \|_U \| v \|_V \qquad \forall \, u \in U \qquad v \in V \tag{3-166}$$

2. There exists a constant $N > 0$ such that

$$\inf_{\substack{u \in U \\ \| u \|_U = 1}} \sup_{\substack{v \in V \\ \| v \|_V \leq 1}} |B(u, v)| \geq N > 0 \tag{3-167}$$

3.

$$\sup_{u \in U} B(u, v) > 0 \qquad v \neq 0 \tag{3-168}$$

Then there exists a unique solution such that

$$B(u, v) = l(v) \qquad \forall \, v \in V \qquad l \in V' \tag{3-169}$$

with the solution u_0 depending continuously on the data such that

$$\| u_0 \|_U \leq \frac{1}{N} \| l \|_{V'} \tag{3-170}$$

The expressions (3-166), (3-167), and (3-168) are known as the Lax–Milgram theorem dealing with existence and uniqueness of the linear elliptic boundary value problems.

Consider now the Galerkin approximation to find $\tilde{u} \in \tilde{U}$ such that

$$B(\tilde{u}, \tilde{v}) = l(\tilde{v}) \qquad \forall \, \tilde{v} \in \tilde{V} \qquad (3\text{-}171)$$

Then we have the following conditions:

1. There exists a unique solution to (3-171) if there exist constants $\tilde{M}$, $m > 0$ such that† $\forall \, \tilde{u} \in \tilde{U}, \forall \, \tilde{v} \in \tilde{V}$

$$B(\tilde{u}, \tilde{v}) \leq \tilde{M} \, \|\tilde{u}\|_{\tilde{v}} \, \|\tilde{v}\|_{\tilde{v}} \qquad (3\text{-}172a)$$

$$\inf_{\|\tilde{u}\|_{\tilde{v}} = 1} \; \sup_{\|\tilde{v}\|_{\tilde{v}} \leq 1} \; |B(\tilde{u}, \tilde{v})| \geq m \qquad (3\text{-}172b)$$

$$\sup_{\tilde{u} \in \tilde{U}} B(\tilde{u}, \tilde{v}) > 0 \qquad \tilde{u} \neq \tilde{v} \qquad (3\text{-}172c)$$

2. $\tilde{U}$ and $\tilde{V}$ are linear subspaces of U and V, respectively.
3. The Galerkin approximation $\tilde{u}_0$ of the solution u_0 of (3-171) is such that

$$\|u_0 - \tilde{u}_0\|_U \leq c \, \|u_0 - \tilde{u}\|_U \qquad \forall \, \tilde{u} \in \tilde{U} \qquad (3\text{-}172d)$$

where c is a constant independent of u, v, or h.

3-4-4 Error Estimates for Elliptic Equation

With the results of Sec. 3-4-2 and Sec. 3-4-3, we are now prepared to derive convergence criteria and a priori error estimates‡ for finite element approximations. We limit our discussions to the linear elliptic boundary value problems of the type

$$Au = f \qquad \text{in } \Omega$$
$$B_k u = g_k \qquad \text{on } \Gamma \qquad 0 \leq k \leq m - 1 \qquad (3\text{-}173)$$

where A is the partial-differential operator of order $2m$ and B_k denote the boundary operators. The associated variational boundary problem is represented by

$$B(u, v) = \int_{\Omega} \hat{f} v \, dx \qquad \forall \, v \in H_0^m(\Omega) \qquad (3\text{-}174)$$

where $\hat{f} = f - Aw$ and $B_k w = g_k$ on Γ. Now the solution can be sought in $H_0^m(\Omega)$.

† Here the subspace $\tilde{U}$ is often referred to as the space of trial functions, and the subspace $\tilde{V}$ as the space of test functions. When $U = V$, we have $\tilde{U} = \tilde{V}$ and the bilinear form $B(\tilde{u}, \tilde{v})$ is symmetric.

‡ An a priori error is meant an error calculated from the data without first determining the approximation itself.

The Galerkin finite element approximation is given by

$$B(\tilde{u}, \tilde{v}) = \int_\Omega \hat{f}\tilde{v} \, dx \qquad \forall \, \tilde{v} \in S_h(\Omega) \tag{3-175}$$

where $S_h(\Omega) \subset H_0^m(\Omega)$. Let u_0 and $\tilde{u}_0$ be the solutions of (3-174) and (3-175), respectively. From the Lax–Milgram theorem (3-166–3-168) and the existence and uniqueness properties (3-172), we are assured of unique solutions for (3-174) and (3-175). Let the error between u_0 and $\tilde{u}_0$ be denoted as $e = u_0 - \tilde{u}_0$. Then from (3-172d), we have

$$\|e\|_{H^m(\Omega)} \leq c\|u_0 - \tilde{u}\|_{H^m(\Omega)} \qquad \forall \, \tilde{u} \in S_h(\Omega)$$

In view of (3-160), we obtain

$$\|e\|_{H^m(\Omega)} \leq ch^\mu \|u_0\|_{H^s(\Omega)} \qquad \mu = \min(k+1-m, s-m) \tag{3-176}$$

Introducing the regularity inequality (3-164) into (3-176) yields

$$\|e\|_{H^m(\Omega)} \leq c\left\{h^{\mu_1}\|f\|_{H^r(\Omega)} + h^{\mu_2}\sum_{j=0}^{m-1}\|g_j\|_{H^{p_j}(\Omega)}\right\} \tag{3-177}$$

where c is a constant > 0, $r = s - 2m$, $p_j = s - q_j - \frac{1}{2}$, and

$$\mu_1 = \min(k+1-m, r+m) \tag{3-177a}$$

$$\mu_2 = \min\left[k+1-m, \min_{0 \leq j \leq m-1}(p_j + q_j + \tfrac{1}{2} - m)\right] \tag{3-177b}$$

with q_j being the order of B_j (boundary operator).

Example 3-1 Discuss an error estimate for the Dirichlet problem of the type

$$-\nabla^2 u + u = f \text{ in } \Omega \qquad \Omega \subset \mathbf{R}^2 \qquad f \in L_2(\Omega)$$

$$u = 0 \quad \text{on } \Gamma$$

Consider a linear triangular element (three nodes). Then $S_h(\Omega) \in H_0^1(\Omega)$ with $m = 1, k = 1, r = 0$. Thus,

$$\|e\|_{H^1(\Omega)} \leq ch\|f\|_{L_2(\Omega)}$$

If a quadratic triangle is used (six nodes), we have $S_h(\Omega) \subset H^1(\Omega)$ with $k = 2$. Thus,

$$\|e\|_{H^1(\Omega)} \leq ch^2\|f\|_{H^0(\Omega)}$$

Example 3-2 Consider the same problem as above except that data are replaced by

$$f = 0 \qquad \text{in } \Omega$$

$$u = g \qquad \text{on } \Gamma, \qquad g \in L_2(\Gamma)$$

The error estimate for both linear and quadratic triangles takes the form

$$\| e \|_{H^1(\Omega)} \le ch^{-1/2} \| g \|_{L_2(\Gamma)}$$

Note that the Galerkin finite element approximations are divergent in the $H^1(\Omega)$-norm.

Example 3-3 Let g belong to $C^\infty(\Gamma)$ in the example above. Then the error estimate is governed by $\mu_1 = k + m - 1 = k$, not $\mu_2 = \infty + 0 + \frac{1}{2} - m = \infty$.

$$\| e \|_{H^1(\Omega)} \le ch^k \| g \|_{H^r(\Gamma)} \qquad k = 1, 2; \ r \ge 2$$

It is sometimes useful to evaluate the error in lower norms. To this end we consider the duality argument of the so-called Aubin–Nitsche method [Aubin, 1967; Nitsche, 1968] which gives

$$\| e \|_{H^\alpha(\Omega)} \le c \left(h^{\sigma_1} \| f \|_{H^r(\Omega)} + h^{\sigma_2} \sum_{j=0}^{m-1} \| g_j \|_{H^{p_j}(\Gamma)} \right) \tag{3-178}$$

where $\alpha \le m$, $\sigma_1 = \mu_1 + \sigma$, $\sigma_2 = \mu_2 + \sigma$, $\sigma = \min(k + 1 - m, \ m - \alpha)$, with μ_1 and μ_2 as defined in (3-177).

To calculate errors in various norms from the results of the finite element solutions, one proceeds as follows: If the error in H^1 norm is desired for a one-dimensional problem $(0 < x < 1)$, we determine

$$\| e \|_{H^1(0,1)} = \left[\int_0^1 \left[|e|^2 + \left| \frac{de}{dx} \right|^2 \right] dx \right]^{1/2}$$

whereas for the $H^0 = L_2$ norm

$$\| e \|_{H^0(0,1)} = \left[\int_0^1 |e|^2 \, dx \right]^{1/2}$$

where $e = u - \hat{u}$ with u and $\hat{u}$ being the exact and finite element solutions, respectively. Here $\hat{u}$ can be replaced by

$$\hat{u} = \Phi_i u_i = \sum_{e=1}^{E} \Phi_N^{(e)} \Delta_{Ni}^{(e)} u_i$$

If the exact solution assumes a complicated form, then the integrals in these norms are cumbersome to calculate.

Finally, we also evaluate the pointwise error, an error at any arbitrary point. Toward this end, let us examine the one-dimensional problem

$$a \frac{d^2 u}{dx^2} + b \frac{du}{dx} + cu = f(x) \qquad 0 < x < 1$$

$$u(0) = u(1) = 0$$

Then the error estimate assumes the form

$$|e(Z_i)| \le \hat{c} h^{2\nu} \| f \|_{H^r(\Omega)} \tag{3-179}$$

where Z_i is a global node number and $v = \min(k, r + 1)$. If the data $f(x)$ is sufficiently smooth, then $r + 1 > k$ and the pointwise error is $0(h^{2k})$.

3-4-5 Parabolic and Hyperbolic Equations

Parabolic equations The mathematical structure of finite element approximations for the parabolic and hyperbolic equations has not been studied as deeply as that in the linear elliptic boundary value problems. However, the basic philosophy introduced in the earlier sections can be extended to other linear boundary-initial value problems or time-dependent problems [Douglas and Dupont, 1970].

Consider the diffusion equation of the form

$$\left.\begin{aligned}
\frac{\partial u(\mathbf{x}, t)}{\partial t} + Au(\mathbf{x}, t) &= f(\mathbf{x}, t) \qquad \mathbf{x} \in \Omega;\ t \in (0, \tau) \\[2mm]
\gamma_j u(\mathbf{x}, t) &= 0 \qquad \mathbf{x} \in \Gamma;\ t \in (0, \tau);\ 0 \le j \le m - 1 \\[2mm]
u(\mathbf{x}, 0) &= u_0(\mathbf{x}) \qquad \mathbf{x} \in \Omega
\end{aligned}\right\} \qquad (3\text{-}180a)$$

where
$$A = \sum_{|\alpha|, |\beta| \le m} (-1)^{|\alpha|} D^{\alpha} a_{\alpha\beta}(\mathbf{x}) D^{\beta}$$

The associated variational problem is represented by the $L_2(\Omega)$ inner products,

$$\left.\begin{aligned}
\left(\frac{\partial u(t)}{\partial t}, v\right) + B(u(t), v) &= (f(t), v) \qquad \forall\, v \in H_0^m(\Omega);\ t \in (0, \tau) \\[2mm]
(u(0), v) &= (u_0, v) \qquad \forall\, v \in H_0^m(\Omega)
\end{aligned}\right\} \qquad (3\text{-}180b)$$

whereas the Galerkin approximation takes the form

$$\left.\begin{aligned}
\left(\frac{\partial \tilde{u}(t)}{\partial t}, \tilde{v}\right) + B(\tilde{u}(t), \tilde{v}) &= (f(t), \tilde{v}) \qquad t \in (0, \tau) \\[2mm]
(\tilde{u}(0), \tilde{v}) &= (\tilde{u}_0, \tilde{v}) \qquad \forall\, \tilde{v} \in S_h(\Omega)
\end{aligned}\right\} \qquad (3\text{-}180c)$$

where u and $\tilde{u}$ are the exact and approximate solutions, respectively, and $B(u, v)$ represents the bilinear form,

$$B(u, v) = \int_{\Omega} \sum_{|\alpha|, |\beta| \le m} a_{\alpha\beta}(\mathbf{x}) D^{\alpha} v D^{\beta} u\, dx$$

Here we assume that there exist constants $M, \gamma > 0$ such that

$$B(u, v) \le M \|u\|_{H^m(\Omega)} \|v\|_{H^m(\Omega)} \qquad \text{and} \qquad B(u, u) \ge \gamma \|u\|_{H^m(\Omega)}^2$$

Each element in $S_h(\Omega)$ is of the form

$$\tilde{v}(\mathbf{x}) = \sum_{i=1}^{r} c_i \Phi_i(\mathbf{x}) \qquad\qquad (3\text{-}181a)$$

and

$$\tilde{u}(\mathbf{x}, t) = \sum_{i=1}^{r} w_i(t) \Phi_i(\mathbf{x}) \qquad\qquad (3\text{-}181b)$$

is assumed to satisfy (3-180). Substituting (3-181) into (3-180) leads to the finite element equations,

$$A_{ij}\dot{w}_j(t) + B_{ij}w_j(t) = f_i(t) \\ A_{ij}w_j(0) = g_i \qquad \left.\right\} \qquad (3\text{-}182)$$

where $A_{ij} = (\Phi_i, \Phi_j)$, $B_{ij} = B(\Phi_i, \Phi_j)$, $f_i = (f, \Phi_i)$ and $g_i = (u_0, \Phi_i)$.

This is known as a semidiscrete Galerkin approximation since $\tilde{u}(\mathbf{x}, t)$ is still dependent on time. Introducing the finite difference approximations to $\dot{w}_j(t)$ as demonstrated in Sec. 3-3-4, the expression (3-182a) is transformed to

$$\left[A_{ij} + \Delta t(1 - \theta)B_{ij}\right]\overset{*}{u}_j^{(n+1)} = \Delta t f_i + (A_{ij} - \Delta t\theta B_{ij})\overset{*}{u}_j^{(n)} \\ A_{ij}\overset{*}{u}_j = g_i \qquad \left.\right\} \qquad (3\text{-}183)$$

where Δt is the time increment $\Delta t = t^{(n+1)} - t^{(n)}$ and θ takes the values

$$\theta = 0 \quad \text{for } w_j(t) = \overset{*}{u}_j^{(n+1)}$$

$$\theta = \tfrac{1}{2} \quad \text{for } w_j(t) = \frac{\overset{*}{u}_j^{(n+1)} + \overset{*}{u}_j^{(n)}}{2}$$

$$\theta = 1 \quad \text{for } w_j(t) = \overset{*}{u}_j^{(n)}$$

Determination of a priori L_2 estimates for semidiscrete Galerkin approximations due to Douglas and Dupont [1970] and others [see Oden and Reddy, 1976] is to find the finite element solution $\overset{*}{u}$ represented by (3-182) as an auxiliary problem,

$$B(\overset{*}{u}, \tilde{v}) = B(u, \tilde{v}) \qquad \forall \, \tilde{v} \in S_h(\Omega); \, t \geq 0$$

Associated with the solution $\overset{*}{u}$ are

$$\left\| u - \overset{*}{u} \right\|_{H^m(\Omega)} \leq c_0 h^{k+1-m} \left\| u(t) \right\|_{H^{k+1}(\Omega)} \qquad (3\text{-}184a)$$

$$\left\| \frac{\partial}{\partial t}(u - \overset{*}{u}) \right\|_{H^m(\Omega)} \leq c_0 h^{k+1-m} \left\| \frac{\partial u(t)}{\partial t} \right\|_{H^{k+1}(\Omega)} \qquad (3\text{-}184b)$$

From the relations (3-179)–(3-184), we obtain

$$\left(\frac{\partial e}{\partial t}, E\right) = \left(\frac{\partial E}{\partial t}, E\right) + B(E, E) \geq \| E \| \frac{d}{dt} \| E \| + \gamma \| E \|^2 \leq \left\| \frac{\partial e}{\partial t} \right\| \| E \| \qquad (3\text{-}185)$$

where e and E are referred to as approximation error and temporal approximation error, respectively

$$e = u - \overset{*}{u}, \qquad E = \tilde{u} - \overset{*}{u}$$

It follows from (3-185) that

$$\frac{d}{dt} \| E \| + \gamma \| E \| \leq \left\| \frac{\partial e}{\partial t} \right\| \qquad (3\text{-}186)$$

This has the solution of the type

$$\| E(t) \| \leq \exp(-\gamma t) \left[\| E(0) \| + \int_0^t \exp(\gamma s) \left\| \frac{\partial e}{\partial t} \right\| ds \right] \qquad (3\text{-}187)$$

Denoting the semidiscrete approximation error between $u(t)$ and $\tilde{u}(t)$ as $\varepsilon(t)$, we have

$$\| \varepsilon(t) \| = \| u(t) - \tilde{u}(t) \| = \| u(t) - \overset{*}{u}(t) + \overset{*}{u}(t) - \tilde{u}(t) \|$$

$$\leq \| e(t) \| + \| E(t) \| \leq ch^{k+1} \| u(t) \|_{H^{k+1}(\Omega)}$$

$$+ \exp(-\gamma t) \left[\| E(0) \| + \int_0^t \exp(\gamma s) \left\| \frac{\partial e}{\partial t} \right\| ds \right] \qquad (3\text{-}188)$$

With these results and in view of (3-184) and (3-159), we have an error estimate for the finite element solution for the diffusion equation as

$$\| \varepsilon(t) \| \leq \exp(-\gamma t) \| \varepsilon(0) \| + ch^{k+1} \left[\| u(t) \|_{H^{k+1}(\Omega)} \right.$$

$$+ \exp(-\gamma t) \| u(0) \|_{H^{k+1}(\Omega)} + \int_0^t \exp(\gamma(s-t)) \left\| \frac{\partial u(s)}{\partial t} \right\|_{H^{k+1}(\Omega)} ds \right] \qquad (3\text{-}189)$$

It is interesting to note that accuracy of the solution here is dictated by h^{k+1} identical to an elliptic problem in $L_2(\Omega)$ norm.

First order hyperbolic equations Let us now examine first order hyperbolic equations of the type

$$\left. \begin{aligned} \frac{\partial u(t)}{\partial t} + Lu(t) &= f(t) \\[2ex] u(0) &= u_0 \end{aligned} \right\} \qquad (3\text{-}190)$$

where L is the first order linear operator

$$L = a_i D^i = \sum_{i=1}^n a_i(\mathbf{x}, t) \frac{\partial}{\partial x^i}$$

Proceeding similarly as in (3-179) through (3-83) leads to

$$\left. \begin{aligned} \left(\frac{\partial u(t)}{\partial t}, v \right) + (Lu(t), v) &= (f(t), v) \qquad \forall \, v \in H_0^1(\Omega); \ t \in (0, \tau) \\[2ex] (u(0), v) &= (u, v) \qquad \forall \, v \in H_0^1(\Omega) \end{aligned} \right\} \qquad (3\text{-}191)$$

$$\left. \begin{aligned} \left(\frac{\partial \tilde{u}(t)}{\partial t}, \tilde{v} \right) + (L\tilde{u}(t), \tilde{v}) &= (f(t), \tilde{v}) \qquad \forall \, \tilde{v} \in S_h(\Omega) \\[2ex] (\tilde{u}(0), \tilde{v}) &= (\tilde{u}_0, \tilde{v}) \qquad \forall \, t \in (0, \tau) \end{aligned} \right\} \qquad (3\text{-}192)$$

$$\left. \begin{aligned} A_{ij} \dot{w}_j + C_{ij} w_j &= f_i(t) \\[2ex] A_{ij} w_j(0) &= g_i \end{aligned} \right\} \qquad (3\text{-}193)$$

and

$$[A_{ij} + \Delta t(1 - \theta)C_{ij}]\overset{*}{u}_j^{(n+1)} = \Delta t f_i + [A_{ij} - \Delta t \theta C_{ij}]\overset{*}{u}_j^{(n)} \left.\right\}$$
$$A_{ij}\overset{*}{u}_j(0) = g_i \qquad\qquad\qquad\qquad\qquad\qquad\qquad\qquad\qquad\qquad\left.\right\} \tag{3-194}$$

Taking the Laplace transforms of (3-191) and (3-192), we obtain

$$s(U(s), v) + (LU(s), v) = (u_0, v) \tag{3-195}$$

$$s(\tilde{U}, \tilde{v}) + (L\tilde{U}(s), \tilde{v}) = (u_0, \tilde{v}) \tag{3-196}$$

where $U(\mathbf{x}, s) = \mathscr{L}[u(\mathbf{x}, t)]$ and $\tilde{U}(\mathbf{x}, s) = \mathscr{L}[\tilde{u}(\mathbf{x}, t)]$. Subtracting (3-196) from (3-195) yields

$$s(\hat{\varepsilon}(s), \tilde{v}) + (L\hat{\varepsilon}(s), \tilde{v}) = 0 \qquad \forall\, v \in S_h(\Omega)$$

where $\hat{\varepsilon} = U(s) - \tilde{U}(s) = \mathscr{L}[u(t) - \tilde{u}(t)]$. If $\beta = 1$, it follows that there exists a constant c such that

$$\| \hat{\varepsilon} \| \le 2 \| \bar{E} \| + \frac{c}{s} \| \bar{E} \|_{H^1(\Omega)}$$

where $\bar{E} = U - \tilde{U}$ with $\tilde{U}$ being the element of $S_h(\Omega)$ that interpolates U. It further follows that the finite element error estimate for the first order hyperbolic equation for $\theta = 1$ takes the form [see Oden and Reddy, 1976]

$$\| \varepsilon(t) \| \le c_1 h^{k+1} \| u \|_{H^{k+1}(\Omega)} + c_2 h^k \int_0^{n\Delta t} \| u \|_{H^{k+1}(\Omega)} \, dt \tag{3-197}$$

Thus the rate of convergence for the first order hyperbolic equation is $0(h^k)$, one power lower than in the case of the second order diffusion equation.

Second order hyperbolic equations The second order hyperbolic equations are given by

$$\frac{\partial^2 u(t)}{\partial t^2} - Au(t) = f(t) \qquad t \in (0, \tau) \tag{3-198a}$$

$$u(0) = 0 \tag{3-198b}$$

$$\frac{\partial u(0)}{\partial t} = 0 \tag{3-198c}$$

with the approximate solution being in a subspace $S_h(\Omega) \subset H_0^1(\Omega)$. Consider the associated variational problem (3-198) in the form

$$\left(\frac{\partial^2 u(t)}{\partial t^2}, v\right) + B(u(t), v) = (f(t), v) \tag{3-199a}$$

$$(u(0), v) = 0 \tag{3-199b}$$

$$\left(\frac{\partial u(0)}{\partial t}, v\right) = 0 \qquad \forall\, v \in H_0^1(\Omega) \tag{3-199c}$$

with
$$B(u, v) = \int_{\Omega} \nabla u \cdot \nabla v \, dx \tag{3-200}$$

Denoting

$$\left(\frac{\partial \tilde{u}}{\partial t}\right)_n = \frac{u^{(n+1)} - u^{(n)}}{\Delta t} \qquad \left(\frac{\partial^2 \tilde{u}}{\partial t^2}\right)_n = \frac{u^{(n+1)} - 2u^{(n)} + u^{(n-1)}}{\Delta t^2}$$

the Galerkin approximation takes the form

$$\left(\frac{\partial^2 \tilde{u}(t)}{\partial t^2}, \tilde{v}\right) + B(\tilde{u}(t), \tilde{v}) = (f(t), \tilde{v}) \qquad \forall \tilde{v} \in S_h(\Omega) \tag{3-201a}$$

$$\left(\frac{\partial \tilde{u}}{\partial t}, \tilde{v}\right) = 0 \qquad \forall \tilde{v} \in S_h(\Omega) \tag{3-201b}$$

$$(\tilde{u}^0, \tilde{v}) = 0 \qquad \forall \tilde{v} \in S_h(\Omega) \tag{3-201c}$$

with
$$\| u \|_{L_2(H(\Omega))} = \int_0^\tau \| u(t) \|_{H(\Omega)}^2 \, dt \tag{3-202}$$

Following Dupont [1973], it is assumed that the solution u of (3-198) exists such that $\partial^4 u/\partial t^4 \in L_2(L_2(\Omega))$. Then it follows that

$$\left\| \left[\left(\frac{\partial^2 \tilde{u}}{\partial t^2}\right)_n - \frac{\partial^2 u(t)}{\partial t^2}\right]_{t=n\Delta t} \right\|_{L_2(\Omega)} \le C\Delta t^3 \int_{t_{n-1}}^{t_{n+1}} \left\| \frac{\partial^4 u(t)}{\partial t^4} \right\|_{L_2(\Omega)} dt \tag{3-203}$$

Let $\varepsilon = u - \tilde{u}$. Then after lengthy algebra it can be shown [see Oden and Reddy, 1976] that there exist positive constants $\hat{C}_1$ and $\hat{C}_2$, independent of h and Δt, such that

$$\left\| \left(\frac{\partial \varepsilon}{\partial t}\right)_n \right\|_{\hat{L}_\infty(L_2(\Omega))} + C_1 \| \varepsilon \|_{\hat{L}_\infty(L_2(\Omega))} \le C_2 \left[\| \varepsilon_0 \|_{H^1(\Omega)} + \| \varepsilon_1 \|_{H^1(\Omega)} \right.$$

$$+ \left\| \left(\frac{\partial \varepsilon_{1/2}}{\partial t}\right)_n \right\| + h^{k+1} \| u \|_{L_\infty(H^{k+1}(\Omega))} + h^{k+1} \left\| \frac{\partial u}{\partial t} \right\|_{L_\infty(H^{k+1}(\Omega))}$$

$$+ h^{k+1} \| u \|_{L_2(H^{k+1}(\Omega))} + \Delta t^2 \left\| \frac{\partial^4 u}{\partial t^4} \right\|_{L_2(L_2(\Omega))} \right] \tag{3-204}$$

It should be noted that the result (3-204) is based on the central difference scheme and we would expect different results if other difference schemes are employed. There are other approaches reported in the literature for derivation of error estimates involved in the second order hyperbolic equations. For example, see Fried [1969], Fujii [1972], and Fix and Nassif [1972], among others.

3-4-6 Various Other Approaches

In the previous sections, we dealt with standard finite element approximations which will be followed in subsequent chapters. For the sake of completeness, however, we briefly mention here various other possibilities; namely, we list H^{-1} methods and hybrid and mixed methods.

The H^{-1} finite element methods were proposed by Rachford and Wheeler [1975]. The idea is to use Galerkin's method in a distributional sense via trial functions not coinciding with the space of test functions ($\tilde{U} \neq \tilde{V}$), thus resulting in unsymmetric matrices. However, the advantage appears to be an effective handling of irregular data.

The hybrid finite element method consists of independent approximations for the dependent variable on the interior of an element and its traces on the boundary. The local element boundary conditions are satisfied by Lagrange multipliers. In this method, we construct independent polynomial approximations for the variable in a local element and normal derivatives along the local boundaries. The purpose of this approach is to relax the requirements on the continuity of approximations of the variable across interelement boundaries. In the mixed finite element, on the other hand, we use independent approximations for both the dependent variable and its derivatives. In this way, for example, the velocity and velocity gradients (or stresses) can be specified at a node. If the hybrid method and mixed method are combined, we refer to such approximations as the mixed hybrid finite element method. These methods have been well developed in solid mechanics and proved to be efficient for calculations of accurate stresses and singularity problems, such as in fracture mechanics. For additional details, see Pian and Tong [1969], Oden and Reddy [1976], and Lee [1976], among others.

FOUR

FLUID DYNAMICS PRELIMINARIES

4-1 COORDINATE SYSTEMS

Descriptions of a variable such as the velocity require a suitable coordinate system to identify magnitudes and directions of the variable in space. The most commonly known coordinate systems are the Eulerian and Lagrangian coordinates. As illustrated in Figs 4-1 and 4-2, the Eulerian coordinates are fixed in space whereas the Lagrangian coordinates are attached to a material point (fluid particle) and expected to translate and rotate through and around the space together with particles of the material. Both orthogonal rectangular cartesian coordinates and arbitrary (not necessarily orthogonal) curvilinear coordinates may be applied to either Eulerian or Lagrangian coordinates. If a coordinate system is not orthogonal rectangular cartesian, then we must invoke the covariant and contravariant components of a vector referenced to the tangent base vectors $\mathbf{g}_i$ and $\mathbf{G}_i$ (Figs 4-1b and 4-2b) rather than the unit vectors $\mathbf{i}_i$. For generality, the discussions are given in terms of curvilinear coordinates and we show that all quantities resulting from the curvilinear coordinates are reduced to the counterparts in cartesian coordinates as a special case.

In general, Eulerian coordinates are preferred in fluid mechanics since it is impractical to trace a fluid particle toward downstream. With an eye fixed at a point in space such as the point P (Figs 4-1a and 4-2a), we observe the fluid particles passing by and wish to measure and calculate the velocity, pressure, etc., of the fluid at this point. The material coordinates ξ_i $(i = 1, 2, 3)$ are fixed permanently at this point, located by the position vector $\mathbf{r} = \hat{x}_m \mathbf{i}_m$ $(m = 1, 2, 3)$, and they do not move with the fluid particles. On the other hand, the Lagrangian coordinates (Figs 4-1*b* and 4-2*b*) are more suitable for solid mechanics because the strains must be measured in terms of the material coordinates moving together with the deforming body. At time $t = t$, the initial position vector $\mathbf{r}$ at P before deformation moves to Q located by the new position vector $\mathbf{R}$ after deformation, thus introducing the displacement $\mathbf{u} = \mathbf{R} - \mathbf{r}$. Note that the initial material coordinates ξ_i coinciding with the reference cartesian coordinates x_i become distorted at $t = t$ (curvilinear, possibly nonorthogonal). Despite the foregoing arguments, some fluid mechanics problems have been solved using the Lagrangian coordinates, particularly for one-dimensional problems. In studies of hyper-

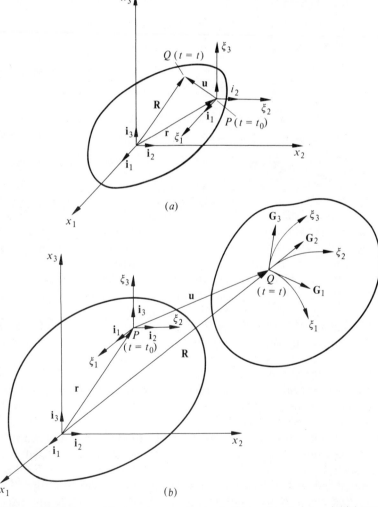

(a)

(b)

Figure 4-1 Eulerian and Lagrangian coordinates (orthogonal cartesian at initial state). (*a*) Eulerian; (*b*) Lagrangian.

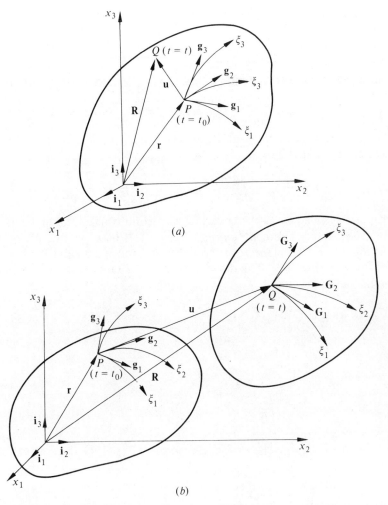

Figure 4-2 Eulerian and Lagrangian coordinates (curvilinear at initial state). (*a*) Eulerian; (*b*) Lagrangian.

velocity impact or fluids sloshing in an elastic container where both the fluid and solid mechanics equations are considered, it may be advisable to use the mixed Eulerian–Lagrangian coordinates.

In the Eulerian coordinates, the coordinates ξ_i placed on the material point P remain stationary in space; thus, they are referred to as the spatial coordinates. In the Lagrangian coordinates, such coordinates ξ_i are fixed on the material and convect along with the material. Therefore they are often called the material (intrinsic) or convective coordinates. The reference coordinates x_i (spatial) and the coordinates on the material (either stationary or convective) are related by

$$x_i = x_i(\xi_i)$$

For the curvilinear coordinates ξ_i, the tangent base vector is given by

$$\mathbf{g}_i = \frac{\partial \mathbf{r}}{\partial \xi_i} = \frac{\partial x_m}{\partial \xi_i} \mathbf{i}_m \qquad (i, m = 1, 2, 3) \tag{4-1a}$$

where $\mathbf{i}_m$ denote the orthonormal unit vectors. The reciprocal base vector is defined as

$$\mathbf{g}^i = \frac{\partial \xi_i}{\partial x_m} \mathbf{i}_m \tag{4-1b}$$

These base vectors are related by

$$\mathbf{g}_i \cdot \mathbf{g}^j = \frac{\partial x_m}{\partial \xi_i} \mathbf{i}_m \cdot \frac{\partial \xi_j}{\partial x_n} \mathbf{i}_n = \frac{\partial x_m}{\partial \xi_i} \frac{\partial \xi_j}{\partial x_n} \delta_{mn} = \frac{\partial x_m}{\partial \xi_i} \frac{\partial \xi_j}{\partial x_m} = \delta_i^j = \delta_{ij} \tag{4-2}$$

where $\mathbf{g}_i$ and $\mathbf{g}^j$ are referred to as the covariant and contravariant components of the tangent base vector. The contravariant (reciprocal) base vectors, $\mathbf{g}^1$, $\mathbf{g}^2$, and $\mathbf{g}^3$ are orthogonal to the planes constructed by $\mathbf{g}_2$ and $\mathbf{g}_3$, $\mathbf{g}_3$ and $\mathbf{g}_1$, and $\mathbf{g}_1$ and $\mathbf{g}_3$, respectively (Fig. 4-3). The dot products of the tangent base vectors are known as the metric tensors,

$$g_{ij} = \mathbf{g}_i \cdot \mathbf{g}_j = \frac{\partial x_m}{\partial \xi_i} \mathbf{i}_m \cdot \frac{\partial x_n}{\partial \xi_j} \mathbf{i}_n = \frac{\partial x_m}{\partial \xi_i} \frac{\partial x_n}{\partial \xi_j} \delta_{mn} = \frac{\partial x_m}{\partial \xi_i} \frac{\partial x_m}{\partial \xi_j} \tag{4-3a}$$

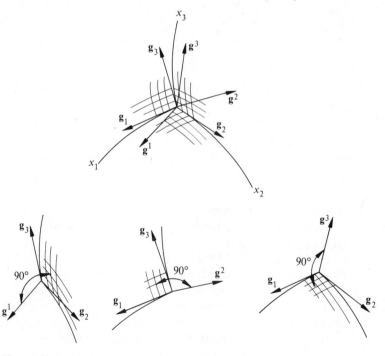

Figure 4-3 Covariant and contravariant components of tangent vectors.

$$g^{ij} = \mathbf{g}^i \cdot \mathbf{g}^j = \frac{\partial \xi_i}{\partial x_m} \mathbf{i}_m \cdot \frac{\partial \xi_j}{\partial x_n} \mathbf{i}_n = \frac{\partial \xi_i}{\partial x_m} \frac{\partial \xi_j}{\partial x_n} \delta_{mn} = \frac{\partial \xi_i}{\partial x_m} \frac{\partial \xi_j}{\partial x_m} \qquad (4\text{-}3b)$$

Here g_{ij} and g^{ij} are referred to as the covariant and contravariant metric tensors, respectively. Additional properties are summarized below.

$$\mathbf{g}_i \times \mathbf{g}_j = \frac{\partial x_m}{\partial \xi_i} \frac{\partial x_n}{\partial \xi_j} \varepsilon_{mnp} \mathbf{i}_p = \frac{\partial x_m}{\partial \xi_i} \frac{\partial x_n}{\partial \xi_j} \frac{\partial x_p}{\partial \xi_k} \varepsilon_{mnp} \mathbf{g}^k$$

$$= \sqrt{|g_{ij}|} \varepsilon_{ijk} \mathbf{g}^k = \sqrt{g} \, \varepsilon_{ijk} \mathbf{g}^k \qquad (4\text{-}4)$$

$$\mathbf{g}_i \times \mathbf{g}_j \cdot \mathbf{g}_k = \frac{\partial x_m}{\partial \xi_i} \frac{\partial x_n}{\partial \xi_j} \frac{\partial x_p}{\partial \xi_k} \varepsilon_{mnp} = \sqrt{g} \, \varepsilon_{ijk} \qquad (4\text{-}5)$$

The infinitesimal volumes in cartesian coordinates and curvilinear coordinates are given by, respectively,

$$dv = \mathbf{i}_1 \, d\xi_1 \times \mathbf{i}_2 \, d\xi_2 \cdot \mathbf{i}_3 \, d\xi_3 = d\xi_1 \, d\xi_2 \, d\xi_3 \qquad (4\text{-}6a)$$

$$d\bar{v} = \mathbf{g}_1 \, d\xi_1 \times \mathbf{g}_2 \, d\xi_2 \cdot \mathbf{g}_3 \, d\xi_3 = \sqrt{g} \, d\xi_1 \, d\xi_2 \, d\xi_3 = \sqrt{g} \, dv \qquad (4\text{-}6b)$$

Obviously, the quantity $\sqrt{g}$ transforms the volume in cartesian coordinates into that in curvilinear coordinates.

4-2 KINEMATICS

4-2-1 Mathematical Operations

In the fluid flow, we are concerned with geometric configurations and changes of any variable through the spatial continuum, generally referred to as kinematics. Let us consider the motion of a continuous media flowing through some region of three-dimensional space. The motion is defined as†

$$x_i = x_i(\xi_i, t)$$

where t is the time. We assume that x_i is differentiable and there exists a mapping such that

$$\left| \frac{\partial x_i}{\partial \xi_j} \right| \geq 0 \qquad (4\text{-}7)$$

The coordinate system considered here is Eulerian, in which we are concerned with what happens at points in space as time elapses. Thus the velocity, density, pressure, and temperature may be assumed to vary with time as well as with respect to the spatial domain. Let us consider a continuous function $f(\xi_i, t)$. The

---

† See Truesdell and Toupin [1960] or Eringen [1967] for complete treatments of mechanics of continua.

time rate of change of this function is given by the material (substantial) derivative,

$$\frac{Df}{Dt} = \frac{\partial f}{\partial t} + \frac{\partial f}{\partial \xi_i}\frac{\partial \xi_i}{\partial x_j}\frac{\partial x_j}{\partial t} = \frac{\partial f}{\partial t} + \frac{\partial f}{\partial \xi_i}\frac{\partial \xi_i}{\partial x_j} V_j = \frac{\partial f}{\partial t} + \frac{\partial f}{\partial x_j} V_j \tag{4-8}$$

where $V_j = \partial x_j/\partial t$ is the velocity. Consider now the velocity vector $\mathbf{V}$ in the arbitrary curvilinear coordinates

$$\mathbf{V} = V_i \mathbf{g}^i = V^i \mathbf{g}_i \tag{4-9}$$

where V_i and V^i are the covariant and contravariant components of the velocity vector $\mathbf{V}$. The derivative of $\mathbf{V}$ with respect to the curvilinear coordinates is given by

$$\frac{\partial \mathbf{V}}{\partial \xi_j} = \frac{\partial V^i \mathbf{g}_i}{\partial \xi_j} = \frac{\partial V^i}{\partial \xi_j}\mathbf{g}_i + \frac{\partial \mathbf{g}_i}{\partial \xi_j}V^i = \left(\frac{\partial V^i}{\partial \xi_j} + \Gamma^i_{jk}V^k\right)\mathbf{g}_i = V^i_{|j}\mathbf{g}_i \tag{4-10}$$

where the stroke "|" represents a covariant derivative defined as

$$V^i_{|j} = \frac{\partial V^i}{\partial \xi_j} + \Gamma^i_{jk}V^k = \frac{1}{\sqrt{g}}\frac{\partial}{\partial \xi_j}\left(\sqrt{g}\,V^i\right) \tag{4-11}$$

and Γ^i_{jk} is the Christoffel symbol of the second kind

$$\Gamma^i_{jk} = \frac{\partial^2 x_m}{\partial \xi_j \partial \xi_k}\frac{\partial \xi^i}{\partial x_m} = g^{il}\Gamma_{jkl} = \mathbf{g}^i \cdot \frac{\partial \mathbf{g}_j}{\partial \xi_k} \tag{4-12a}$$

with Γ_{jkl} denoting the Christoffel symbol of the first kind

$$\Gamma_{jkl} = \frac{\partial^2 x_m}{\partial \xi_j \partial \xi_k}\frac{\partial x_m}{\partial \xi_l} = \tfrac{1}{2}(g_{lj,k} + g_{lk,j} - g_{jk,l}) \tag{4-12b}$$

Note also that

$$\mathbf{V}\phi = \mathbf{g}^i \frac{\partial \phi}{\partial \xi^i}$$

$$\nabla^2 \phi = \mathbf{V} \cdot \mathbf{V}\phi = \phi_{|ij}g^{ij} = \frac{1}{\sqrt{g}}\frac{\partial}{\partial \xi^j}\left(\sqrt{g}\,g^{ij}\phi_{,i}\right) \tag{4-13}$$

$$\phi_{|ij} = \frac{\partial^2 \phi}{\partial \xi_i \partial \xi_j} + \Gamma^k_{kj}\frac{\partial \phi}{\partial \xi_i} \tag{4-14}$$

$$\mathbf{V} \cdot \mathbf{V} = \frac{1}{\sqrt{g}}\frac{\partial}{\partial \xi_i}\left(\sqrt{g}\,V^i\right) = V^i_{|i} \tag{4-15}$$

$$\mathbf{V} \times \mathbf{V} = V_{j|i}\varepsilon^{ijk}\mathbf{g}_k \tag{4-16}$$

For cylindrical coordinates (Fig. 4-4a), we may set $\xi_1 = r$ (radial), $\xi_2 = \theta$ (tangential), and $\xi_3 = z$ (axial); $x_1 = r\cos\theta$, $x_2 = r\sin\theta$, and $x_3 = z_3$. Thus the position vector is

$$\mathbf{r} = x_1\mathbf{i}_1 + x_2\mathbf{i}_2 + x_3\mathbf{i}_3$$

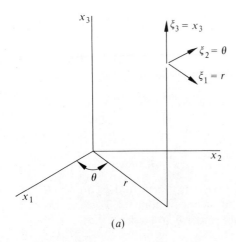

(a)

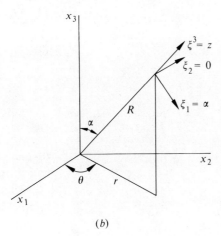

(b)

Figure 4-4 Curvilinear coordinates. (a) Cylindrical coordinates; (b) spherical coordinates.

or
$$\mathbf{r} = r\cos\theta\mathbf{i}_1 + r\sin\theta\mathbf{i}_2 + z\mathbf{i}_3$$

The covariant metric tensors are

$$g_{11} = \frac{\partial x_1}{\partial\xi_1}\frac{\partial x_1}{\partial\xi_1} + \frac{\partial x_2}{\partial\xi_1}\frac{\partial x_2}{\partial\xi_1} + \frac{\partial x_3}{\partial\xi_1}\frac{\partial x_3}{\partial\xi_1} = \cos^2\theta + \sin^2\theta = 1$$

We find similarly $g_{22} = r^2$ and all other $g_{ij} = 0$. Using the relations (4-10) and (4-11) yields

$$\Gamma^1_{22} = -r \qquad \Gamma^2_{21} = \Gamma^2_{12} = \frac{1}{r}$$

with all other $\Gamma^i_{jk} = 0$. Therefore, the Laplace equation (4-13) assumes the form

$$\nabla^2\phi = \phi_{|11} + \frac{1}{r^2}\phi_{|22} + \phi_{|33}$$

or
$$\nabla^2 \phi = \frac{\partial^2 \phi}{\partial r^2} + \frac{1}{r} \frac{\partial \phi}{\partial r} + \frac{1}{r^2} \frac{\partial^2 \phi}{\partial \theta^2} + \frac{\partial^2 \phi}{\partial z^2} \tag{4-17}$$

For spherical coordinates with $\xi_1 = R$, $\xi_2 = \alpha$, $\xi_3 = \theta$, $x_1 = R \sin \alpha \cos \theta$, $x_2 = R \sin \alpha \sin \theta$, and $x_3 = R \cos \alpha$ (Fig. 4-4b), we obtain

$$g_{11} = 1 \qquad g_{22} = R^2 \qquad g_{33} = R^2 \sin^2 \alpha$$

and

$$\nabla^2 \phi = \frac{\partial^2 \phi}{\partial R^2} + \frac{2}{R} \frac{\partial \phi}{\partial R} + \frac{1}{R^2} \frac{\partial^2 \phi}{\partial \alpha^2} + \frac{\cot \alpha}{R^2} \frac{\partial \phi}{\partial \alpha} + \frac{1}{R^2 \sin^2 \alpha} \frac{\partial^2 \phi}{\partial \theta^2} \tag{4-18}$$

To determine the divergence $\nabla \cdot \mathbf{V}$, we first note that V_i and V^i are the tensor components and must be reduced to the physical components. Toward this end, we express the tangent base vectors as unit vectors $\overset{*}{\mathbf{g}}_i$ in the form,

$$\overset{*}{\mathbf{g}}_i = \frac{\mathbf{g}_i}{|\mathbf{g}_i|} = \frac{\mathbf{g}_i}{\sqrt{g_{(ii)}}} = \sqrt{g^{(ii)}} \, \mathbf{g}_i \tag{4-19a}$$

and
$$\overset{*}{\mathbf{g}}{}^i = \sqrt{g_{(ii)}} \, \mathbf{g}^i \tag{4-19b}$$

Using these relationships, we write

$$\mathbf{V} = V^i \mathbf{g}_i = V^i \sqrt{g_{(ii)}} \, \overset{*}{\mathbf{g}}_i = \overset{*}{V}{}^i \overset{*}{\mathbf{g}}_i$$

where

$$\overset{*}{V} = V^i \sqrt{g_{(ii)}} \tag{4-20a}$$

and similarly,

$$\overset{*}{V}_i = V_i \sqrt{g^{(ii)}} \tag{4-20b}$$

Here, $\overset{*}{V}{}^i$ and $\overset{*}{V}_i$ are the physical components of the velocity vector based now on the unit vectors instead of the arbitrary tangent base vectors. The indices inside the parentheses are not to be summed.

Now consider the divergence of the velocity vector in which we must determine the physical vector components. From (4-12) and (4-15), we write

$$\nabla \cdot \mathbf{V} = V^i_{|i} = \frac{\partial V^i}{\partial \xi_i} + \Gamma^i_{ij} V^j$$

or

$$\nabla \cdot \mathbf{V} = \frac{\partial V^1}{\partial \xi_1} + \frac{\partial V^2}{\partial \xi_2} + \frac{\partial V^3}{\partial \xi_3} + \Gamma^1_{11} V^1 + \Gamma^1_{12} V^2 + \cdots + \Gamma^3_{33} V^3 \tag{4-21}$$

For cylindrical coordinates, we have

$$V^1 = \sqrt{g^{(11)}} \, \overset{*}{V}{}^1 = \overset{*}{V}{}^1 \qquad V^2 = \sqrt{g^{(22)}} \, \overset{*}{V}{}^2 = \frac{1}{r} \overset{*}{V}{}^2 \qquad V^3 = \sqrt{g^{(33)}} \, \overset{*}{V}{}^3 = \overset{*}{V}{}^3$$

Letting $\overset{*}{V}{}^1 = u$, $\overset{*}{V}{}^2 = v$, $\overset{*}{V}{}^3 = w$, $\xi_1 = r$, $\xi_2 = \theta$, and $\xi_3 = z$, we obtain

$$\nabla \cdot \mathbf{V} = \frac{\partial u}{\partial r} + \frac{1}{r}\frac{\partial v}{\partial \theta} + \frac{\partial w}{\partial z} + \frac{u}{r} \tag{4-22}$$

It is a simple matter to extend the expression for divergence to spherical coordinates. For cartesian coordinates, we simply substitute the covariant derivative to a partial derivative; thus the terms with the Christoffel symbols are no longer present.

4-2-2 Rate-of-Deformation Tensor and Strain Rates

Let us now examine the rate-of-deformation tensor which is the basic kinematic quantity in fluid mechanics. Let ds_0 and ds denote a material line element in some reference configuration $t = t_0$ and the same element at $t = t$, respectively. Then the time rate of the measure of the deformation is given by the substantial derivative of the difference of the squares of the line elements ds_0 and ds before and after deformation, respectively. Thus for cartesian coordinates we have

$$\frac{D(ds^2 - ds_0^2)}{Dt} = \frac{D\,ds^2}{Dt} = \frac{D(dx_i\,dx_i)}{Dt} = 2\,dx_i\,\frac{D\,dx_i}{Dt} \tag{4-23}$$

but

$$\frac{D\,dx_i}{Dt} = \frac{D}{Dt}\frac{\partial x_i}{\partial \xi_j}\,d\xi_j = \frac{\partial}{\partial \xi_j}\left(\frac{\partial x_i}{\partial t}\right)d\xi_j = \frac{\partial V_i}{\partial \xi_j}\,d\xi_j$$

$$= \frac{\partial V_i}{\partial x_j}\frac{\partial x_j}{\partial \xi_k}\,d\xi_k = \frac{\partial V_i}{\partial x_j}\,dx_j = V_{i,j}\,dx_j$$

Since the velocity gradient may be written in the form

$$V_{i,j} = \tfrac{1}{2}(V_{i,j} + V_{j,i}) + \tfrac{1}{2}(V_{i,j} - V_{j,i}) = d_{ij} + \omega_{ij} \tag{4-24}$$

where d_{ij} and ω_{ij} are referred to as the rate of deformation tensor (symmetric) and rotation tensor (antisymmetric), respectively

$$d_{ij} = \tfrac{1}{2}(V_{i,j} + V_{j,i}) \tag{4-25}$$

$$\omega_{ij} = \tfrac{1}{2}(V_{i,j} - V_{j,i}) \tag{4-26}$$

Combining (4-23) and (4-24) yields

$$\frac{D\,ds^2}{Dt} = 2\,dx_i(d_{ij} + \omega_{ij})\,dx_j = 2d_{ij}\,dx_i\,dx_j$$

Or alternatively, we may write

$$\frac{D\,ds^2}{Dt} = \frac{D}{Dt}\,dx_i\,dx_j\,\delta_{ij} = \delta_{ij}\,dx_j\,\frac{D\,dx_i}{Dt} + \delta_{ij}\,dx_i\,\frac{D\,dx_j}{Dt}$$

$$= \delta_{ij}(dx_j V_{i,k}\,dx_k + dx_i V_{j,k}\,dx_k) = dx_i V_{i,j}\,dx_j + dx_j V_{j,i}\,dx_i$$

$$= 2d_{ij}\,dx_i\,dx_j$$

Thus we obtain the same result as before

$$d_{ij} = \tfrac{1}{2}(V_{i,j} + V_{j,i})$$

For curvilinear coordinates, it can be shown that the rate of deformation tensor assumes the form

$$d_{ij} = \tfrac{1}{2}(V_{i|j} + V_{j|i}) = \frac{1}{2}\left(\frac{\partial V_i}{\partial x_j} + \frac{\partial V_j}{\partial x_i} - 2\Gamma_{ij}^m V_m\right) \tag{4-27}$$

To relate the rate-of-deformation tensor to the strain rates, we first note that, in the cartesian Eulerian coordinates (see Fig. 4-1a),

$$ds^2 - ds_0^2 = dx_i\, dx_i - d\xi_i\, d\xi_i = (\delta_{ij} - F_{ij})\, dx_i\, dx_j = 2E_{ij}\, dx_i\, dx_j \tag{4-28}$$

with

$$E_{ij} = \tfrac{1}{2}(\delta_{ij} - F_{ij}) \qquad \text{and} \qquad F_{ij} = \frac{\partial \xi_m}{\partial x_i}\frac{\partial \xi_m}{\partial x_j} \tag{4-29}$$

Here E_{ij} is known as Eulerian strain tensor or Almansi strain tensor. Using the relation $\xi_m = x_m - u_m$, u_m being the displacement components, we obtain

$$E_{ij} = \frac{1}{2}\left(\frac{\partial u_i}{\partial x_j} + \frac{\partial u_j}{\partial x_i} - \frac{\partial u_k}{\partial x_i}\frac{\partial u_k}{\partial x_j}\right) \tag{4-30}$$

Similarly, for the cartesian Lagrangian coordinates, we have (Fig. 4-1),

$$ds^2 - ds_0^2 = dx_i\, dx_i - d\xi_i\, d\xi_i = (G_{ij} - \delta_{ij})\, d\xi_i\, d\xi_j = 2\gamma_{ij}\, d\xi_i\, d\xi_j \tag{4-31}$$

with

$$\gamma_{ij} = \tfrac{1}{2}(G_{ij} - \delta_{ij}) = \frac{1}{2}\left(\frac{\partial u_i}{\partial \xi_j} + \frac{\partial u_j}{\partial \xi_i} + \frac{\partial u_m}{\partial \xi_i}\frac{\partial u_m}{\partial \xi_j}\right) \tag{4-32}$$

$$G_{ij} = \frac{\partial x_m}{\partial \xi_i}\frac{\partial x_m}{\partial \xi_j}$$

Here γ_{ij} is the Lagrangian strain tensor or Green–St. Venant strain tensor. The time rate of change of the Eulerian strain tensor assumes the form

$$\frac{D(ds^2 - ds_0^2)}{Dt} = 2\dot{E}_{ij}\, dx_i\, dx_j + 2E_{ij}\frac{\partial V_i}{\partial x_k}\, dx_k\, dx_j + 2E_{ij}\frac{\partial V_j}{\partial x_k}\, dx_k\, dx_i$$

$$= 2\left(\dot{E}_{ij} + E_{kj}\frac{\partial V_k}{\partial x_i} + E_{ik}\frac{\partial V_k}{\partial x_j}\, dx_i\, dx_j\right) \tag{4-33}$$

Equating (4-24) and (4-33) yields

$$\dot{E}_{ij} = \frac{dE_{ij}}{dt} = d_{ij} - \left(E_{kj}\frac{\partial V_k}{\partial x_i} + E_{ik}\frac{\partial V_k}{\partial x_j}\right) \tag{4-34}$$

whereas the Lagrangian strain rate is of the form

$$\dot{\gamma}_{ij} = \frac{d\gamma_{ij}}{dt} = \frac{1}{2}\left(\frac{\partial V_i}{\partial \xi_j} + \frac{\partial V_j}{\partial \xi_i} + \frac{\partial V_m}{\partial \xi_i}\frac{\partial u_m}{\partial \xi_j} + \frac{\partial u_m}{\partial \xi_i}\frac{\partial V_m}{\partial \xi_j}\right) \tag{4-35}$$

where the Lagrangian velocity is defined as $V_i = \partial u_i / \partial t$ in contrast with the Eulerian velocity $V_i = \partial x_i / \partial t$. Note also that taking the material time derivative of (4-31) and equating it with (4-24) yields

$$\frac{d\gamma_{ij}}{dt} = d_{mn} \frac{\partial x_m}{\partial \xi_i} \frac{\partial x_n}{\partial \xi_j} \tag{4-36}$$

Here the expressions (4-28–4-36) apply only to the cartesian coordinates.

4-3 KINETICS

4-3-1 Conservation Laws

The axiom of global mass conservation states that the total mass is invariant under the motion. Global mass conservation when applied to an infinitesimal neighborhood of a material point implies local mass conservation. These statements are mathematically expressed as

$$\int_{\Omega_0} \rho_0 \, d\Omega_0 = \int \rho \, d\Omega \tag{4-37}$$

where ρ is the mass density and the subscript 0 denotes the initial state. The volumes $d\Omega_0$ at $t = t_0$ and the $d\Omega$ at $t = t$ are given by

$$d\Omega_0 = d\xi_1 \mathbf{i}_1 \times d\xi_2 \mathbf{i}_2 \cdot d\xi_3 \mathbf{i}_3 = d\xi_1 \, d\xi_2 \, d\xi_3$$

$$d\Omega = dx_1 \mathbf{G}_1 \times dx_2 \mathbf{G}_2 \cdot dx_3 \mathbf{G}_3 = \sqrt{G} \, d\xi_1 \, d\xi_2 \, d\xi_3$$

or
$$d\Omega = \sqrt{G} \, d\Omega_0 \tag{4-38}$$

In view of (4-37) and (4-38), we write

$$\int_{\Omega_0} \rho_0 \, d\Omega_0 = \int_{\Omega_0} \sqrt{G} \rho \, d\Omega_0 \tag{4-39}$$

From the law of conservation, the substantial derivative of (4-37) is written as

$$\frac{D}{Dt} \int_{\Omega} \rho \, d\Omega = \int_{\Omega} \frac{D\rho}{Dt} \, d\Omega + \int_{\Omega} \rho \frac{D \, d\Omega}{Dt} = 0 \tag{4-40}$$

It can be shown that

$$\int_{\Omega} \rho \frac{D \, d\Omega}{Dt} = \int_{\Omega} \rho V^i_{|i} \, d\Omega$$

Thus the expression (4-40) assumes the form

$$\int_{\Omega} \left(\frac{\partial \rho}{\partial t} + \rho_{|i} V^i + \rho V^i_{|i} \, d\Omega \right) = 0$$

or
$$\int_{\Omega} \left[\frac{\partial \rho}{\partial t} + (\rho V^i)_{|i} \right] d\Omega = 0 \tag{4-41}$$

For all infinitesimal volumes ($d\Omega$), the global mass conservation given by (4-41) must be satisfied. Therefore, the integrand must vanish and we obtain the local mass conservation of the form

$$\frac{\partial \rho}{\partial t} + (\rho V^i)_{|i} = 0 \tag{4-42a}$$

For cartesian coordinates, the covariant derivative may be replaced by the partial derivative so that

$$\frac{\partial \rho}{\partial t} + (\rho V_i)_{,i} = 0 \tag{4-42b}$$

Let us now consider the first law of thermodynamics in its global form as applied to the curvilinear Eulerian coordinates

$$\frac{D}{Dt}(K + U) = R + Q \tag{4-43}$$

where K is the kinetic energy, U is the internal energy, R is the mechanical power and Q is the heat energy given by

$$K = \frac{1}{2}\int_\Omega \rho \mathbf{V} \cdot \mathbf{V} \, d\Omega = \frac{1}{2}\int_\Omega \rho V^i V_i \, d\Omega \tag{4-44a}$$

$$U = \int_\Omega \rho \varepsilon \, d\Omega \tag{4-44b}$$

$$R = \int_\Omega \rho \mathbf{F} \cdot \mathbf{V} \, d\Omega + \int_\Gamma \mathbf{S} \cdot \mathbf{V} \, d\Gamma = \int_\Omega \rho F^i V_i \, d\Omega + \int_\Gamma S^i V_i \, d\Gamma \tag{4-44c}$$

$$Q = \int_\Omega \rho h \, d\Omega + \int_\Gamma \mathbf{q} \cdot \mathbf{n} \, d\Gamma = \int_\Omega \rho h \, d\Omega + \int_\Gamma q^i n_i \, d\Gamma \tag{4-44d}$$

Here, ρ is the mass density per unit volume, ε is the internal energy density per unit mass, F^i are the components of body force per unit mass, S^i are the components of surface pressure, h is the heat supply per unit mass, q^i are the components of the surface heat flux, and n_i refers to the components of the unit vector normal to the surface. It follows from (4-44a), (4-13), and the Green–Gauss theorem that

$$\int_\Omega \rho \frac{DV^i}{Dt} V_i \, d\Omega + \int_\Omega \rho \frac{D\varepsilon}{Dt} \, d\Omega = \int_\Omega \rho F^i V_i \, d\Omega + \int_\Omega \sigma^{ij}_{|j} V_i \, d\Omega + \int_\Omega \sigma^{ij} V_{i|j} \, d\Omega$$
$$+ \int_\Omega \rho h \, d\Omega + \int_\Omega q^i_{|i} \, d\Omega \tag{4-45}$$

It should be noted that the surface pressure S^i was replaced by

$$S^i = \sigma^{ij} n_j \tag{4-46}$$

where σ^{ij} is known as the Cauchy stress tensor and its covariant derivative is given by

$$\sigma_{|i}^{ij} = \sigma_{,i}^{ij} + \Gamma_{im}^j \sigma^{im} + \Gamma_{im}^m \sigma^{ij} \tag{4-47}$$

We further define the components of acceleration a^i as

$$a^i = \frac{DV^i}{Dt} = \dot{V}^i + V^i_{|j} V^j \tag{4-48}$$

Similarly, the material time derivative of the internal energy density becomes

$$\frac{D\varepsilon}{Dt} = \dot{\varepsilon} + \varepsilon_{|i} V^i \tag{4-49}$$

It follows from (4-44) that

$$\cdot \int_\Omega (\sigma_{|j}^{ij} + \rho F^i - \rho a^i) V_i \, d\Omega - \int_\Omega \left(\rho \frac{D\varepsilon}{Dt} - \sigma^{ij} V_{i|j} - \rho h - q_{|i}^i \right) d\Omega = 0 \tag{4-50}$$

In view of conservation of linear momentum and for (4-50) to be satisfied for all arbitrary volumes, the integrands in (4-50) must vanish. Thus we have

$$\sigma_{|j}^{ij} + \rho F^i - \rho a^i = 0 \tag{4-51}$$

$$\rho \frac{D\varepsilon}{Dt} - \sigma^{ij} V_{i|j} - \rho h - q_{|i}^i = 0 \tag{4-52}$$

Here σ^{ij} is symmetric and we may write (4-52) in the form

$$\rho \dot{\varepsilon} + \rho \varepsilon_{|i} V^i - \sigma^{ij} d_{ij} - \rho h - q_{|i}^i = 0 \tag{4-53}$$

The equations (4-51) and (4-53) represent the local forms of the conservation of linear momentum and conservation of energy, respectively. The conservation of angular momentum can be shown as

$$\sigma^{ij} = \sigma^{ji} \tag{4-54}$$

Finally, we require an equation of state for complete description of a fluid

$$f(\rho, P, T) = 0$$

in which P is the hydrostatic pressure and T is the temperature.

For cartesian coordinates, the foregoing equations may be summarized as†

1. *Momentum equation*

$$\rho \frac{\partial V_i}{\partial t} + \rho \frac{\partial V_i}{\partial x_j} V_j - \frac{\partial \sigma_{ij}}{\partial x_j} - \rho F_i = 0 \tag{4-55}$$

2. *Continuity equation*

$$\frac{\partial \rho}{\partial t} + \frac{\partial (\rho V_i)}{\partial x_i} = 0 \tag{4-56}$$

---

† Note that all indices raised for the contravariant components are meaningless in cartesian coordinates and they may be lowered for convenience.

3. *Energy equation*

$$\rho \frac{\partial \varepsilon}{\partial t} + \rho \frac{\partial \varepsilon}{\partial x_i} V_i - \sigma_{ij} d_{ij} - \rho h - q_{i,i} = 0 \tag{4-57}$$

4. *Equation of state*

$$f(\rho, P, T) = 0 \tag{4-58}$$

4-3-2 Constitutive Equations

Various conservation laws have been discussed in the previous section without specific constitutive relations between the kinematics and kinetics of the fluid. If we assume that fluids cannot sustain a shear stress, then we may write for cartesian coordinates,

$$\sigma_{ij} = -P\delta_{ij} \tag{4-59}$$

where P is a static pressure. It is possible, however, that for some fluids, the pressure depends on density, so that

$$\sigma_{ij} = -\pi(\rho)\delta_{ij} = -R\rho\gamma\delta_{ij} \tag{4-60}$$

Here π is the thermodynamic pressure, R is the gas constant, $\gamma = c_p/c_v$, and c_p and c_v are the specific heats at constant pressure and constant volume, respectively. The relationship characterized by (4-60) is called the barotropic flow. In 1687 Newton proposed a relation

$$\sigma_{ij} = -\pi\delta_{ij} + T_{ij}(d_{ij}) \tag{4-61}$$

in which T_{ij} is the linear function of d_{ij}, commonly known as Newtonian flow. If T_{ij} is arbitrary, not necessarily the linear function of d_{ij}, we define such fluid as the non-Newtonian flow. Recently, Reiner [1945], Rivlin [1948], and Rivlin–Erickson [1955], among others, have proposed more complicated and general laws as functions of invariants of d_{ij}. For example, the Reiner–Rivlin fluid may be written in the form

$$T_{ij} = F_1 d_{ij} + F_2 d_{ik} d_{kj}$$

where F_1 and F_2 are arbitrary functions of the invariants

$$F_1 = \hat{F}_1(d_{ij}d_{ij}) \qquad F_2 = \hat{F}_2(d_{ij}d_{jk}d_{ki})$$

For isotropic Newtonian fluids, the stress tensor may be written as

$$\sigma_{ij} = -\pi\delta_{ij} + \lambda d_{kk}\delta_{ij} + 2\mu d_{ij} \tag{4-62}$$

where λ and μ are viscosity constants. Stokes proposed the relation,

$$3\lambda + 2\mu = 0$$

which reduces (4-38) to

$$\sigma_{ij} = -\pi\delta_{ij} + 2\mu d'_{ij} \tag{4-63}$$

where $d_{ij} = d_{ij} - \frac{1}{3}d_{kk}\delta_{ij}$. For incompressible flow, we set $\pi = P$, and $d_{kk} = 0$.

If the fluid is compressible, we must define the internal energy as a function of pressure and temperature. For a perfect (ideal) gas, we write

$$\varepsilon = H - \frac{P}{\rho} \qquad (4\text{-}64a)$$

or

$$\varepsilon = c_v T \qquad (4\text{-}64b)$$

Here $H = c_p T$ is the enthalpy. For a real gas, it is postulated that

$$\frac{Pv}{RT} = Z(P, \rho)$$

where $Z(P, \rho)$ is called the compressibility factor. Thus $Z = 1$ refers to the perfect gas.

If the constitutive relation can be established for entropy, one may use the second law of thermodynamics characterized by

$$D + \frac{1}{\theta} q_i \theta_{,i} \geq 0 \qquad (4\text{-}65)$$

where

$$D = \rho \theta \frac{D\eta}{Dt} - q_{i,i} - \rho h \geq 0 \qquad (4\text{-}66)$$

with θ the absolute temperature and η the entropy density per unit mass. Thus it is possible to use (4-66) instead of (4-53) as the energy equation. The heat flux q_i may be given by the linear Fourier law

$$q_i = \kappa T_{,i} \qquad (4\text{-}67)$$

where κ is the coefficient of thermal conductivity. Note that the familiar notion of negative sign on the right-hand side of (4-67) is already incorporated as a negative gradient of q_i in (4-66).

4-3-3 Navier–Stokes Equation

The Navier–Stokes equation, well known in fluid mechanics, is obtained from the momentum equation (4-51) and the constitutive equation (4-62). Since all essential ingredients for curvilinear coordinates have been presented, the reader may be able to verify that the Navier–Stokes equation in the cylindrical coordinates takes the form

$$\rho \left(\frac{\partial V_r}{\partial t} + V_r \frac{\partial V_r}{\partial r} + \frac{V_\theta}{r} \frac{\partial V_r}{\partial \theta} + V_z \frac{\partial V_r}{\partial z} - \frac{V_\theta^2}{r} \right)$$

$$= \rho F_r - \frac{\partial \pi}{\partial r} + \mu \left(2 \frac{\partial^2 V_r}{\partial r^2} + \frac{1}{r^2} \frac{\partial^2 V_r}{\partial \theta^2} + \frac{1}{r} \frac{\partial^2 V_\theta}{\partial \theta \partial r} - \frac{3}{r^2} \frac{\partial V_\theta}{\partial \theta} + \frac{\partial^2 V_z}{\partial z \partial r} + \frac{\partial^2 V_r}{\partial z^2} \right.$$

$$\left. - 2 \frac{V_r}{r^2} + \frac{2}{r} \frac{\partial V_r}{\partial r} \right) + \lambda \left[\frac{\partial^2 V_r}{\partial r^2} + \frac{1}{r} \frac{\partial^2 V_\theta}{\partial r \partial \theta} - \frac{1}{r^2} \frac{\partial V_\theta}{\partial \theta} + \frac{\partial}{\partial r} \left(\frac{V_r}{r} \right) + \frac{\partial^2 V_z}{\partial r \partial z} \right]$$

for the radial direction r. Similar forms may be derived for the tangential and axial directions.

For cartesian coordinates, the above equation may be considerably simplified. For the x component we have

$$\rho\left(\frac{\partial V_x}{\partial t} + V_x\frac{\partial V_x}{\partial x} + V_y\frac{\partial V_x}{\partial y} + V_z\frac{\partial V_x}{\partial z}\right)$$

$$= \rho F_x - \frac{\partial \pi}{\partial x} + \mu\left(2\frac{\partial^2 V_x}{\partial x^2} + \frac{\partial^2 V_x}{\partial y^2} + \frac{\partial^2 V_x}{\partial z^2} + \frac{\partial^2 V_y}{\partial x\,\partial y} + \frac{\partial^2 V_z}{\partial x\,\partial z}\right)$$

$$+ \lambda\left(\frac{\partial^2 V_x}{\partial x^2} + \frac{\partial^2 V_y}{\partial x\,\partial y} + \frac{\partial^2 V_z}{\partial x\,\partial z}\right)$$

with the equations for the y and z components derived similarly. Note that the Stokes hypothesis $3\lambda + 2\mu = 0$ leads to simplified forms. For incompressible fluids, $d_{kk} = 0$, further simplifications result.

The reader should realize that the mathematical exercises given in this book are not intended for derivation of fluid mechanics equations but are designed for use in derivation of finite element equations in the most convenient and rigorous manner. This fact will become apparent in later chapters.

4-4 VELOCITY POTENTIAL AND STREAM FUNCTION

For two-dimensional flow, it is sometimes convenient to express the velocity in terms of either the velocity potential ϕ or stream function ψ. From the equation of continuity for incompressible fluid in cartesian coordinates, we have

$$\nabla \cdot \mathbf{V} = V_{i,i} = \frac{\partial V_1}{\partial x_1} + \frac{\partial V_2}{\partial x_2} = 0 \tag{4-68}$$

It is possible to write the velocity components in the forms,

$$V_1 = V_x = \frac{\partial \psi}{\partial y} = \psi_{,2} \tag{4-69a}$$

$$V_2 = V_y = \frac{\partial \psi}{\partial x} = -\psi_{,1} \tag{4-69b}$$

which satisfy the equation of continuity, automatically. On the other hand, for two-dimensional irrotational flow or potential flow, the vorticity is zero. Thus we obtain the relation,

$$\nabla \times \mathbf{V} = (V_{2,1} - V_{1,2})\mathbf{i}_3 = 0$$

or

$$V_{2,1} - V_{1,2} = 0 \tag{4-70}$$

Substituting (4-69) into (4-70) yields

$$\frac{\partial^2 \psi}{\partial x^2} + \frac{\partial^2 \psi}{\partial y^2} = \psi_{,11} + \psi_{,22} = 0$$

or
$$\nabla^2 \psi = \psi_{,ii} = 0 \qquad (i = 1, 2) \tag{4-71}$$

which is the Laplace equation.

For analyses involving surface profiles, it is convenient to use the velocity potential related by

$$V_1 = \phi_{,1} \qquad V_2 = \phi_{,2} \tag{4-72}$$

In view of (4-68) and (4-72), we obtain

$$\frac{\partial^2 \phi}{\partial x^2} + \frac{\partial^2 \phi}{\partial y^2} = 0$$

or
$$\nabla^2 \phi = \phi_{,ii} = 0 \tag{4-73}$$

which is the Laplace equation in terms of the velocity potential.

FIVE

INCOMPRESSIBLE FLOW

5-1 GENERAL

In this chapter, we are concerned with the mechanics of incompressible fluids and numerical solutions via finite elements. Both inviscid and viscous fluids will be considered.

We examine the ideal flow (incompressible inviscid flow) to include details of the finite element solutions of the Laplace equation for a two-dimensional flow and the Stokesian equation for an axisymmetrical flow. Example problems are presented for flows around a cylinder and sphere and out of an orifice. We then show the applications to incompressible viscous fluids with velocities and pressures as unknowns. We further consider the problems of wave motion and the eigenvalue solutions. Rotational and boundary layer flows, three-dimensional analysis, and boundary singularities are also discussed.

As we have observed in the first three chapters, the finite element equations can be developed with the governing equations and interpolation functions written directly in the global form. We noted that the same results can be obtained by the local formulation and assembly into a global form. For simplicity, in this chapter and subsequent chapters, we will discuss the finite element formulations in a local form only.

5-2 EQUATIONS OF MOTION AND CONTINUITY OF AN IDEAL FLUID

The ideal fluid motion may be described by an expression referred to as the Euler equation

$$a_i = \frac{DV_i}{Dt} = \dot{V}_i + V_{i,j}V_j = \frac{1}{\rho}(G_i - P_{,i}) \tag{5-1}$$

where the gravitational body force G_i is defined as

$$G_i = \rho F_i = -(\rho g H)_{,i} = -J_{,i} \tag{5-2}$$

with g the gravitational acceleration and H the vertical distance relative to earth above some datum. Substituting (5-2) in (5-1) yields

$$\dot{V}_i + V_{i,j}V_j = -\frac{1}{\rho}(P + J)_{,i} \tag{5-3a}$$

or

$$\frac{\partial \mathbf{V}}{\partial t} + (\mathbf{V} \cdot \nabla)\mathbf{V} = -\nabla\left(\frac{P}{\rho} + \frac{J}{\rho}\right) \tag{5-3b}$$

It follows from (1-10b) and (5-3) that

$$\mathbf{a} = \frac{\partial \mathbf{V}}{\partial t} + \nabla\left(\frac{V^2}{2}\right) - \nabla \times \boldsymbol{\omega} = -\nabla\left(\frac{P}{\rho} + \frac{J}{\rho}\right)$$

For steady state motion, this becomes

$$\nabla \times \boldsymbol{\omega} = \nabla\left(\frac{P}{\rho} + \frac{V^2}{2} + gH\right)$$

For irrotational motion $\omega = 0$, we have

$$\nabla\left(\frac{P}{\rho} + \frac{V^2}{2} + gH\right) = 0$$

Consequently upon integration, we obtain

$$\frac{P}{\rho} + \frac{V^2}{2} + gH = \Gamma \tag{5-4}$$

This is known as Bernoulli's equation. Here the quantity Γ is constant when $\omega = 0$ or $\mathbf{V} = 0$. Bernoulli's equation may be generalized to include an unsteady motion in the form

$$\frac{\partial \mathbf{V}}{\partial t} + \nabla\left(\frac{P}{\rho} + \frac{V^2}{2} + gH\right) = 0$$

Using the relation $V = \nabla\phi$, we obtain

$$\nabla\left(\frac{\partial \phi}{\partial t} + \frac{P}{\rho} + \frac{V^2}{2} + gH\right) = 0 \tag{5-5}$$

or

$$\frac{\partial \phi}{\partial t} + \frac{P}{\rho} + \frac{V^2}{2} + gH = \Gamma(t) \tag{5-6}$$

5-3 TWO-DIMENSIONAL INVISCID FLOW

5-3-1 Finite Element Formulations

In potential flow problems, either stream functions or velocity potentials are calculated. Velocity distributions then follow from

$$V_i = \varepsilon_{ij}\psi_{,j}$$

or
$$V_i = \phi_{,i}$$

In general, the choice between velocity potential and stream function in the finite element formulation depends on boundary conditions, whichever is easier to specify. No advantage of one over the other is claimed if the geometry is simple.

Consider the Laplace equation (4-71) in which the stream function ψ may be approximated within a finite element in the form

$$\psi = \Phi_N \psi_N$$

with $N = 1, 2, \ldots, r$ (r is the total number of nodes in the element), Φ_N are the interpolation functions, and ψ_N are the nodal values of ψ. Let the residual of (4-71) be equated to a residual ε. Then we have

$$\nabla^2 \psi = \varepsilon \tag{5-7}$$

Consider now an orthogonal projection of the residual space (5-7) onto a subspace spanned by the interpolation function Φ_N acting as weighting function in the sense of Galerkin. This process is represented by

$$(\varepsilon, \Phi_N) = \int_\Omega \psi_{,ii} \Phi_N \, d\Omega = 0 \tag{5-8a}$$

Use of the Green–Gauss theorem in (5-8a) leads to

$$\int_\Gamma \psi_{,i} n_i \overset{*}{\Phi}_N \, d\Gamma - \int_\Omega \psi_{,i} \Phi_{N,i} \, d\Omega = 0$$

or
$$\left\{ \int_\Omega \Phi_{N,i} \Phi_{M,i} \, d\Omega \right\} \psi_M = \int_\Gamma \psi_{,i} n_i \overset{*}{\Phi}_N \, d\Gamma \tag{5-8b}$$

Here $\overset{*}{\Phi}_N$ is the interpolation function corresponding to the boundary flux or the normal gradient of the stream function.

Using the simplified notations, we write

$$A_{NM} \psi_M = F_N \tag{5-9}$$

where A_{NM} and F_N are called the coefficient matrix and the flux vector, respectively

$$A_{NM} = \int_\Omega \Phi_{N,i} \Phi_{M,i} \, d\Omega \tag{5-10}$$

$$F_N = \int_\Gamma \psi_{,i} n_i \overset{*}{\Phi}_N \, d\Gamma \tag{5-11}$$

The flux vector may be written alternatively

$$F_N = \int_\Gamma \varepsilon_{ji} V_j n_i \overset{*}{\Phi}_N \, d\Gamma \tag{5-12}$$

Consider the inclined boundary nodes 1 and 2 as shown in Fig. 5-1. Assume that the velocity distribution between the two nodes is linear. We have for unit thickness,

$$F_N = \int_0^l \psi_{,i} n_i \overset{*}{\Phi}_N \, ds = \int_0^l \left\{ \frac{\partial \psi}{\partial x} \overset{*}{\Phi}_N \cos(n, x) + \frac{\partial \psi}{\partial y} \overset{*}{\Phi}_N \cos(n, y) \right\} ds \qquad (5\text{-}13)$$

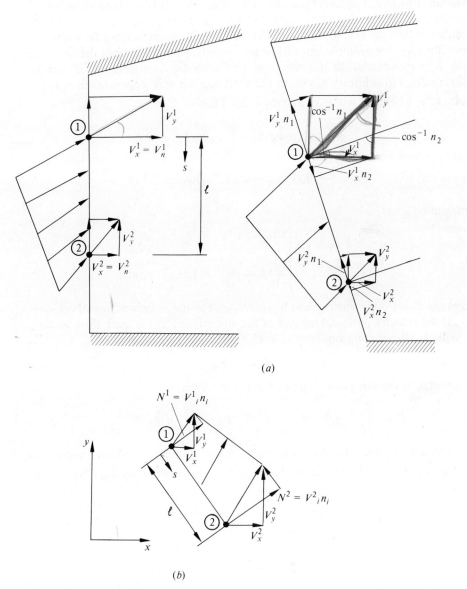

(a)

(b)

Figure 5-1 Boundary velocities. (a) Stream function boundary condition; (b) potential function boundary condition.

or
$$F_N = \int_0^l \varepsilon_{ij} V_j n_i \overset{*}{\Phi}_N \; ds = \int_0^l [V_x \cos(n, y) - V_y \cos(n, x)] \overset{*}{\Phi}_N \; ds \qquad (5\text{-}14)$$

Note that $\varepsilon_{ji} V_j n_i$ represent the components of the boundary velocity parallel to the boundary surface as shown in Fig. 5-1a. If the incoming velocity vector is perpendicular to the boundary surface, then such boundary represents the potential line on which $\varepsilon_{ji} V_j n_i$ must lie. In this case, F_N vanishes regardless of the orientations of the boundary entrance surface. Physically, F_N represents the amount of flow parallel to the boundary surface. Obviously such a flow quantity does not exist, for example, parallel to the boundary entrance when the incoming flow is perpendicular to the entrance. For boundary velocities with variable magnitudes and arbitrary angles to the entrance, we may approximate $\varepsilon_{ji} V_j n_i = \varepsilon_{ji} \overset{*}{\Phi}_M V_j^M n_i$ where $\overset{*}{\Phi}_1 = 1 - s/l$ and $\overset{*}{\Phi}_2 = s/l$. Then

$$F_N = \int_0^l \varepsilon_{ji} V_j n_i \overset{*}{\Phi}_N \; ds = \int_0^l \varepsilon_{ji} \overset{*}{\Phi}_M V_j^M n_i \overset{*}{\Phi}_N \; ds$$

or
$$F_N = \int_0^l (\overset{*}{\Phi}_M V_x^M n_2 - \overset{*}{\Phi}_M V_y^M n_1) \overset{*}{\Phi}_N \; ds$$

Integrating gives

$$F_N = \begin{bmatrix} F_1 \\ F_2 \end{bmatrix} = \frac{l}{3} \begin{bmatrix} V_x^1 n_2 - V_y^1 n_1 + \dfrac{V_x^2 n_2}{2} - \dfrac{V_y^2 n_1}{2} \\[2mm] \dfrac{V_x^1 n_2}{2} - \dfrac{V_y^1 n_1}{2} + V_x^2 n_2 - V_y^2 n_1 \end{bmatrix} \qquad (5\text{-}15)$$

Here the direction cosines n_1 and n_2 correspond to the associated nodal velocity.

If the velocity potential instead of the stream function is used, then we have $\phi = \Phi_N \phi_N$. Proceeding similarly as in (5-7) through (5-9), we obtain

$$A_{NM} \phi_M = F_N \qquad (5\text{-}16)$$

where A_{NM} is the same as (5-10) but F_N is of the form

$$F_N = \int_\Gamma \phi_{,i} n_i \overset{*}{\Phi}_N \; d\Gamma \qquad \text{or} \qquad F_N = \int_\Gamma V_i n_i \overset{*}{\Phi}_N \; d\Gamma \qquad (5\text{-}17)$$

Note that $V_i n_i$ represents the component of the boundary velocities normal to the surface (Fig. 5-1b). Therefore, $V_i n_i$ may be approximated by $\overset{*}{\Phi}_M V_i^N n_i$ and we have

$$F_N = \int_0^l \overset{*}{\Phi}_M V_i^M n_i \overset{*}{\Phi}_N \; ds$$

or

$$F_N = \begin{bmatrix} F_1 \\ F_2 \end{bmatrix} = \frac{l}{3} \begin{bmatrix} V_x^1 n_1 + V_y^1 n_2 + \dfrac{V_x^2 n_1}{2} + \dfrac{V_y^2 n_2}{2} \\[2mm] \dfrac{V_x^1 n_1}{2} + \dfrac{V_y^1 n_2}{2} + V_x^2 n_1 + V_y^2 n_2 \end{bmatrix} \qquad (5\text{-}18)$$

Instead of the Galerkin method, we may use the Rayleigh–Ritz method to obtain the finite element analogue of the Laplace equation. We first require the functional of the form

$$I(\psi) = \int_\Omega f(x_i, \psi, \psi_{,i}) \, d\Omega \qquad (5\text{-}19)$$

in which the competing functions $\psi(x_i)$ assume on the boundary Γ of the region Ω preassigned continuous values $\psi = \psi(s)$. The variational functional is of the form

$$f(x_i, \psi, \psi_{,i}) = \tfrac{1}{2}\psi_{,i}\psi_{,i} \qquad (5\text{-}20)$$

where ψ is written as

$$\psi = \Phi_N \psi_N$$

In view of (5-19) and (5-20), we have

$$I = \frac{1}{2} \int_\Omega \psi_{,i}\psi_{,i} \, d\Omega \qquad (5\text{-}21)$$

The variation of the integral with respect to the nodal values of the stream function leads to

$$\delta I = \frac{\partial I}{\partial \psi_N} \, \delta\psi_N = \left\{ \int_\Omega \psi_{,i} \frac{\partial \psi_{,i}}{\partial \psi_N} \, d\Omega \right\} \delta\psi_N$$

or

$$\delta I = \left\{ \int_\Omega \psi_{,i}\Phi_{N,i} \, d\Omega \right\} \delta\psi_N = 0 \qquad (5\text{-}22)$$

From Green–Gauss theorem we have

$$\delta I = \left(\int_\Gamma \psi_{,i}n_i\overset{*}{\Phi}_N \, d\Gamma - \int_\Omega \psi_{,ii}\Phi_N \, d\Omega \right) \delta\psi_N = \left(\int_\Gamma \psi_{,i}n_i\overset{*}{\Phi}_N \, d\Gamma \right) \delta\psi_N \qquad (5\text{-}23)$$

Equating (5-22) and (5-23), we obtain

$$\left\{ \int_\Omega \Phi_{N,i}\Phi_{M,i} \, d\Omega \right\} \psi_M = \int_\Gamma \psi_{,i}n_i\overset{*}{\Phi}_N \, d\Gamma$$

or

$$A_{NM}\psi_M = F_N \qquad (5\text{-}24)$$

which is identical to (5-9) as derived from the Galerkin method. It is obvious that determination of the form of the variational functional as in (5-21) is not easily available if the governing differential equation is complicated. The Galerkin method which does not require such functionals is therefore more convenient.

Let us now examine the treatment of boundary conditions for a two-dimensional flow over a circular cylinder between the parallel plates as shown in Fig. 5-2.

Stream function formulation (Fig. 5-2b, c) Because of symmetry, the quadrant $a\text{-}b\text{-}d\text{-}c\text{-}g$ is used. By inspection we note that the boundaries $a\text{-}g\text{-}c$ and $b\text{-}d$ are

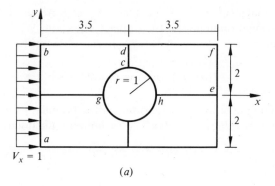

(a)

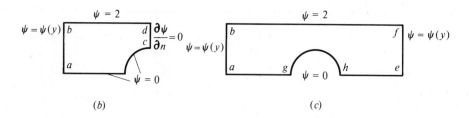

(b)   (c)

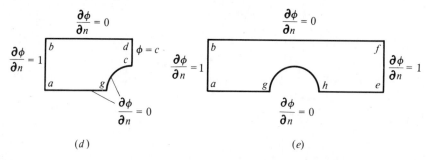

(d)   (e)

Figure 5-2 Flow around a cylinder. (a) Entire domain; (b) quadrant ψ boundaries; (c) upper-half ψ boundaries; (d) quadrant ϕ boundaries; (e) upper-half ϕ boundaries.

streamlines and they are constants. For the purpose of reference, let $\psi = 0$ along a-g-c. Since the input free stream velocity is constant along a-b, we may set

$$\int_{\psi_a}^{\psi_b} d\psi = \int_{y_a}^{y_b} V_x \, dy \tag{5-25}$$

Integrating between the limits indicated, we obtain

$$V_x = \frac{\psi_b - \psi_a}{y_b - y_a} \tag{5-26}$$

Since ψ_a is referenced as zero, we have $\psi_b = 2$ for the free stream velocity $V_x = 1$, with values of ψ linearly varying from 0 to 2 between a-b and $\psi = 2$ between b-d. Thus the ψ values can be prescribed along the boundaries g-a-b-d to be imposed as Dirichlet boundary conditions. All Neumann boundary conditions as calculated from (5-15) are zero ($n_1 = 1$, $n_2 = 0$) along the boundaries a-b and c-d.

If the free stream velocity is not constant but still perpendicular to the entrance face, then the integral (5-25) must be evaluated with V_x as a function of y. If the free stream velocity is not perpendicular to the entrance face, then F_N representing the Neumann boundary conditions is nonzero and should be calculated from (5-15). In both of these cases, the symmetry is no longer maintained and the entire domain must be analyzed. Thus the Dirichlet boundary conditions such as $\psi = 0$ can be specified along either the top or the bottom plate.

Velocity potential formulation (Fig. 5-2d, e) If the velocity potential function is used in the finite element equations, the apparent boundary condition is that ϕ is constant along a-b and d-c (only a quadrant is used). The reference values ϕ may be specified as Dirichlet conditions along d-c. However, the entrance face is subject to the Neumann condition of the type (5-18). The condition $\phi = $ constant along a-b need not be applied because the Neumann condition must be applied here to introduce the input information. Along the boundaries b-d and a-g-c, the Neumann conditions $\phi_{,i}n_i = V_i n_i = 0$ are automatically satisfied by setting $F_N = 0$. If the entire upper-half domain is used, then $\partial\phi/\partial n = v_x$ must also be applied along e-f.

5-3-2 Calculations by Triangular Elements

Consider the flow around an infinitely long cylinder positioned symmetrically between the two flat plates of infinite dimension. Let us choose the equivalent finite domain as shown in Fig. 5-2. We note that with proper treatment of boundary conditions, it is possible to use only one quarter of the entire domain.

> **Example 5-1** (*Stream function formulation*) Choice of dimensions in the direction of flow is arbitrary, but the free stream velocity is considered to prevail at distances sufficiently far from the cylinder. In this example, we use the triangular elements and nodes are numbered in the shorter direction rather than longer. Moreover, the numbering does not zigzag but moves toward the same direction (Fig. 5-3). These schemes reduce the adjacent node number differences to a minimum and contribute to a narrowly banded matrix such that more stable and accurate solutions may be obtained. Since the realistic discretization as shown in Fig. 5-3 yields a large number of equations, let us consider a crude approximation represented in Fig. 5-4 for numerical demonstration purposes. The stream function is considered to vary linearly within an element and is related by
>
> $$\psi = \Phi_N \psi_N \qquad (N = 1, 2, 3) \tag{5-27}$$

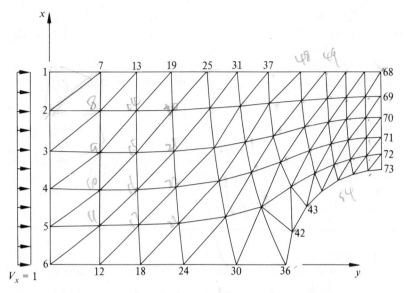

Figure 5-3 Finite element discretization (73 nodes, 111 elements) for flow around the cylinder in Fig. 5-2 (smaller amounts are used for higher velocity gradients and/or for rapid changes of geometry).

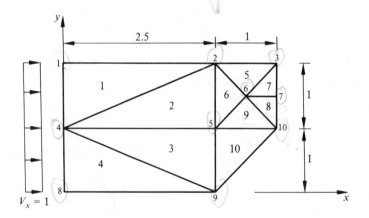

Element node no.	Element no.									
	Global node number									
	1	2	3	4	5	6	7	8	9	10
1	1	4	4	4	2	2	6	6	5	5
2	4	5	9	8	6	5	7	10	10	9
3	2	2	5	9	3	6	3	7	6	10

Figure 5-4 Crude discretization of Fig. 5-3.

From (2-17) we obtain

$$\Phi_N = a_N + b_N x + c_N y \tag{5-28}$$

$$a_1 = (x_2 y_3 - x_3 y_2)/2A \qquad a_2 = (x_3 y_1 - x_1 y_3)/2A \qquad a_3 = (x_1 y_2 - x_2 y_1)/2A$$

$$b_1 = (y_2 - y_3)/2A \qquad b_2 = (y_3 - y_1)/2A \qquad b_3 = (y_1 - y_2)/2A$$

$$c_1 = (x_3 - x_2)/2A \qquad c_2 = (x_1 - x_3)/2A \qquad c_3 = (x_2 - x_1)/2A$$

$$\tag{5-29}$$

with A the area of triangular element. The finite element equation is of the form

$$A_{NM} \psi_M = F_N \tag{5-30}$$

On expanding

$$\begin{bmatrix} A_{11} & A_{12} & A_{13} \\ A_{21} & A_{22} & A_{23} \\ A_{31} & A_{32} & A_{33} \end{bmatrix} \begin{bmatrix} \psi_1 \\ \psi_2 \\ \psi_3 \end{bmatrix} = \begin{bmatrix} F_1 \\ F_2 \\ F_3 \end{bmatrix} \tag{5-31}$$

where

$$A_{11} = \int_\Omega \Phi_{1,i} \Phi_{1,i} \, d\Omega = \int_\Omega (\Phi_{1,1} \Phi_{1,1} + \Phi_{1,2} \Phi_{1,2}) \, d\Omega$$

$$= \int_\Omega (b_1^2 + c_1^2) \, d\Omega = \int_\Omega (b_1^2 + c_1^2) \, d\dot{y} \, dx \, dz$$

For unit thickness perpendicular to the x-y plane, we obtain

$$A_{11} = A(b_1^2 + c_1^2) \tag{5-32}$$

All other coefficients are determined similarly and we have

$$A_{NM} = A \begin{bmatrix} b_1^2 + c_1^2 & b_2 b_1 + c_2 c_1 & b_3 b_1 + c_3 c_1 \\ b_2 b_1 + c_2 c_1 & b_2^2 + c_2^2 & b_3 b_2 + c_3 c_2 \\ b_3 b_1 + c_3 c_1 & b_3 b_2 + c_3 c_2 & b_3^2 + c_3^2 \end{bmatrix} \tag{5-33}$$

To calculate A_{NM} for all elements, it is necessary first to number the element nodes arbitrarily but counterclockwise. The counterclockwise numbering of

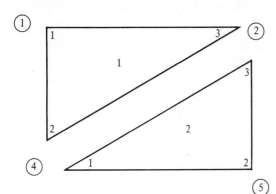

Figure 5-5 Local and global nodes for elements 1 and 2 as assigned in Fig. 5-4.

the element nodes is required as the area A becomes negative in (5-28), otherwise. Since the coordinate values of element 1 are (Figs. 5-4, 5-5)

$$x_1 = 0 \qquad y_1 = 2$$
$$x_2 = 0 \qquad y_2 = 1$$
$$x_3 = 2.5 \qquad y_3 = 2$$

we calculate b's and c's from these coordinate values and obtain the coefficient matrix for element 1 as

$$(A_{NM})_1 = \begin{bmatrix} 1.45 & -1.25 & -0.2 \\ -1.25 & 1.25 & 0 \\ -0.2 & 0 & 0.2 \end{bmatrix} \begin{matrix} 1 \\ 4 \\ 2 \end{matrix}$$

Similarly, for all other elements we have

$$(A_{NM})_2 = \begin{bmatrix} 0.2 & -0.2 & 0 \\ -0.2 & 1.45 & -1.25 \\ 0 & -1.25 & 1.25 \end{bmatrix} \qquad (A_{NM})_3 = \begin{bmatrix} 0.2 & 0 & -0.2 \\ 0 & 1.25 & -1.2 \\ -0.2 & -1.25 & 1.45 \end{bmatrix}$$

$$(A_{NM})_4 = \begin{bmatrix} 1.25 & -1.25 & 0 \\ -1.25 & 1.45 & -0.2 \\ 0 & -0.2 & 0.2 \end{bmatrix} \qquad (A_{NM})_5 = \begin{bmatrix} 0.5 & -0.5 & 0 \\ -0.5 & 1 & -0.5 \\ 0 & -0.5 & 0.5 \end{bmatrix}$$

$$(A_{NM})_6 = \begin{bmatrix} 0.5 & 0 & -0.5 \\ 0 & 0.5 & -0.5 \\ -0.5 & -0.5 & 1 \end{bmatrix} \qquad (A_{NM})_7 = \begin{bmatrix} 0.5 & -0.5 & 0 \\ -0.5 & 1 & -0.5 \\ 0 & -0.5 & 0.5 \end{bmatrix}$$

$$(A_{NM})_8 = \begin{bmatrix} 0.5 & 0 & -0.5 \\ 0 & 0.5 & -0.5 \\ -0.5 & -0.5 & 1 \end{bmatrix} \qquad (A_{NM})_9 = \begin{bmatrix} 0.5 & 0 & -0.5 \\ 0 & 0.5 & -0.5 \\ -0.5 & -0.5 & 1 \end{bmatrix}$$

$$(A_{NM})_{10} = \begin{bmatrix} 1 & -0.5 & -0.5 \\ -0.5 & 0.5 & 0 \\ -0.5 & 0 & 0.5 \end{bmatrix}$$

These element matrices are now to be assembled as described in Sec. 3-1,

$$A_{\alpha\beta}\psi_\beta = F_\alpha \tag{5-34}$$

$$A_{\alpha\beta} = \sum_{e=1}^{e=10} A_{NM}^{(e)} \Delta_{N\alpha}^{(e)} \Delta_{M\beta}^{(e)} \qquad F_\alpha = \sum_\Gamma F_N^{(e)} \Delta_{N\alpha}^{(e)}$$

with $N, M = 1, 2, 3$ and $\alpha, \beta = 1, 2, \dots, 10$. Since the line 8–9–10 is the constant streamline, we may set $\psi = 0$ along this line. Thus we must have $\psi = 1$ at node 4 and $\psi = 2$ at nodes 1, 2, and 3, as calculated from (5-26). These are the Dirichlet boundary conditions. Note that all Neumann boundary conditions are satisfied by setting $\psi_{,i} n_i = 0$ or $F_N = 0$ for the entire vertical boundary nodes.

The assembled global finite element equations (5-34) must now be modified for the boundary conditions by any one of the methods as described in Sec. 3-2. For example, using the method of (3-15), we obtain

$$
\begin{bmatrix}
1 & 0 & 0 & 0 & 0 & 0 & 0 \\
0 & 1 & 0 & 0 & 0 & 0 & 0 \\
0 & 0 & 1 & 0 & 0 & 0 & 0 \\
0 & 0 & 0 & 1 & 0 & 0 & 0 \\
0 & 0 & 0 & 0 & 4.9 & -1 & 0 \\
0 & 0 & 0 & 0 & -1 & 4 & -1 \\
0 & 0 & 0 & 0 & 0 & -1 & 2
\end{bmatrix}
\begin{bmatrix}
\psi_1 \\ \psi_2 \\ \psi_3 \\ \psi_4 \\ \psi_5 \\ \psi_6 \\ \psi_7
\end{bmatrix}
=
\begin{bmatrix}
2 \\ 2 \\ 2 \\ 1 \\ 2.5+0.4 \\ 2+1 \\ 1
\end{bmatrix}
\qquad (5\text{-}35)
$$

Discarding the first four equations, we only need to solve

$$
\begin{bmatrix}
4.9 & -1 & 0 \\
-1 & 4 & -1 \\
0 & -1 & 2
\end{bmatrix}
\begin{bmatrix}
\psi_5 \\ \psi_6 \\ \psi_7
\end{bmatrix}
=
\begin{bmatrix}
2.9 \\ 3 \\ 1
\end{bmatrix}
\qquad (5\text{-}36)
$$

from which we obtain

$$\psi_5 = 0.845$$

$$\psi_6 = 1.241$$

$$\psi_7 = 1.120$$

Since the stream function is assumed to be linear between the nodes, the velocity is constant. Therefore, we may use the formula (5-26) to calculate the velocities between the nodes. For example, along the vertical line 10–7–3, we obtain

$$(V_x)_{7-10} = \frac{\psi_7 - \psi_{10}}{y_7 - y_{10}} = \frac{1.12 - 0}{1.5 - 1} = 2.24$$

$$(V_x)_{3-7} = \frac{\psi_3 - \psi_7}{y_3 - y_7} = \frac{2 - 1.12}{2 - 1.5} = 1.76$$

These results may be compared with an approximate analytical solution via the method of images. Considering that line 9–10 is a circular segment, the stream function and horizontal velocity are given by

$$\psi = U\left\{ y - \frac{H}{2\pi}\sinh^2(\pi b/H)\sin(2\pi y/H)/[\cosh^2(\pi x/H) - \cos^2(\pi y/H)] \right\}$$

$$V_x = \frac{\partial \psi}{\partial y} = U\left\{ 1 - \frac{\sinh^2(\pi b/H)\cos(2\pi y/H)}{\cosh^2(\pi x/H) - \cos^2(\pi y/H)} \right. \qquad (5\text{-}37a)$$

$$\left. + \frac{\tfrac{1}{2}\sinh^2(\pi b/H)\sin^2(2\pi y/H)}{[\cosh^2(\pi x/H) - \cos^2(\pi y/H)]^2} \right\} \qquad (5\text{-}37b)$$

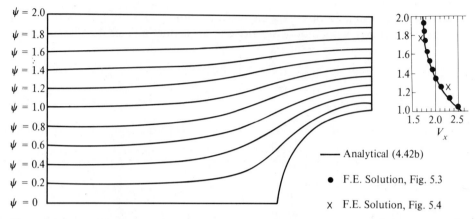

Figure 5-6 Stream function contours and velocity distributions along the vertical line above the crest of cylinder—based on Fig. 5-3 and Fig. 5-4. Streamline contours are interpolated from nodal stream functions and velocities are calculated from the streamline contour values.

where x, y are the coordinates with origin at the center of the cylinder, b is the radius, and H is the vertical distance between the two plates. The velocities along the vertical line above the crest of the cylinder calculated from (5-37b) are given in Fig. 5-6. It is unfair to compare the results of the present example because we solved for the flow around the pointed wedge rather than a circular cylinder with unreasonably crude mesh sizes. Therefore, returning to the finite element geometry of Fig. 5-3 (73 nodes, 111 elements), the results obtained by computer programs (Appendix B-1) are compared with those of the analytical solution in Fig. 5-6. The average deviation for the velocity profiles between the two cases is less than one percent. However, it should be warned that the analytical solution loses accuracy for large values of b/H ratio. (The error built in (5-37) for $b/H = 0.25$ is 2.34%.)†

In general, accuracy of the finite element solution is expected to improve with finer mesh and higher order interpolation functions. Irregular mesh configurations with needle-like elements contribute to deterioration of the accuracy. These topics are discussed in Sec. 5-3-3.

Example 5-2 (*Velocity potential formulation*) Calculations via velocity potential formulation differ very slightly from those of stream function formulation. The globally assembled coefficient matrix is identical. Only the boundary conditions are different. For the problem of Fig. 5-4, we may set $\phi_3 = \phi_7 = \phi_{10} = 0$, or any convenient number for a reference ϕ value. We now have the Neumann boundary condition $F_N \neq 0$

$$F_N = \int_\Gamma \phi_{,i} n_i \overset{*}{\Phi}_N \, d\Gamma = \int_\Gamma v_i n_i \overset{*}{\Phi}_N \, d\Gamma$$

† C. C. Shih (Private Communication).

From the formula (5-18) with $V_x = 1$, $V_y = 0$, $n_1 = 1$, $n_2 = 0$, and $l = 1$ for boundaries 1–4 and 4–8, we obtain

$$F_N = \frac{lV_x}{2}\begin{bmatrix}1\\1\end{bmatrix} = \frac{1}{2}\begin{bmatrix}1\\1\end{bmatrix}$$

For all other boundaries $F_N = 0$ and thus the assembled input vector is of the form

$$F_\alpha = \sum_{e=1}^{E} F_N^{(e)} \Delta_{N\alpha}^{(e)} \qquad \begin{pmatrix}\alpha = 1,2,\ldots,10\\ E = 1,2,\ldots,10\end{pmatrix}$$

or

$$F_\alpha = [\tfrac{1}{2}\ \ 0\ \ 0\ \ 1\ \ 0\ \ 0\ \ 0\ \ \tfrac{1}{2}\ \ 0\ \ 0]^T$$

Thus the global equations take the form

$$\begin{bmatrix}
1.45 & -0.2 & 0 & -1.25 & 0 & 0 & 0 & 0 & 0 & 0\\
 & 2.45 & 0 & 0 & -1.25 & -1.0 & 0 & 0 & 0 & 0\\
 & & 1.0 & 0 & 0 & -0.5 & -0.5 & 0 & 0 & 0\\
 & & & 2.9 & -0.4 & 0 & 0 & -1.25 & 0 & 0\\
 & & & & 4.9 & -1.0 & 0 & 0 & -1.75 & -0.5\\
 & & & & & 4.0 & -1.0 & 0 & 0 & -0.5\\
 & & & & & & 2.0 & 0 & 0 & -0.5\\
\text{Symm.} & & & & & & & 1.45 & -0.2 & 0\\
 & & & & & & & & 1.95 & 0\\
 & & & & & & & & & 1.5
\end{bmatrix}
\begin{bmatrix}\phi_1\\\phi_2\\\phi_3\\\phi_4\\\phi_5\\\phi_6\\\phi_7\\\phi_8\\\phi_9\\\phi_{10}\end{bmatrix}
=
\begin{bmatrix}0.5\\0\\0\\1\\0\\0\\0\\0.5\\0\\0\end{bmatrix}$$

$$(5\text{-}38)$$

Because $\phi_3 = \phi_7 = \phi_{10} = 0$, we may either zero out the 3rd, 7th, and 10th rows and columns with one's at the diagonal or leave them out completely (see Eq. 3-14) and solve the remaining 7×7 equations. Thus we obtain

$$\phi_1 = 3.787, \quad \phi_2 = 1.204, \quad \phi_3 = 0, \quad \phi_4 = 3.841, \quad \phi_5 = 1.261$$

$$\phi_6 = 6.161, \quad \phi_7 = 0, \quad \phi_8 = 3.827, \quad \phi_9 = 1.491, \quad \phi_{10} = 0$$

From these, the average x-velocity between the nodes 6 and 7 is calculated

$$(V_x)_{6-7} = \frac{\phi_6 - \phi_7}{x_6 - x_7} = \frac{6.161 - 0}{0.5} = 1.2322$$

Because of the crude mesh used in this example, we cannot expect to have reasonable results.

The velocities may be calculated at the centroid of each element from

$$V_x = \frac{\partial \psi}{\partial y} = \frac{\partial \Phi_N}{\partial y}\psi_N \qquad (5\text{-}39a)$$

$$V_y = -\frac{\partial \psi}{\partial x} = -\frac{\partial \Phi_N}{\partial x} \psi_N \qquad (5\text{-}39b)$$

for the ψ formulation, and

$$V_x = \frac{\partial \phi}{\partial x} = \frac{\partial \Phi_N}{\partial x} \phi_N \qquad (5\text{-}40a)$$

$$V_y = \frac{\partial \phi}{\partial y} = \frac{\partial \Phi_N}{\partial y} \phi_N \qquad (5\text{-}40b)$$

for the ϕ formulation. If we use four-node isoparametric elements, then it is possible to calculate the velocities at nodes along the line d–c by substituting the isoparametric coordinates $(1, -1)$ corresponding to the nodes. Detailed descriptions of the use of isoparametric elements are presented in the next section for axisymmetric flow. Nevertheless, for the benefit of the reader, a computer program using the isoparametric elements (110 elements, 132 nodes) is listed in Appendix B-2. The results are presented in Fig. 5-7. In this program, the pressures on the cylinder are calculated from

$$P = \tfrac{1}{2}\rho q^2 \qquad (5\text{-}41)$$

with

$$q = (V_x^2 + V_y^2)^{1/2}$$

Verification of accuracy of the solutions discussed here is an important part of the analysis. This subject is taken up in the following section.

5-3-3 Error Analysis

The finite element solutions must be evaluated as to their accuracies. The basic criteria for accuracy of the finite element analysis were presented in Sec. 3-4. Let

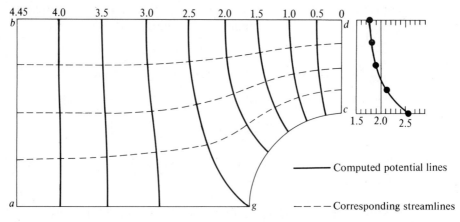

Figure 5-7 Velocity potential function contours and velocity distributions. (Geometric configuration same as Fig. 5-2, 88 Isoparametric elements.)

us now examine the governing partial differential equations together with the boundary conditions discussed in Sec. 5-3-2.

For the stream function formulation

$$\nabla^2 \psi = 0 \qquad \text{in } \Omega$$

$$\psi = \text{constant} \qquad \text{on } \Gamma_1$$

$$\frac{\partial \psi}{\partial n} = 0 \qquad \text{on } \Gamma_2$$

Obviously, this example represents nonhomogeneous mixed boundary conditions. The error estimate follows from (3-177),

$$\| e \|_{H^m(\Omega)} \leq c \left\{ h^{\mu_1} \| f \|_{H^r(\Omega)} + h^{\mu_2} \sum_{j=0}^{m-1} \| g_j \|_{H^{p_j}(\Gamma)} \right\}$$

$$\mu_1 = \min(k + 1 - m, r + m)$$

$$\mu_2 = \min\left[k + 1 - m, \quad \min_{0 \leq j \leq m-1} (p_j + q_j + \tfrac{1}{2} - m) \right]$$

Here Γ_1 represents the boundaries a–b–d and a–g–c, and Γ_2 is the boundary c–d for Fig. 5-2b; whereas Γ_1 covers the entire boundaries and there is no Γ_2 boundary for Fig. 5-2c. Thus Fig. 5-2b is the mixed (Dirichlet–Neumann) boundary value problem and Fig. 5-2c represents the Dirichlet boundary value problem. Note that this formula has been derived for essential boundary conditions, not for mixed boundary conditions. Unfortunately, at present, no error estimate formula for the mixed boundary conditions is available. Nevertheless, in this particular example problem, a great deal is known about the behavior of the solution on the boundary thanks to physical arguments. Since ψ is assumed to be zero along a–g–c, ψ varies linearly on a–b reaching a value of 2 at b and remains constant on b–d with $\partial \psi / \partial n = 0$ along c–d. In other words, ψ is continuous on all of the boundaries $\Gamma = \Gamma_1 + \Gamma_2$. It makes sense, therefore, to regard the boundary data g as an element of a space of functions which have piece wise constant first derivatives on Γ. By using Fourier transforms, it is possible to show that such a function belongs to $H^{3/2}(\Gamma)$. Hence, we take $p_j = p_o = 3/2$.

Let us first evaluate the rate of convergence for the Dirichlet boundary value problem of Fig. 5-2c. In this case, $k = 1$ for linear triangles, $m = 1$, $f = 0$, $r \geq 2$ for Laplace equation, $g = C^\infty$ for $p_j = \infty$, $q_j = 0$ for Dirichlet boundary operator. Thus $\mu_1 = 0$, and

$$\| e \|_{H^1(\Omega)} \leq c h^{\mu_2} \| g \|_{H^{p_j}(\Gamma)}$$

with

$$\mu_2 = \min \begin{cases} k + 1 - m = 1 \\ p_j + q_j + \tfrac{1}{2} - m = \infty \end{cases}$$

This gives $\mu_2 = 1$ as governed by the minimum, and

$$\| e \|_{H^1(\Omega)} \leq c h \| g \|_{C^\infty(\Gamma)}$$

We conclude that the solution is in $H^1(\Omega)$ and the rate of convergence is $0(h)$. It is apparent that the rate of convergence becomes $0(h^2)$ if quadratic triangles $(k = 2)$ are used.

On the other hand, if we use Fig. 5-2b, we have $g \in H^{3/2}$, $p_j = 3/2$, and $q_j = 1$. Thus,

$$\mu_2 = \min \begin{cases} k + 1 - m = 1 \\ p_j + q_j + \tfrac{1}{2} - m = 1 \end{cases}$$

This implies that μ_2 is governed by the second of the above criteria and that no improvement arises even if higher order elements $(k > 1)$ are used.

We may also calculate the error in $H^0(\Omega) = L_2(\Omega)$ norm (3-178) and find that the error estimate is $0(h^{k+1})$. The pointwise error (3-179), on other hand, shows $0(h^{2k})$.

Next, let us look at the velocity potential formulation,

$$\nabla^2 \phi = 0 \qquad \text{in } \Omega$$

$$\phi = \text{constant} \qquad \text{on } \Gamma_1$$

$$\frac{\partial \phi}{\partial n} = \begin{cases} 0 \\ \text{constant} \end{cases} \qquad \begin{matrix} \text{on } \Gamma_{21} \\ \text{on } \Gamma_{22} \end{matrix} \qquad \Gamma_2 = \Gamma_{21} \cup \Gamma_{22}$$

This is the mixed boundary value problem for both Fig. 5-2d and Fig. 5-2e. Therefore, the error estimate for the velocity potential formulation is the same as in the stream function formulation (Fig. 5-2c).

Having confirmed the mathematical convergence properties, let us now examine numerical experiments for the convergence as a function of various mesh sizes. Since we have shown two solutions for the stream function formulation with 10 nodes and 86 nodes, let us choose additional discretizations for 19, 33, and 46 nodes as shown in Fig. 5-8. Calculations of velocities above the crest at two points, $c(x = 3.5, y = 1)$ and $d(x = 3.5, y = 2)$ are plotted in Fig. 5-9a. Stream functions are expected to converge upper-bound. However, the velocities calculated from stream functions may be either upper-bound or lower-bound, depending on lengths between adjacent nodes. It is seen that convergence appears to have been achieved at 33 node discretization.

The errors in $H^1(\Omega)$ and $H^0(\Omega)$ norms are calculated, respectively, from

$$\| e \|_{H^1(\Omega)} = \left\{ \int \int \left[e^2 + \left(\frac{\partial e}{\partial x} \right)^2 + \left(\frac{\partial e}{\partial y} \right)^2 \right] dx \, dy \right\}^{1/2}$$

$$\| e \|_{H^0(\Omega)} = \left\{ \int \int e^2 \, dx \, dy \right\}^{1/2}$$

where $e = \psi - \hat{\psi} = \psi - \Phi_i \psi_i$ with Φ_i and ψ_i being the finite element interpolation functions and the nodal values of ψ as determined from the solution, respectively. Considering (5-37a) to be the exact solution ψ, the integration can be performed as implied above. The pointwise error is determined at any arbitrary point as the difference between the exact and finite element solutions.

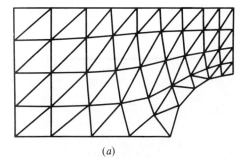

(a)

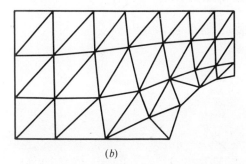

(b)

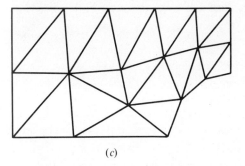

(c)

Figure 5-8 Various mesh configurations. (a) 46 nodes, 74 elements; (b) 33 nodes, 44 elements; (c) 19 nodes, 22 elements.

For any two sets of data $\| e_1 \|$ and $\| e_2 \|$ corresponding to the mesh sizes h_1 and h_2, we have the relation

$$\left(\frac{h_1}{h_2} \right)^p = \frac{\| e_1 \|}{\| e_2 \|}$$

or

$$p = \frac{\ln \| e_1 \| - \ln \| e_2 \|}{\ln h_1 - \ln h_2}$$

where p is the exponent denoting the order of error related to mesh size. Thus the slope of the errors versus the mesh sizes in the natural log—log scale determines

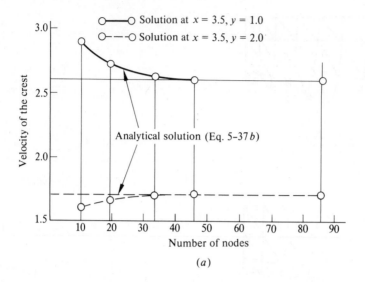

(a)

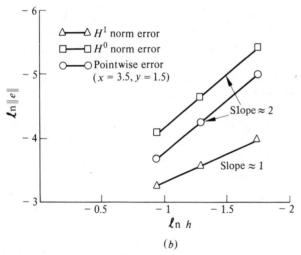

(b)

Figure 5-9 Error analysis—ψ formulation. (a) Convergence of velocities above the crest; (b) errors at various norms.

the order of the error estimate. The mesh size h for a given discretization is determined by h_{max}/H where h_{max} and H are the maximum dimensions of the element and the domain, respectively.

For the linear triangular element ($k = 1$), we confirm the theoretical predictions of $O(h)$ for the $H^1(\Omega)$ norm and $O(h^2)$ for the $H^0(\Omega)$ norm as shown in Fig. 5-9b. The pointwise error is evaluated at any arbitrary point. An evaluation at $x = 3.5$ and $y = 1.5$ is shown in Fig. 5-9b. The pointwise error turns out to be of $O(h^2)$ as expected (3-179).

Accuracy of finite element solutions can be affected adversely when shapes of the elements are irregular. Let the domain be partitioned into elements e_i with the maximum dimension of any element $h = \max h_i$. In the case of a triangular element depicted in Fig. 5-10, the variation of the interpolation function across the element in the x-direction has the following value at node 2,

$$\frac{\partial \Phi_N(z_M)}{\partial x} = \frac{1}{h \tan \theta}$$

Here the interpolation function Φ_N is of the order m and there exists constant c such that

$$\max \left| D^\alpha \Phi_N(z_M) \right| \leq ch^{k+1-m}$$

For $\alpha = 1$, $k = 1$, and $m = 1$, c is found to be

$$c \geq \frac{1}{\tan \theta}$$

It is clear that for $\theta = 0$, with the triangle degenerated into a vertical line, the interpolation function will not be uniform, leading to serious errors in the finite element solution.

It is also important to examine errors due to the approximated curved boundaries such as around the circle being replaced by straight lines. If data are applied along the approximated boundaries, then additional errors would occur because of possible perturbation of the data. Estimates for these errors have been studied by Ciarlet and Raviart [1972] and Zlamal [1973].

The effects of corners or geometric singularities along the boundaries have been studied by Babuska [1970] and Whiteman [1975], among others. For example, the point g in Fig. 5-2 represents a singularity with an abrupt change

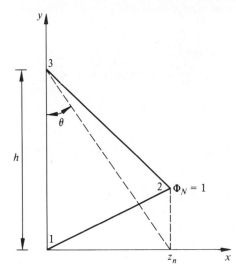

Figure 5-10 Linear interpolation function in a triangular element.

in geometry. Moreover, the points a, b, c, and d signify abrupt changes in boundary conditions. These are intriguing mathematical questions. In practice, however, the remedy is to use refined mesh or interpolation functions with singular terms. See Sec. 5-11 for additional discussions.

If the finite element equations are not exactly integrated, such as in the case of Gaussian quadrature integration, this contributes to still another source of errors. This subject will be discussed in Sec. 5-4-5.

5-4 AXISYMMETRIC INVISCID FLOW

5-4-1 Governing Equations

A flow around an axisymmetric body requires a special treatment in the analysis. All governing equations must be transformed into appropriate curvilinear coordinates: polar, cylindrical, or spherical. The Laplace equation in terms of potential function for the curvilinear coordinates assumes the form (see Sec. 4-2)

$$\nabla^2\phi = \mathbf{V}\cdot\mathbf{V} = \mathbf{V}\cdot\nabla\phi = \mathbf{g}^i\frac{\partial}{\partial\xi_i}\cdot\left(\mathbf{g}^j\frac{\partial\phi}{\partial\xi_j}\right) = \phi_{|ij}g^{ij} = 0 \qquad (5\text{-}42)$$

with

$$\phi_{|ij} = \phi_{,ij} - \Gamma^k_{ij}\phi_{,k} = \phi_{,ij} + \Gamma^k_{ki}\phi_{,j} \qquad (5\text{-}43)$$

Thus, for cylindrical coordinates ($\xi_1 = r$, $\xi_2 = \theta$, $\xi_3 = z$, $g_{11} = 1$, $g_{22} = r^2$, $g_{33} = 1$, $\Gamma^1_{22} = -r$, $\Gamma^2_{12} = 1/r$), we have

$$\nabla^2\phi = \frac{\partial^2\phi}{\partial r^2} + \frac{1}{r^2}\frac{\partial^2\phi}{\partial\theta^2} + \frac{\partial^2\phi}{\partial z^2} + \frac{1}{r}\frac{\partial\phi}{\partial r} = 0 \qquad (5\text{-}44)$$

For axisymmetric flow, ϕ is constant in the circumferential direction (θ). Therefore

$$\nabla^2\phi = \frac{\partial^2\phi}{\partial r^2} + \frac{\partial^2\phi}{\partial z^2} + \frac{1}{r}\frac{\partial\phi}{\partial r} = 0 \qquad (5\text{-}45)$$

For irrotational flow, the curl of the velocity assumes the form (Fig. 5-11)

$$\nabla\times\mathbf{V} = \mathbf{g}^i\frac{\partial}{\partial\xi_i}\times(V_j\mathbf{g}^j) = V_{j|i}\frac{1}{\sqrt{g}}e_{kij}\mathbf{g}_k = 0$$

$$= \left(\frac{\partial V_j}{\partial\xi_i} - \Gamma^m_{ji}V_m\right)\frac{1}{\sqrt{g}}e_{kij}\mathbf{g}_k = 0 \qquad (5\text{-}46)$$

Expanding (5-46) and replacing the tensor components of V_j and $\mathbf{g}_k$ by appropriate physical components for cylindrical coordinates, we obtain

$$\left(\frac{\partial V_\theta}{\partial z} - \frac{1}{r}\frac{\partial V_z}{\partial\theta}\right)\mathbf{e}_r + \left(\frac{\partial V_z}{\partial r} - \frac{\partial V_r}{\partial z}\right)\mathbf{e}_\theta + \left(\frac{1}{r}\frac{\partial V_r}{\partial\theta} - \frac{1}{r}\frac{\partial(rV_\theta)}{\partial r}\right)\mathbf{e}_z = 0 \qquad (5\text{-}47)$$

Here $V_1 = V_r\sqrt{g_{11}}$, $\mathbf{g}_1 = \mathbf{e}_r\sqrt{g_{11}}$, etc. If the flow is axisymmetric, the terms with

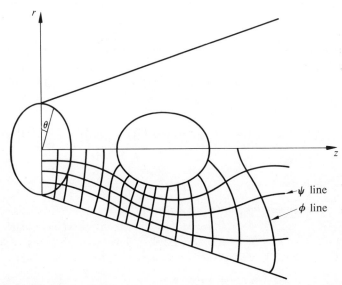

Figure 5-11 Axisymmetric flow.

the unit vectors $\mathbf{e}_r$ and $\mathbf{e}_z$ vanish, suggesting the existence of the stream functions related by

$$V_z = \frac{1}{r} \frac{\partial \psi}{\partial r} \tag{5-48a}$$

and

$$V_r = -\frac{1}{r} \frac{\partial \psi}{\partial z} \tag{5-48b}$$

Here ψ is called the Stokesian stream function having the unit of ft^3/s as opposed to the two-dimensional stream function often called Lagrangian stream function having the unit of ft^2/s. Thus we obtain the curl of the velocity vector with the terms associated with the unit vector $\mathbf{e}_\theta$ perpendicular to the plane of the flow

$$\frac{1}{r}\left(\frac{\partial^2 \psi}{\partial r^2} + \frac{\partial^2 \psi}{\partial z^2} - \frac{1}{r}\frac{\partial \psi}{\partial r}\right)\mathbf{e}_\theta = 0$$

and the expression known as the Stokesian equation

$$D^2\psi = \frac{\partial^2 \psi}{\partial r^2} + \frac{\partial^2 \psi}{\partial z^2} - \frac{1}{r}\frac{\partial \psi}{\partial r} = 0 \tag{5-49}$$

For finite element applications, either (5-45) or (5-49) may be used, leading to a velocity potential formulation or stream function formulation, respectively.

5-4-2 Derivatives of Isoparametric Interpolation Functions

The axisymmetric element geometry may be generated from integration around the circumference of a two-dimensional plane element (either three-node or four-node), which provides a ring element. Since we used a triangular element in two-dimensional flow, we intend to demonstrate the use of four-node isoparametric element in this section. The two-dimensional linear isoparametric interpolation functions as represented in Fig. 5-12 are given by (see Chap. 2 for details)

$$\Phi_N(\xi_i) = \tfrac{1}{4}[(1 + \xi_{N1}\xi_1)(1 + \xi_{N2}\xi_2)] \tag{5-50}$$

Thus the stream function may be expressed as

$$\psi = \Phi_N(\xi_i)\psi_N \qquad (N = 1, 2, 3, 4; i = 1, 2) \tag{5-51}$$

The cartesian coordinates are related by

$$x_i = \Phi_N(\xi_i)x_{Ni} \tag{5-52}$$

or
$$x_i = \tfrac{1}{4}[a_i + b_i\xi_1 + c_i\xi_2 + d_i\xi_1\xi_2]$$

where

$$a_i = x_{1i} + x_{2i} + x_{3i} + x_{4i}$$
$$b_i = -x_{1i} + x_{2i} + x_{3i} - x_{4i}$$
$$c_i = -x_{1i} - x_{2i} + x_{3i} + x_{4i}$$
$$d_i = x_{1i} - x_{2i} + x_{3i} - x_{4i}$$

$$\tag{5-53}$$

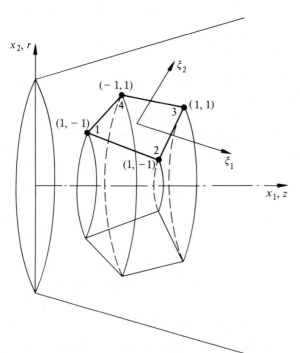

Figure 5-12 Linear isoparametric element.

The partial derivatives with respect to ξ_i are of the form

$$\frac{\partial \Phi_N}{\partial \xi_i} = \frac{\partial \Phi_N}{\partial x_k} \frac{\partial x_k}{\partial \xi_i} = \frac{\partial \Phi_N}{\partial x_k} J_{ki} \qquad (i, k = 1, 2) \tag{5-54}$$

Solving for $\partial \Phi_N / \partial x_k$ gives

$$\frac{\partial \Phi_N}{\partial x_k} = (J_{ki})^{-1} \frac{\partial \Phi_N}{\partial \xi_i} \tag{5-55}$$

where J_{ki} is a Jacobian. It can easily be shown that

$$\frac{\partial \Phi_N}{\partial x_k} = \hat{J}(A_{Nk} + B_{Nk}^i \xi_i) \tag{5-56}$$

with

$$\hat{J} = \frac{1}{8|J_{ki}|} = \frac{1}{8|\mathbf{J}|}$$

$$
\begin{array}{lll}
A_{11} = x_{22} - x_{42} & B_{11}^1 = x_{42} - x_{32} & B_{11}^2 = x_{32} - x_{22} \\
A_{21} = x_{32} - x_{12} & B_{21}^1 = x_{32} - x_{42} & B_{21}^2 = x_{12} - x_{42} \\
A_{31} = x_{42} - x_{22} & B_{31}^1 = x_{12} - x_{22} & B_{31}^2 = x_{42} - x_{12} \\
A_{41} = x_{12} - x_{32} & B_{41}^1 = x_{22} - x_{12} & B_{41}^2 = x_{22} - x_{32}
\end{array}
\tag{5-57a}
$$

$$
\begin{array}{lll}
A_{12} = x_{41} - x_{21} & B_{12}^1 = x_{31} - x_{41} & B_{12}^2 = x_{21} - x_{31} \\
A_{22} = x_{11} - x_{31} & B_{22}^1 = x_{41} - x_{31} & B_{22}^2 = x_{41} - x_{11} \\
A_{32} = x_{21} - x_{41} & B_{32}^1 = x_{21} - x_{11} & B_{32}^2 = x_{11} - x_{41} \\
A_{42} = x_{31} - x_{11} & B_{42}^1 = x_{11} - x_{21} & B_{42}^2 = x_{31} - x_{21}
\end{array}
\tag{5-57b}
$$

$$|\mathbf{J}| = \frac{\partial x_1}{\partial \xi_1} \frac{\partial x_2}{\partial \xi_2} - \frac{\partial x_2}{\partial \xi_1} \frac{\partial x_1}{\partial \xi_2} = \tfrac{1}{8}(\alpha_0 + \alpha_1 \xi_1 + \alpha_2 \xi_2) \tag{5-58}$$

$$\alpha_0 = (x_{41} - x_{21})(x_{12} - x_{32}) - (x_{11} - x_{31})(x_{42} - x_{22})$$

$$\alpha_1 = (x_{31} - x_{41})(x_{12} - x_{22}) - (x_{11} - x_{21})(x_{32} - x_{42})$$

$$\alpha_2 = (x_{41} - x_{11})(x_{22} - x_{32}) - (x_{21} - x_{31})(x_{42} - x_{12})$$

5-4-3 Finite Element Formulations

Consider now the Stokesian equation in terms of stream function,

$$\frac{\partial^2 \psi}{\partial z^2} + \frac{\partial^2 \psi}{\partial r^2} = \frac{1}{r} \frac{\partial \psi}{\partial r} \tag{5-59}$$

As discussed in Sec. 2-3-3, we may use the same interpolation function for the stream function as used for defining the geometry. Thus we write

$$\psi = \Phi_N \psi_N$$

The Galerkin integral assumes the form

$$\int_\Omega \left(\frac{\partial^2 \psi}{\partial z^2} + \frac{\partial^2 \psi}{\partial r^2} - \frac{1}{r} \frac{\partial \psi}{\partial r} \right) \Phi_N \, d\Omega = 0$$

where $dv = r \, d\theta \, dr \, dz$ with $z = x_1$, $r = x_2$. Integrating by parts or using the Green–Gauss theorem yields

$$\int_\Gamma \frac{\partial \psi}{\partial z} \overset{*}{\Phi}_N r \, dr \, d\theta - \int_\Omega \frac{\partial \psi}{\partial z} \frac{\partial \Phi_N}{\partial z} r \, d\theta \, dr \, dz + \int_\Gamma \frac{\partial \psi}{\partial r} \overset{*}{\Phi}_N r \, d\theta \, dz$$

$$- \int_\Omega \frac{\partial \psi}{\partial r} \frac{\partial \Phi_N}{\partial r} r \, d\theta \, dr \, dz - 2 \int_\Omega \frac{\partial \psi}{\partial r} \Phi_N \, d\theta \, dr \, dz = 0$$

where $\overset{*}{\Phi}_N$ is the interpolation function denoting the distribution of the normal gradient of stream function on the boundary. Recall that the integrals of physical coordinates can be transformed into isoparametric coordinates in the form (2-61),

$$\int \int dr \, dz = \int_{-1}^{1} \int_{-1}^{1} |\mathbf{J}| \, d\xi \, d\eta$$

with $\xi = \xi_1$, $\eta = \xi_2$. Therefore, the local finite element equation assumes the form

$$D_{NM} \psi_M = F_N \tag{5-60}$$

where

$$D_{NM} = 2\pi \int_{-1}^{1} \int_{-1}^{1} \left(\frac{\partial \Phi_N}{\partial z} \frac{\partial \Phi_M}{\partial z} + \frac{\partial \Phi_N}{\partial r} \frac{\partial \Phi_M}{\partial r} \right) r \, |\mathbf{J}| \, d\xi \, d\eta$$

$$+ 4\pi \int_{-1}^{1} \int_{-1}^{1} \Phi_N \frac{\partial \Phi_M}{\partial r} |\mathbf{J}| \, d\xi \, d\eta \tag{5-61}$$

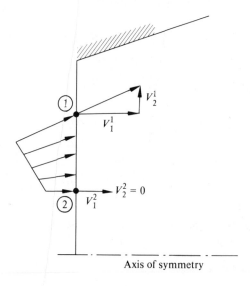

Figure 5-13 Boundary entrance velocity distribution in axisymmetry geometry.

Axis of symmetry

and

$$F_N = 2\pi \int_0^l \frac{\partial \psi}{\partial z} \overset{*}{\Phi}_N \cos(n, z) r \, ds + 2\pi \int_0^l \frac{\partial \psi}{\partial r} \overset{*}{\Phi}_N \cos(n, r) r \, ds \qquad (5\text{-}62)$$

$$\frac{\partial \Phi_N}{\partial z} \frac{\partial \Phi_M}{\partial z} = (a_{NM}^{(2)} + b_{NM}^{(2)} \xi + c_{NM}^{(2)} \eta + d_{NM}^{(2)} \xi\eta + e_{NM}^{(2)} \xi^2 + f_{NM}^{(2)} \eta^2) \hat{J}^2 \qquad (5\text{-}63)$$

$$\frac{\partial \Phi_N}{\partial r} \frac{\partial \Phi_M}{\partial r} = (a_{NM}^{(1)} + b_{NM}^{(1)} \xi + c_{NM}^{(1)} \eta + d_{NM}^{(1)} \xi\eta + e_{NM}^{(1)} \xi^2 + f_{NM}^{(1)} \eta^2) \hat{J}^2 \qquad (5\text{-}64)$$

$$\Phi_N \frac{\partial \Phi_M}{\partial r}$$
$$= (g_{NM}^{(1)} + h_{NM}^{(1)} \xi + k_{NM}^{(1)} \eta + p_{NM}^{(1)} \xi\eta + q_{NM}^{(1)} \xi^2 + s_{NM}^{(1)} \eta^2 + t^{(1)} \xi^2 \eta + u_{NM}^{(1)} \xi\eta^2) \frac{\hat{J}}{4} \quad (5\text{-}65)$$

with

$$a_{NM}^{(2)} = A_{M2} A_{N2} \qquad\qquad c_{NM}^{(2)} = A_{M2} B_{N2}^2 + B_{M2}^2 A_{N2}$$

$$b_{NM}^{(2)} = A_{M2} B_{N1}^1 + B_{M2}^1 A_{N2} \qquad d_{NM}^{(2)} = B_{M2}^1 B_{N2}^2 + B_{M2}^2 B_{N2}^1, \text{ etc.}$$

The input vector F_N as pointed out for the two-dimensional problem represents the component of the flow rate in the direction parallel to the entrance boundary surface. Thus for the incoming velocity perpendicular to the entrance surface, F_N vanishes. For the purpose of generality, however, let us consider an arbitrary incoming velocity distribution as shown in Fig. 5-13. The input vector F_N (5-62) may be written as

$$F_N = 2\pi \int_0^l \varepsilon_{ji} V_j n_i \overset{*}{\Phi}_N r^2 \, ds$$

As shown in the two-dimensional case, we approximate V_j by $\overset{*}{\Phi}_N V_j^N$. We also set $r = \overset{*}{\Phi}_N r_N$ between the two boundary nodes. Thus we obtain

$$F_N = 2\pi \int_0^l \varepsilon_{ji} \overset{*}{\Phi}_M V_j^M n_i \overset{*}{\Phi}_P r_P \overset{*}{\Phi}_Q r_Q \overset{*}{\Phi}_N \, ds$$

or

$$F_N = \frac{2\pi l}{5} \begin{bmatrix} (V_a + \tfrac{1}{4}V_b)r_1^2 + (\tfrac{1}{2}V_a + \tfrac{1}{3}V_b)r_1 r_2 + (\tfrac{1}{6}V_a + \tfrac{1}{4}V_b)r_2^2 \\ (\tfrac{1}{4}V_a + \tfrac{1}{6}V_b)r_1^2 + (\tfrac{1}{3}V_a + \tfrac{1}{2}V_b)r_1 r_2 + (\tfrac{1}{4}V_a + V_b)r_2^2 \end{bmatrix}$$

with

$$V_a = V_1^1 n_2 - V_2^1 n_1 \qquad V_b = V_1^2 n_2 - V_2^2 n_1$$

Note that for the vertical entrance such as shown in Fig. 5-13, we have $n_1 = 1$ and $n_2 = 0$. In general engineering applications, the boundary entrance velocity distribution of the type in Fig. 5-13 would not occur but our formulas provide a capability of accommodating any type of boundary input.

Recall that integration of the equation such as (5-61) may be performed effectively by means of Gaussian quadrature, as discussed in Sec. 2-3-3. Let us consider the integral of the first term in (5-61),

$$H_{NM} = 2\pi \int_{-1}^{1} \int_{-1}^{1} \frac{\partial \Phi_N}{\partial z} \frac{\partial \Phi_M}{\partial z} r |J| \, d\xi \, d\eta$$

$$= 2\pi \int_{-1}^{1} \int_{-1}^{1} \frac{\hat{J}}{8} (a_{NM}^{(2)} + b_{NM}^{(2)} \xi + c_{NM}^{(2)} \eta + d_{NM}^{(2)} \xi\eta + e_{NM}^{(2)} \xi^2 + f_{NM}^{(2)} \eta^2) r \, d\xi \, d\eta$$

$$= 2\pi \int_{-1}^{1} \int_{-1}^{1} \frac{1}{32} \frac{(a_{NM}^{(2)} + b_{NM}^{(2)} \xi + \cdots)(a_2 + b_2\xi + c_2\eta + d_2\xi\eta)}{\alpha_0 + \alpha_1\xi + \alpha_2\eta} \, d\xi \, d\eta$$

$$= \frac{\pi}{16} \int_{-1}^{1} \int_{-1}^{1} f_{NM}(\xi, \eta) \, d\xi \, d\eta$$

or

$$H_{NM} = \frac{\pi}{16} \sum_{i=1}^{n} \sum_{j=1}^{n} w_i w_j f_{NM}(\xi_i, \eta_j) \tag{5-66}$$

where n is the number of Gaussian points. Suppose that we choose to use the three-point Gaussian quadrature ($n = 3$). Then we have

$$H_{NM} = \frac{\pi}{16} \sum_{i=1}^{3} \sum_{j=1}^{3} w_i w_j f_{NM}(\xi_i, \eta_j)$$

Referring to Table 2-1 for $n = 3$, the values of abscissae and weight coefficients are as shown in Table 5-1. For the purpose of illustration, consider a simple example of the form ($M = 1$, $N = 1$)

$$H_{11} = \int_{-1}^{1} \int_{-1}^{1} (1 + \xi + \eta)^{-1}(1 + \xi^2 + \eta^2 + \xi\eta) \, d\xi \, d\eta$$

$$= \sum_{i=1}^{3} \sum_{j=1}^{3} w_i w_j f_{11}(\xi_i, \eta_j) \tag{5-67}$$

Table 5-1 Three-point Gaussian quadrature constants

	w_i	ξ_i, η_i
$i = 3$	0.5555 →	← 0.77459
$i = 2$	0.8888 →	← 0
$i = 1$	0.5555 →	← −0.77459

Let us first calculate $f_{11}(\xi_i, \eta_j)$ for $i, j = 1, 2, 3$. Then

$$f_{11}(\xi_1, \eta_1) = (1 - 0.77459 - 0.77459)^{-1}[1 + (-0.77459)^2$$
$$+ (-0.77459)^2 - (-0.77459)^2] = -5.098454$$

$$f_{11}(\xi_1, \eta_2) = f_{11}(\xi_2, \eta_1) = 7.09813$$

$$f_{11}(\xi_1, \eta_3) = f_{11}(\xi_3, \eta_1) = 1.599989$$

$$\vdots$$

$$f_{11}(\xi_3, \eta_2) = f_{11}(\xi_2, \eta_3) = 0.90161$$

$$f_{11}(\xi_3, \eta_3) = 1.09838$$

Therefore,

$$H_{11} = w_1 w_1 f_{11}(\xi_1, \eta_1) + w_1 w_2 f_{11}(\xi_1, \eta_2) + \cdots + w_3 w_3 f_{11}(\xi_3, \eta_3)$$
$$= (0.5555)^2(-5.098454) + (0.5555)(0.8888)(7.09813)$$
$$+ \cdots + (0.5555)^2(1.09838) = 8.4424$$

For more accurate integration, one may choose to use additional Gaussian points.

5-4-4 Calculations by Isoparametric Elements

Example 5-3 (*Stream function formulation*) To illustrate the solution procedure, let us choose a simple geometry as shown in Fig. 5-14 with six elements and twelve nodes. The coordinates z and r of each node and element node number are given in Tables 5-2 and 5-3, respectively. For element 1, we have

$z_1 = 0$	$r_1 = 1$	$A_{11} = -2.5$	$B_{11}^1 = 2.5$	$B_{11}^2 = 0$
$z_2 = 2.5$	$r_2 = 1$	$A_{21} = -2.5$	$B_{21}^1 = -2.5$	$B_{21}^2 = 0$
$z_3 = 2.5$	$r_3 = 2$	$A_{31} = 2.5$	$B_{31}^1 = 2.5$	$B_{31}^2 = 0$
$z_4 = 0$	$r_4 = 2$	$A_{41} = 2.5$	$B_{41}^1 = -2.5$	$B_{41}^2 = 0$
$a_1 = 5.0$	$a_2 = 6.0$	$A_{12} = -1$	$B_{12}^1 = 0$	$B_{12}^2 = 1$
$b_1 = 5.0$	$b_2 = 0.0$	$A_{22} = 1$	$B_{22}^1 = 0$	$B_{22}^2 = -1$
$c_1 = 0.0$	$c_2 = 2.0$	$A_{32} = 1$	$B_{22}^1 = 0$	$B_{32}^2 = 1$
$d_1 = 0.0$	$d_2 = 0.0$	$A_{42} = -1$	$B_{42}^1 = 0$	$B_{42}^2 = -1$

$$r = \Phi_N r_N = x_2 = \tfrac{1}{4}(a_2 + b_2\xi + c_2\eta + d_2\xi\eta) = \tfrac{1}{4}(6 + 2\eta)$$

$$|\mathbf{J}| = \tfrac{1}{8}(\alpha_0 + \alpha_1\xi + \alpha_2\eta)$$
$$= \tfrac{1}{8}[\{(2.5)(1) - (2.5)(-1)\} + \{(-2.5)(0) - (2.5)(0)\}\xi$$
$$+ \{(0)(1) - (0)(-1)\}\eta] = 0.625$$

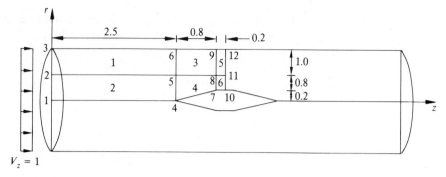

Figure 5-14 Flow around a pointed axisymmetric body in a cylinder.

$$\hat{J} = \frac{1}{8|\mathbf{J}|} = \frac{1}{8(0.625)} = 0.2$$

$$D_{NM} = 2\pi \int_{-1}^{1} \int_{-1}^{1} \frac{\hat{J}}{8} (a_{NM}^{(2)} + b_{NM}^{(2)} \xi + c_{NM}^{(2)} \eta + d_{NM}^{(2)} \xi\eta$$

$$+ e_{NM}^{(2)} \xi^2 + f_{NM}^{(2)} \eta^2)r \, d\xi \, d\eta + 2\pi \int_{-1}^{1} \int_{-1}^{1} \frac{\hat{J}}{8} (a_{NM}^{(1)} + b_{NM}^{(1)} \xi + c_{NM}^{(1)} \eta$$

$$+ d_{NM}^{(1)} \xi\eta + e_{NM}^{(1)} \xi^2 + f_{NM}^{(1)} \eta^2)r \, d\xi \, d\eta + 4\pi \int_{-1}^{1} \int_{-1}^{1} \frac{1}{32} (g_{NM}^{(1)} + h_{NM}^{(1)} \xi$$

$$+ k_{NM}^{(1)} \eta + p_{NM}^{(1)} \xi\eta + q_{NM}^{(1)} \xi^2 + s_{NM}^{(1)} \eta^2 + t_{NM}^{(1)} \xi^2\eta + u_{NM}^{(1)} \xi\eta^2) \, d\xi \, d\eta$$

Table 5-2 Nodal coordinates

	Global node											
	1	2	3	4	5	6	7	8	9	10	11	12
z	0	0	0	2.5	2.5	2.5	3.3	3.3	3.3	3.5	3.5	3.5
r	0	1.0	2.0	0	1.0	2.0	0.2	1.0	2.0	0.2	1.0	2.0

Table 5-3 Element and global node number incidence

Element node no.	Element no.					
	1	2	3	4	5	6
1	2	1	5	4	8	7
2	5	4	8	7	11	10
3	6	5	9	8	9	11
4	3	2	6	5	12	8

In view of various constants corresponding to element geometries, we obtain

$$D_{11} = 2\pi \int_{-1}^{1} \int_{-1}^{1} (0.025)(1 - 2\eta + \eta^2)(0.25)(6 + 2\eta) \, d\xi \, d\eta$$

$$+ 2\pi \int_{-1}^{1} \int_{-1}^{1} (0.025)(6.25 - 12.5\xi + 6.25\xi^2)(0.25)(6 + 2\eta) \, d\xi \, d\eta$$

$$+ 4\pi \int_{-1}^{1} \int_{-1}^{1} (0.0313)(-2.5 + 5\xi + 2.5\eta - 5\xi\eta - 2.5\xi^2 + 2.5\xi^2\eta) \, d\xi \, d\eta$$

$$= 2\pi(0.582) = 3.65681$$

Note here that the explicit simple integration is possible because the Jacobian is constant as opposed to the sample calculation in (5-67). It is seen that if the element is rectangular, the Jacobian becomes constant. It is interesting to compare the accuracy of the Gaussian quadrature integration of A_{11} with that of exact integration. Using the six-point Gaussian quadrature, we obtain

$$D_{11} = \sum_{i=1}^{6} \sum_{j=1}^{6} w_i w_j f_{11}(\xi_i, \eta_i) = 2\pi(0.5833)$$

in contrast to $2\pi(0.582)$, an error of 0.22 percent. Understanding now that the six-point Gaussian quadrature integration is quite accurate, we present below the coefficient matrix integrated by the computer:

$$(D_{NM})_1 = 2\pi \begin{bmatrix} 0.583 & 0.042 & -0.308 & -0.317 \\ 0.042 & 0.583 & -0.317 & -0.308 \\ -1.142 & -1.983 & 2.317 & 0.808 \\ -1.983 & -1.142 & 0.808 & 2.316 \end{bmatrix}$$

$$(D_{NM})_2 = 2\pi \begin{bmatrix} -0.383 & -0.242 & 0.175 & 0.450 \\ -0.242 & -0.383 & 0.450 & 0.175 \\ -0.658 & -1.217 & 1.350 & 0.525 \\ -1.217 & -0.658 & 0.525 & 1.350 \end{bmatrix}$$

$$(D_{NM})_3 = 2\pi \begin{bmatrix} 0.654 & -0.454 & -0.379 & 0.179 \\ -0.454 & 0.654 & 0.179 & -0.379 \\ -0.646 & -0.354 & 1.396 & -0.396 \\ -0.354 & -0.646 & -0.396 & 1.396 \end{bmatrix}$$

$$(D_{NM})_4 = 2\pi \begin{bmatrix} -0.021 & -0.173 & -0.027 & 0.221 \\ -0.173 & 0.667 & 0.183 & -0.077 \\ -0.293 & -0.350 & 0.717 & -0.073 \\ -0.312 & -0.343 & -0.073 & 0.729 \end{bmatrix}$$

$$(D_{NM})_5 = 2\pi \begin{bmatrix} 2.117 & -2.067 & -1.267 & 1.217 \\ -2.067 & 2.117 & 1.217 & -1.267 \\ -1.333 & 1.083 & 3.083 & -2.833 \\ 1.083 & -1.333 & -2.833 & 3.083 \end{bmatrix}$$

$$(D_{NM})_6 = 2\pi \begin{bmatrix} 0.517 & -0.542 & -0.392 & 0.412 \\ -0.542 & 0.517 & 0.417 & -0.392 \\ -0.458 & 0.283 & 1.183 & -1.008 \\ 0.283 & -0.458 & -1.008 & 1.183 \end{bmatrix}$$

The stream function along the entrance face is determined by integrating (5-48a) for uniform flow V_z. Thus

$$\psi = \tfrac{1}{2}r^2 V_z$$

This gives $\psi_3 = \tfrac{1}{2}(2^2)(1) = 2$, $\psi_2 = \tfrac{1}{2}$, with $\psi_1 = 0$. With these boundary conditions imposed similarly as in the two-dimensional flow, we obtain the global finite element equations in the form

$$2\pi \begin{bmatrix} 3.3162 & -0.5275 & 0.0000 \\ -0.5275 & 4.6709 & -3.0750 \\ 0.0000 & -3.0750 & 3.3000 \end{bmatrix} \begin{bmatrix} \psi_5 \\ \psi_8 \\ \psi_{11} \end{bmatrix} = 2\pi \begin{bmatrix} 1.366 \\ 0.500 \\ 0.100 \end{bmatrix}$$

Solving these equations gives

$$\psi_5 = 0.4868 \qquad \psi_8 = 0.4708 \qquad \psi_{11} = 0.4690$$

The corresponding axial velocities along the line passing through the nodes 10, 11, and 12 are calculated using the formula

$$V_z = \frac{2(\psi_B - \psi_A)}{r_A^2 - r_B^2}$$

Thus,

$$(V_z)_{10-11} = \frac{2(\psi_{10} - \psi_{11})}{r_{10}^2 - r_{11}^2} = \frac{2(0.469 - 0)}{1^2 - (0.2)^2} = 0.977$$

$$(V_z)_{11-12} = \frac{2(2 - 0.469)}{2^2 - 1^2} = 1.021$$

To demonstrate the solution with finer meshes, let us now consider the geometry shown in Fig. 5-15. This geometry is so chosen that the finite element discretization may be done in the same area as in the case of two-dimensional flow of Fig. 5-3, except that here we use the isoparametric element instead of triangular element. The results are shown in Fig. 5-16. Since the flow area is reduced at the crest of the sphere to $\tfrac{3}{4}$ of the cylindrical area, the velocity increases to a mean value of approximately 1.33 times the free stream velocity, with a maximum of about 1.65 at the crest of the sphere and a minimum of approximately 1.25 on the wall directly above the crest. The computer program for this problem is listed in Appendix B-3.

$$\| e \|_{H^1(\Omega)} = 0(h^6)$$

Note that a higher order polynomial in the interpolation functions must be balanced by the increase of Gaussian points.

5-5 FREE SURFACE FLOW

As a last example for an ideal fluid, let us consider a free surface flow subjected to atmosphere. The pressure at free surface in two-dimensional flow is related by

$$\frac{1}{2} V_i V_i + \frac{P}{\rho} + gy = \frac{1}{2} V_i^{(r)} V_i^{(r)} + \frac{P_{(r)}}{\rho} + gy_{(r)} \tag{5-69}$$

in which (r) refers to the downstream reference point on the free surface. We note that the pressure is constant on the free surface. For an nth node, this requires that

$$q_{(n)} = [q_{(r)}^2 - 2g(y_{(n)} - y_{(r)})]^{1/2} \tag{5-70}$$

where

$$q = \sqrt{V_i V_i} = \nabla \phi = \frac{\partial \phi}{\partial s} \tag{5-71}$$

along the line s on free surface. If the velocity between adjacent nodes m and n is assumed to be linear, we have

$$\phi_{(n)} = \phi_{(m)} - \frac{q_{(m)} + q_{(n)}}{2} \Delta s \tag{5-72}$$

Here $q_{(m)}$ and $q_{(n)}$ may be determined by (5-70). The requirement of zero normal velocity along the free surface is satisfied by

$$V_i n_i = 0$$

or

$$V_x \cos \theta + V_y \sin \theta = 0 \tag{5-73}$$

where θ is defined in Fig. 5-17. We may rewrite (5-73) in the form

$$\frac{dy}{dx} = \frac{1}{\tan \theta} = -\frac{V_y}{V_x} = \frac{\Delta y}{\Delta x}$$

or

$$\Delta y = -\frac{V_y}{V_x} \Delta x \tag{5-74}$$

The relation in (5-74) may be used to calculate $y_{(n)}$ in (5-70). To initiate computation, assume a free surface downstream. We may assume $P_{(r)}$ as atmospheric pressure and determine $q_{(r)}$ from (5-69). We specify the potential function ϕ at each node of free surface from (5-72) and solve the entire system of equations to calculate the velocity components through an iterative procedure involved in

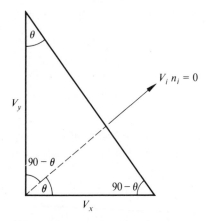

Figure 5-17 Zero normal velocity along the free surface.

(5-70), (5-72), and (5-74). Example problems of a free surface flow for two-dimensional geometries are presented by Chan and Larock [1972].

To illustrate, let us consider an orifice flow with entrance velocity V_e from the face AD as shown in Fig. 5-18. The computational steps are as follows:

Step 1. Assume a free surface CF and a uniform flow along FE. Calculate V_x at F from entrance velocity V_e.

Step 2. Calculate the velocities from (5-70) at each node along the free surface CF upstream starting from the calculated velocity at F.

Step 3. Calculate the potential function ϕ_i from (5-72).

Step 4. Solve the Laplace equation (5-17) subject to boundary conditions (5-73), (5-72) and Step 3 to calculate ϕ at all nodes.

Step 5. Calculate the velocities along the free surface from $V_x = \Delta\phi/\Delta x$ and $V_y = \Delta\phi/\Delta y$.

Step 6. Calculate the new elevations of free surface from (5-74).

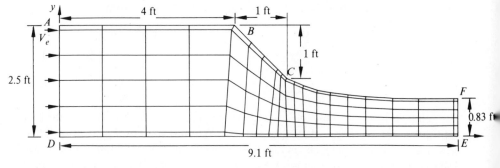

Figure 5-18 Discretization of free surface flow. Free stream entrance velocity $V_e = 4$ ft/s, 140 nodes, 114 elements, 13 nodes on a free surface.

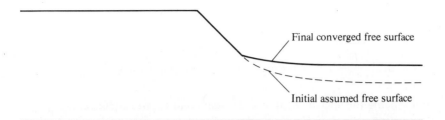

Figure 5-19 Converged free surface of the problem in Fig. 5-14.

Step 7. Using the new velocities at free surface, repeat Step 3 through Step 6 until the weighted average of new and old Δy values at all nodes becomes negligible.

In this problem, several iterations were required for final converged free surface, as shown in Fig. 5-19.

5-6 INCOMPRESSIBLE VISCOUS FLOW

5-6-1 Basic Equations

Most of the real fluids exhibit viscous behavior. This causes energy dissipation, which affects the motion of fluids and leads to thermodynamic irreversibility. In incompressible fluids, density is constant, and thus the continuity equation becomes

$$\mathbf{V} \cdot \mathbf{V} = V_{i,i} = 0 \qquad (5\text{-}75)$$

Also, the constitutive relation (4-62) is reduced to

$$\sigma_{ij} = -P\delta_{ij} + 2\mu \, d_{ij} \qquad (5\text{-}76)$$

where μ is the dynamic viscosity. The momentum equation (4-55) with (5-76) leads to the well-known Navier–Stokes equation,

$$\dot{V}_i + V_{i,j}V_j - F_i + \frac{1}{\rho}P_{,i} - \nu V_{i,jj} = 0 \qquad (5\text{-}77a)$$

or

$$\frac{\partial \mathbf{V}}{\partial t} + (\mathbf{V} \cdot \mathbf{V})\mathbf{V} - \mathbf{F} + \frac{1}{\rho}\nabla P - \nu\nabla^2\mathbf{V} = 0 \qquad (5\text{-}77b)$$

Here $\nu = \mu/\rho$ is called the kinematic viscosity. In incompressible fluids, the kinematic viscosity is considered constant. For steady state, $\dot{V}_i$ is zero but the convective term $V_{i,j}V_j$ remains

$$V_{i,j}V_j - F_i + \frac{1}{\rho}P_{,i} - \nu V_{i,jj} = 0 \qquad (5\text{-}78)$$

If nonisothermal condition is considered, then we add the energy equation (4-57) for incompressible fluids

$$\rho c_v \left(\frac{\partial T}{\partial t} + T_{,i} V_i \right) + P \delta_{ij} - 2\mu \, d_{ij} d_{ij} - \kappa T_{,ii} - \rho h = 0 \qquad (5\text{-}79)$$

with c_v and κ being the specific heat and thermal conductivity coefficients.

In this section, we study the finite element equations for the Navier–Stokes equation (5-77) and the continuity equation (5-75), along with the simultaneous solution of these equations.†

5-6-2 Formulations and Solutions

Velocity pressure formulation The unknowns sought in the Navier–Stokes equation and the continuity equation are the velocities and pressures at all nodes. The arbitrary element velocity and pressure are approximated as

$$V_i(\mathbf{x}, t) = \Phi_N(\mathbf{x}) V_{Ni}(t) \qquad (5\text{-}80a)$$

$$P(\mathbf{x}, t) = \Psi_N(\mathbf{x}) P_N(t) \qquad (5\text{-}80b)$$

It is considered reasonable to choose the interpolation functions Φ_N and Ψ_N as linear functions of spatial domain although a quadratic approximation of velocity may be required, particularly for flow problems of high Reynolds number. This argument is based on the fact that ideal momentum transfer due to pressure is of first order whereas the viscous momentum transfer due to rate of deformation is of second order; and for high speed flow, the velocity distribution within the finite element may well be of higher order function of spatial coordinates.

In what follows, the finite element equations are derived in various alternate forms depending on the boundary conditions to be imposed. The method of numerical solution of the resulting nonlinear equations of unsteady flow will be described.

Consider the residual space of the momentum and continuity equations in cartesian coordinates

$$\rho \dot{V}_i + \rho V_{i,j} V_j + \rho F_i - \sigma_{ij,j} = \varepsilon_i^{(1)}$$

$$V_{i,i} = \varepsilon^{(2)}$$

Now construct an orthogonal projection of each of these residual spaces onto a subspace spanned by appropriate weighting functions. The subspace consists of a function suitable to construct an inner product with the residual of the differential

---

† Earlier contributions were made by Oden [1973], Baker [1973], and Taylor and Hood [1973], among others.

equation. This inner product must yield a spatial invariant. For the momentum equation, the obvious variable is the velocity so that we obtain the invariant.†

$$I_1 = (\varepsilon_i^{(1)}, V_i) = (\varepsilon_i^{(1)}, \Phi_N(\mathbf{x})V_{Ni}(t)) = \int_\Omega \varepsilon_i^{(1)}\Phi_N(\mathbf{x})V_{Ni}(t)\, d\Omega = 0 \qquad (5\text{-}81)$$

For the continuity equation, however, the velocity is no longer qualified as subspace because the inner product $(\varepsilon^{(2)}, V_i)$ does not yield a spatial invariant with i appearing as a free index. It is seen that since the residual $\varepsilon^{(2)}$ is a scalar, the subspace must also be a scalar. The pressure P, the only remaining variable, is eligible for this purpose. Thus we have†

$$I_2 = (\varepsilon^{(2)}, P) = (\varepsilon^{(2)}, \Psi_N(\mathbf{x})P_N(t)) = \int_\Omega \varepsilon^{(2)}\Psi_N(\mathbf{x})P_N(t)\, d\Omega = 0 \qquad (5\text{-}82)$$

It is clear now that the weighting functions for the equations of momentum and continuity are the velocity interpolation function $\Phi_N(\mathbf{x})$ and the pressure interpolation function $\Psi_N(\mathbf{x})$, respectively. Physically, the invariant I_1 is the energy (ft-lb/s) due to the velocity field whereas the invariant I_2 represents the energy required to maintain the incompressibility of the fluid.

For all arbitrary nodal values of velocity and pressure, we obtain the local Galerkin finite element equations for the momentum and continuity, respectively,

$$\int_\Omega \varepsilon_i^{(1)}\Phi_N\, d\Omega = 0 \qquad \text{and} \qquad \int_\Omega \varepsilon^{(2)}\Psi_N\, d\Omega = 0$$

The finite element equation of momentum takes the form

$$\int_\Omega (\rho\dot{V}_i + \rho V_{i,j}V_j - \rho F_i - \sigma_{ij,j})\Phi_N\, d\Omega = 0 \qquad (5\text{-}83)$$

Noting that

$$\int_\Omega \sigma_{ij,j}\Phi_N\, d\Omega = \int_\Gamma \sigma_{ij}n_j\overset{*}{\Phi}_N\, d\Gamma - \int_\Omega \sigma_{ij}\Phi_{N,j}\, d\Omega$$

$$= \int_\Gamma S_i\overset{*}{\Phi}_N\, d\Gamma - \int_\Omega \{-P\delta_{ij} + \mu(V_{i,j} + V_{j,i})\}\Phi_{N,j}\, d\Omega$$

$$\int_\Omega \rho V_{i,j}V_j\Phi_N\, d\Omega = \int_\Omega \rho\frac{1}{2}(V_jV_j)_{,i}\, d\Omega = \int_\Gamma \frac{\rho}{2} V_jV_jn_i\overset{*}{\Phi}_N\, d\Gamma - \int_\Omega \frac{\rho}{2} V_jV_j\Phi_{N,i}\, d\Omega$$

---

† This argument stems from a physical viewpoint. However, the mathematical nature of the test functions (weighting functions) versus trial functions (interpolation or basis functions) are not related to the physical reasoning presented here. In fact, with terms of first order derivatives (convective terms or pressure gradients) present, the test functions of order higher than the trial functions may contribute to significant improvements of the solutions. See Sec. 5-12 for the notion of upwind finite elements in diffusion equations. Studies of this kind have not been carried out for Navier–Stokes equation.

for irrotational flow, we obtain

$$A_{NM} \dot{V}_{Mi} + B_{NiRM} V_{Mj} V_{Rj} + C_{NiM} P_M + D_{NMik} V_{Mk} = E_{Ni}^{(b)} + E_{Ni}^{(s)} \qquad (5\text{-}84)$$

with $N, M, R = 1, 2, \dots, r$ (r is the total number of element nodes), i, j is the spatial dimension, and

$$A_{NM} = \int_{\Omega} \rho \Phi_N \Phi_M \, d\Omega \qquad \text{(mass matrix)}$$

$$B_{NjRM} = -\int_{\Omega} \frac{\rho}{2} \Phi_{N,i} \Phi_R \Phi_M \, d\Omega \qquad \text{(convective matrix)}$$

$$C_{NiM} = -\int_{\Omega} \Phi_{N,i} \Psi_M \, d\Omega \qquad \text{(pressure matrix)}$$

$$D_{NMik} = \int_{\Omega} \mu(\Phi_{N,j} \Phi_{M,j} \delta_i^k + \Phi_{N,j} \Phi_{M,i} \delta_j^k) \, d\Omega \qquad \text{(dissipation matrix)}$$

$$E_{Ni}^{(b)} = \int_{\Gamma} \rho F_i \Phi_N \, d\Gamma \qquad \text{(body force vector)}$$

$$E_{Ni}^{(s)} = E_{Ni}^{(s_1)} + E_{Ni}^{(s_2)} \qquad \text{(surface force vector)}$$

$$E_{Ni}^{(s_1)} = \int_{\Gamma} \sigma_{ij} n_j \overset{*}{\Phi}_N \, d\Gamma \qquad E_{Ni}^{(s_2)} = -\int_{\Gamma} \frac{\rho}{2} V_j V_j n_i \overset{*}{\Phi}_N \, d\Gamma$$

Here the sum $E_{Ni}^{(s)} = E_{Ni}^{(s_1)} + E_{Ni}^{(s_2)}$ is equivalent to the forces exerted at the ends of the conduit or those exerted by the walls on the fluid,

$$f_i = \int_{\Gamma} (S_i - P n_i) \, d\Gamma$$

where $S_i = \sigma_{ij} n_j$ and $P n_i = \frac{1}{2} \rho V_j V_j n_i$. Note also that $\overset{*}{\Phi}_N$ represents the interpolation for the variation of boundary forces.

In the same manner, the local form of Galerkin finite element equation for continuity is derived as

$$\int_{\Omega} V_{i,i} \Psi_N \, d\Omega = 0 \qquad (5\text{-}85)$$

or

$$C_{NiM} V_{Mi} = G_N \qquad (5\text{-}86)$$

with $G_N = -\int_{\Gamma} V_i n_i \overset{*}{\Psi}_N \, d\Gamma$ (boundary velocity vector) and C_{NiM} is identified as a continuity matrix which is a transpose of the pressure matrix in the momentum equation.

For the same order approximation for both velocity and pressure, the matrix C_{NiM} contributes to ill-conditioning. However, this may be remedied by prescribing the normal pressure gradient on the rigid wall boundary. An alternative which has been preferred is to approximate the pressure with one order lower than the

velocity approximation (e.g., quadratic velocity and linear pressure approximations).

Additionally, difficulty may arise from nonlinearity of the convective or
advective term $B_{NiRM} V_{Mj} V_{Rj}$, particularly for high Reynolds number, a problem
long noted in the finite difference literature. Furthermore, the presence of the
first derivative associated with diffusion problems [Spalding, 1972] has led to a
procedure called *upwind* finite element scheme in which test functions are chosen
with orders higher than trial (basis) functions [Heinrich, Huyakorn, Zienkiewicz,
and Mitchell, 1977]. See further discussions in Sec. 5-12.

It is often considered convenient to write the Navier–Stokes equation in
nondimensional form in terms of Reynolds number

$$\dot{V}_i + V_{i,j} V_j + P_{,i} - \frac{1}{Re} V_{i,jj} = 0 \tag{5-87}$$

where Re is the Reynolds number

$$Re = \frac{U_\infty L}{v}$$

and $V_i = \bar{V}_i/U_\infty$, $x_i = \bar{x}_i/L$, $t = \bar{t}/(L/U_\infty)$, and $P = \bar{P}/(\bar{\rho}U_\infty^2)$ with U_∞ the free stream
velocity and L the characteristic length. The bars indicate physical quantities.
With this nondimensional form, it is clear that for high Reynolds number ($Re \gg 1$),
the convective term in (5-87) dominates the viscous diffusion term. This will
cause the nonlinear character to be more dominant, making the convergence of
finite element solutions a key issue.

We are now left with the problem of solving the finite element equations
for unsteady conditions ((5-84), (5-86)), or for steady state without the time-
dependent term. We discuss the solution procedures below.

Steady state solution The assembled finite element equations for (5-84) and (5-86)
are of the form

$$B_{\alpha i \beta \gamma} V_{\beta j} V_{\gamma j} + D_{\alpha \beta} V_{\beta i} + C_{\alpha i \lambda} P_{\lambda} = E_{\alpha i} \tag{5-88a}$$

$$C_{\alpha i \beta} V_{\beta i} = G_{\alpha} \tag{5-88b}$$

where

$$B_{\alpha i \beta \gamma} = \sum_{e=1}^{E} B_{NiRM}^{(e)} \Delta_{N\alpha}^{(e)} \Delta_{R\beta}^{(e)} \Delta_{M\gamma}^{(e)}, \text{ etc.}$$

and $E_{\alpha i}$ represents the global form of the body force $E_{Ni}^{(b)}$ and the surface pressure
$E_{Ni}^{(s)}$. Note also that $C_{\alpha i \beta}$ is the transpose of $C_{\alpha i \lambda}$ if $\Phi_N = \Psi_N$ and signs are changed
on both sides of (5-86). Note also that $i, j = 1, 2$ for two-dimensional problems,
and $\alpha, \beta, \gamma, \lambda$ represent the global node numbers. The equations (5-88a,b) can
be written in matrix form as

$$\begin{bmatrix} B_{\alpha i \beta \gamma} V_{\gamma j} + D_{\alpha \beta} \delta_{ij} & C_{\alpha i \lambda} \\ C_{\alpha j \beta} & 0 \end{bmatrix} \begin{bmatrix} V_{\beta j} \\ P_{\lambda} \end{bmatrix} = \begin{bmatrix} E_{\alpha i} \\ G_{\alpha} \end{bmatrix} \tag{5-89}$$

It is interesting to note that the continuity equation representing the incompressibility condition acts as a constraint with the pressure playing the role of Lagrange multipliers. Since the expression (5-89) is nonliniear, it is now necessary to use any one of the iterative methods described in Sec. 3-3-2.

Typical boundary conditions are

$$\sigma_{ij} n_j = f_i \qquad \text{on } \Gamma_1$$

$$V_i n_i = g \qquad \text{on } \Gamma_2$$

On the solid walls of boundary layers, it is possible to prescribe $P_{,i} n_i = 0$, although such a boundary condition does not occur *naturally* (natural boundary condition) from integration of the governing equation by parts.

The solution of simultaneous equations (5-89) is obtained using the iterative methods described in Sec. 3-3-2. Let (5-89) be written as

$$A_{mnp} X_n X_p = Y_m$$

Here the ranges of indices m, n, p represent the total number of equations, and X_n, X_p are the unknowns for nonlinear terms, but we set $A_{mnp} X_p = \hat{A}_{mn}$ for linear terms, with summing at X_p suppressed to a unity. If the Newton–Raphson method is used, an $r + 1$th iterative cycle takes the form

$$X_m^{(r+1)} = X_m^{(r)} - [J_{mn}^{(r)}]^{-1} R_n^{(r)} \tag{5-90}$$

where

$$R_m^{(r)} = A_{mnp} X_n^{(r)} X_p^{(r)} - Y_m$$

and

$$J_{mn}^{(r)} = \frac{\partial R_m^{(r)}}{\partial X_n}$$

The Jacobian may be determined for local elements and assembled into a global form

$$J_{mn}^{(r)} = (A_{mnp} + A_{mpn}) X_p^{(r)}$$

or, using the notation of the matrix on the left-hand side of (5-89), we may express the Jacobian in the form

$$\mathbf{J} = \begin{bmatrix} B_{\alpha i \beta \gamma} V_{\gamma j}^{(r)} + B_{\alpha i \gamma \beta} V_{\gamma j}^{(r)} + D_{\alpha \beta} \delta_{ij} & C_{\alpha i \lambda} \\ C_{\alpha j \beta} & 0 \end{bmatrix} \tag{5-91}$$

Computations begin with initial values of $X_m^{(r)}$. If the convergence is reached, then the term $[J_{mn}^{(r)}]^{-1} R_n^{(r)}$ in (3-88) becomes negligibly small so that

$$|X_n^{(r+1)} - X_n^{(r)}| / |X_n^{(r)}| \le \varepsilon$$

where ε is a permissible error.

Unsteady solution Let us return to the unsteady finite element equations of the type (5-84). The global form for (5-85) is written as

$$A_{\alpha \beta} V_{\beta i} + B_{\alpha i \beta \gamma} V_{\gamma j} V_{\beta j} + D_{\alpha \beta} V_{\beta i} + C_{\alpha i \lambda} P_\lambda = E_{\alpha i} \tag{5-92}$$

The temporal operators for time-dependent problems to solve equations of this type are discussed in Sec. 3-3-3. The rate of change of velocity may be put in the forward difference

$$\dot{V}_{\beta i} = \frac{V_{\beta i}^{(n+1)} - V_{\beta i}^{(n)}}{\Delta t}$$

Then the expression (5-92) becomes

$$\left[A_{\alpha\beta}\delta_{ij} + \Delta t(1 - \theta)(B_{\alpha i\beta\gamma}V_{\gamma j}^{(n+1)} + D_{\alpha\beta}\delta_{ij})\right] V_{\beta j}^{(n+1)} + \Delta t(1 - \theta)C_{\alpha i\lambda}P_{\lambda}^{(n+1)}$$
$$= \Delta t E_{\alpha i}^{(n+1)} + \left[A_{\alpha\beta}\delta_{ij} - \Delta t\theta(B_{\alpha i\beta\gamma}V_{\gamma j}^{(n)} + D_{\alpha\beta}\delta_{ij})\right] V_{\beta j}^{(n)} - \Delta t\theta C_{\alpha i\lambda}P_{\lambda}^{(n)}$$

$$\theta = \begin{cases} 0 \text{ for } V_{\beta j} = V_{\beta j}^{(n+1)} \\ \dfrac{1}{2} \text{ for } V_{\beta j} = \dfrac{V_{\beta j}^{(n+1)} + V_{\beta j}^{(n)}}{2} \\ 1 \text{ for } V_{\beta j} = V_{\beta j}^{(n)} \end{cases}$$

with similar definitions being applied for the pressure. Combining this with the continuity equation, we write

$$\begin{bmatrix} K_{\alpha\beta ij} & \Delta t(1 - \theta)C_{\alpha i\lambda} \\ C_{\alpha j\beta} & 0 \end{bmatrix} \begin{bmatrix} V_{\beta j} \\ P_{\lambda} \end{bmatrix}^{(n+1)} = \begin{bmatrix} \Delta t E_{\alpha i} \\ G_{\alpha} \end{bmatrix}^{(n+1)} + \begin{bmatrix} F_{\alpha i} \\ 0 \end{bmatrix}^{(n)} \tag{5-93}$$

with

$$K_{\alpha\beta ij} = A_{\alpha\beta}\delta_{ij} + \Delta t(1 - \theta)(B_{\alpha i\beta\gamma}V_{\gamma j}^{(n+1)} + D_{\alpha\beta}\delta_{ij})$$
$$F_{\alpha i}^{(n)} = \left[A_{\alpha\beta}\delta_{ij} - \Delta t\theta(B_{\alpha i\beta\gamma}V_{\gamma j}^{(n)} + D_{\alpha\beta}\delta_{ij})\right] V_{\beta j}^{(n)} - \Delta t\theta C_{\alpha i\lambda}P_{\lambda}^{(n)}$$

The expression (5-93) is similar to (5-89). If $K_{\alpha\beta ij}$ is linear (absence of convective term), then initially $F_{\alpha i}^{(n)} = 0$ and the response can be calculated immediately from the initial and boundary conditions. Indeed this is the way one would start the problem. The results obtained in this manner (first time increment) are then used to calculate $F_{\alpha i}^{(n)}$ as well as the convective term. We are now ready to combine the Newton–Raphson method or any other iterative method to obtain convergent solutions. Toward this end we write (5-93) in the form

$$A_{mnp}X_n^{(n+1)}X_p^{(n+1)} = Y_m^{(n)}$$

This leads to an iterative solution,

$$X_m^{(n+1),(r+1)} = X_m^{(n+1),(r)} - \left[J_{mn}^{(n+1),(r)}\right]^{-1} R_n^{(n+1),(r),(n),(0)} \tag{5-94}$$

with the Jacobian defined as

$$\mathbf{J} = \begin{bmatrix} A_{\alpha\beta}\delta_{ij} + \Delta t(1 - \theta)(B_{\alpha i\beta\gamma}V_{\gamma j}^{(r)} + B_{\alpha i\gamma\beta}V_{\gamma j}^{(r)} + D_{\alpha\beta}\delta_{ij}) & C_{\alpha i\lambda} \\ C_{\alpha j\beta} & 0 \end{bmatrix} \tag{5-95}$$

When the converged state is reached, we have

$$X_m^{(n+1),(r+1)} \simeq X_m^{(n+1),(r)}$$

For example, let us consider one-dimensional Burger's equation of the form

$$\frac{\partial u}{\partial t} + u\frac{\partial u}{\partial x} - v\frac{\partial^2 u}{\partial x^2} = 0 \qquad 0 < x < 1 \left.\begin{array}{c} \\ \\ \\ \end{array}\right\}$$

$$u(x, 0) = u_0(x) \qquad\qquad\qquad\qquad (5\text{-}96)$$

$$u(0, t) = u(1, t) = 0$$

Using a linear function for $u = \Phi_N u_N$, we obtain the finite element equation

$$\int_0^h \left(\frac{\partial u}{\partial t} + u\frac{\partial u}{\partial x} - v\frac{\partial^2 u}{\partial x^2}\right)\Phi_N \, dx = 0$$

or

$$A_{NM}\dot{u}_M + B_{NRM}u_R u_M + C_{NM}u_M = E_N \qquad (5\text{-}97a)$$

where

$$A_{NM} = \frac{h}{6}\begin{bmatrix} 2 & 1 \\ 1 & 2 \end{bmatrix}, \qquad C_{NM} = \frac{1}{h}\begin{bmatrix} 1 & -1 \\ -1 & 1 \end{bmatrix}$$

$$B_{NRM}u_R u_M = -\frac{1}{12}\begin{bmatrix} -2u_1 - u_2 & -u_1 - 2u_2 \\ 2u_1 + u_2 & u_1 + 2u_2 \end{bmatrix}\begin{bmatrix} u_1 \\ u_2 \end{bmatrix}$$

$$E_N = -\left(\frac{u^2}{2}\overset{*}{\Phi}_N\right)_0^h + \left(v\frac{\partial u}{\partial x}\overset{*}{\Phi}_N\right)_0^h = 0$$

The global form of (5-97a) becomes

$$A_{\alpha\beta}\dot{u}_\beta + B_{\alpha\beta\gamma}u_\beta u_\gamma + C_{\alpha\beta}u_\beta = 0 \qquad (5\text{-}97b)$$

In accordance with (5-92), we write

$$\left[A_{\alpha\beta} + \Delta t(1 - \theta)(B_{\alpha\beta\gamma}u_\gamma^{(n+1)} + C_{\alpha\beta})\right]u_\beta^{(n+1)} = \left[A_{\alpha\beta} - \Delta t\theta(B_{\alpha\beta\gamma}u_\gamma^{(n)} + C_{\alpha\beta})\right]u_\beta^{(n)}$$

$$(5\text{-}98)$$

The Newton–Raphson iteration at $(r + 1)$ for the time step $(n + 1)$ is performed,

$$u_\alpha^{(n+1),(r+1)} = u_\alpha^{(n+1),(r)} - (J_{\alpha\beta}^{(n+1),(r)})^{-1} R_\beta^{(n+1),(r);(n),(0)}$$

with

$$J_{\alpha\beta}^{(n+1),(r)} = A_{\alpha\beta} + \Delta t(1 - \theta)(B_{\alpha\beta\gamma}u_\gamma^{(n+1),(r)} + B_{\alpha\gamma\beta}u_\gamma^{(n+1),(r)} + C_{\alpha\beta})$$

$$R_\beta^{(n+1),(r);(n),(0)} = \left[A_{\alpha\beta} + \Delta t(1 - \theta)(B_{\alpha\beta\gamma}u_\gamma^{(n+1),(r)} + C_{\alpha\beta})\right]u_\beta^{(n+1),(r)}$$

$$- \left[A_{\alpha\beta} - \Delta t\theta(B_{\alpha\beta\gamma}u_\gamma^{(n),(0)} + C_{\alpha\beta})\right]u_\beta^{(n),(0)}$$

Notice that $u_\beta^{(n)}$ in the last two terms of $R_\beta^{(n+1),(r),(n)}$ represents the converged response of the previous time step and remains constant during the Newton–Raphson iterations, and that $u_\alpha^{(n+1),(r)}$ is calculated from

$$u_\alpha^{(n+1),(r)} = \left[A_{\alpha\beta} + \Delta t(1 - \theta)(B_{\alpha\beta\gamma}u_\gamma^{(n),(0)})\right]^{-1}\left[A_{\alpha\beta} - \Delta t\theta(B_{\alpha\beta\gamma}u_\gamma^{(n),(0)} + C_{\alpha\beta})\right]u_\beta^{(n),(0)}$$

To solve, we begin with initial conditions $u_\beta^{(0),(0)}$ and calculate

$$u_\alpha^{(1),(0)} = [A_{\alpha\beta} + \Delta t(1 - \theta)(B_{\alpha\beta\gamma}u_\gamma^{(0),(0)} + C_{\alpha\beta})]^{-1}[A_{\alpha\beta} - \Delta t\theta(B_{\alpha\beta\gamma}u_\gamma^{(0),(0)}$$
$$+ C_{\alpha\beta})]u_\beta^{(0),(0)}$$

$$u_\alpha^{(1),(1)} = u_\alpha^{(1),(0)} - (J_{\alpha\beta}^{(1),(0)})^{-1} R_\beta^{(1),(0),(0)}$$

$$u_\alpha^{(1),(2)} = u_\alpha^{(1),(1)} - (J_{\alpha\beta}^{(1),(1)})^{-1} R_\beta^{(1),(1),(0)}$$

$$\vdots$$

In this manner, we arrive at the converged solution at the rth cycle for the first time step

$$u_\alpha^{(1),(r+1)} \simeq u_\alpha^{(1),(r)}$$

Then the calculations move on to the next time increment and continue similarly until the desired time has been reached.

At the expense of accuracy, we may set the Jacobian equal to

$$J_{\alpha\beta} \simeq A_{\alpha\beta} + \Delta t(1 - \theta)(B_{\alpha\beta\gamma}u_\gamma + C_{\alpha\beta}) \tag{5-99}$$

Then we observe that

$$u_\alpha^{(n+1)} = [A_{\alpha\beta} + \Delta t(1 - \theta)(B_{\alpha\beta\gamma}u_\gamma^{(n)} + C_{\alpha\beta})]^{-1}[A_{\beta\mu} - \Delta t\theta(B_{\beta\mu\nu}u_\nu^{(n)} + C_{\beta\mu})]u_\mu^{(n)}$$
$$\tag{5-100}$$

This is equivalent to solving (5-98) directly as a form of piecewise linearization.

Other solution techniques such as perturbation method (Sec. 3-3-2) have been used successfully [Kawahara, et al., 1976] in both steady and unsteady problems. In the perturbation method, the nodal values of both velocity and pressure are expanded in the form

$$V_{\beta i} = V_{\beta i}^{(0)} + V_{\beta i}^{(1)} + V_{\beta i}^{(2)} + \cdots \tag{5-101a}$$

$$P_\lambda = P_\lambda^{(0)} + P_\lambda^{(1)} + P_\lambda^{(2)} + \cdots \tag{5-101b}$$

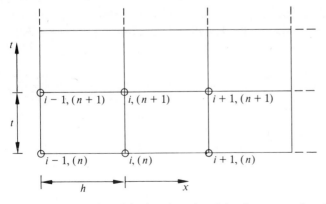

Figure 5-20 Expansion of the time-dependent finite element equations for solution at i and $(n + 1)$.

Following the procedure outlined in (3-50), the nth perturbation responses are calculated from the recursive formula identical to (5-93) except for the term $E_{\alpha i}^{(n+1)}$ which must be replaced by

$$\hat{E}_{\alpha i}^{(n+1)} = E_{\alpha i}^{(n+1)} - \sum_{r=1}^{n} B_{\alpha i \beta \gamma} V_{\beta j}^{(r)} V_{\gamma j}^{(n+1-r)}$$

The input velocity is applied in small increments in (5-93) and the response due to each of these increments are to be summed according to (5-101).

It is of interest to note that the finite difference analog can be derived directly from (5-98). For example, if we choose $\theta = \frac{1}{2}$, then the ith equation is expanded as (see Fig. 5-20)

$$\frac{1}{6\Delta t}(u_{i+1}^{(n+1)} + 4u_i^{(n+1)} + u_{i-1}^{(n+1)} - u_{i+1}^{(n)} - 4u_i^{(n)} - u_{i-1}^{(n)}) + \frac{1}{12h}[(u_{i+1}^{(n+1)})^2 + u_{i+1}^{(n+1)}u_i^{(n+1)}$$

$$- u_i^{(n+1)}u_{i-1}^{(n+1)} - (u_{i-1}^{(n+1)})^2 + (u_{i+1}^{(n)})^2 + u_{i+1}^{(n)}u_i^{(n)} - u_i^{(n)}u_{i-1}^{(n)} - (u_{i-1}^{(n)})^2]$$

$$- \frac{v}{2h^2}(u_{i-1}^{(n+1)} - 2u_i^{(n+1)} + u_{i+1}^{(n+1)} + u_{i-1}^{(n)} - 2u_i^{(n)} + u_{i+1}^{(n)}) = 0$$

$$(5\text{-}102a)$$

Solving for $u_i^{(n+1)}$ of the linear terms we get†

$$(u_i^{(n+1)})^{(r+1)} = \frac{1}{(2h/3) + vk}\left[\frac{\Delta t}{12}[(u_{i-1}^{(n)})^2 + u_i^{(n)}(u_{i-1}^{(n)} - u_{i+1}^{(n)}) - (u_{i+1}^{(n)})^2]\right.$$

$$+ \left(\frac{vk}{2} + \frac{h}{6}\right)u_{i-1}^{(n)} + \left(\frac{h}{6} - vk\right)u_i^{(n)} + \left(\frac{vk}{2} + \frac{h}{6}\right)u_{i+1}^{(n)} + \frac{\Delta t}{12}\{[(u_{i-1}^{(n+1)})2]^{(r)}$$

$$+ (u_i^{(n+1)})^{(r)}[(u_{i-1}^{(n+1)})^{(r)} - (u_{i+1}^{(n+1)})^{(r)}] - [(u_{i+1}^{(n+1)})^2]^{(r)}\}$$

$$\left. + \left(\frac{vk}{2} - \frac{h}{6}\right)(u_{i-1}^{(n+1)})^{(r)} + \left(\frac{vk}{2} - \frac{h}{6}\right)(u_{i+1}^{(n+1)})^{(r)}\right]$$

$$(5\text{-}102b)$$

where $k = \Delta t/h$. For the $r + 1$ iteration step, the terms with the superscript (r) may be taken as the nth time step initially. The Picard iteration method [Crandall, 1956] can be used to carry out the solution, marching in time.

Stream function formulation In this approach, we consider steady state problems. The desired finite element equations are obtained by substituting the velocity components in terms of the stream function. We also assume that the pressure is constant. The basic definitions are

$$\psi = \Phi_N(\mathbf{x})\psi_N \qquad (5\text{-}103a)$$

---

† The reader will note that (5-102a) is similar to the standard finite difference or well-known Crank–Nicholson scheme [1947].

and

$$V_i = \varepsilon_{ij}\psi_{,j} = \varepsilon_{ij}\Phi_{N,j}\psi_N \tag{5-103b}$$

From the general form of finite element equation (5-81) for steady state, we obtain

$$\int_{\Omega} (\rho V_{i,j} V_j - \sigma_{ij,j} - \rho F_i) V_i \, d\Omega = 0$$

Integrating by parts leads to

$$\int_{\Gamma} \frac{1}{2} \rho V_j V_j n_i V_i \, d\Gamma - \int_{\Omega} \frac{1}{2} \rho V_j V_j V_{i,i} \, d\Omega - \int_{\Gamma} \sigma_{ij} n_j V_i \, d\Gamma$$

$$+ \int_{\Omega} \sigma_{ij} V_{i,j} \, d\Omega - \int_{\Omega} \rho F_i V_i \, d\Omega = 0$$

Substituting (5-103) into the above expression and for all arbitrary nodal values of stream function which do not vanish, it is necessary that the following integral must vanish:

$$\int_{\Gamma} \frac{1}{2} \rho \varepsilon_{ik} V_j V_j n_i \overset{*}{\Phi}_{N,k} \, d\Gamma - \left(\int_{\Omega} \frac{1}{2} \rho \varepsilon_{jl} \varepsilon_{jm} \varepsilon_{ik} \Phi_{R,l} \Phi_{M,m} \Phi_{N,ki} \, d\Omega \right) \psi_R \psi_M$$

$$- \int_{\Gamma} \varepsilon_{ik} \sigma_{ij} n_j \overset{*}{\Phi}_{N,k} \, d\Gamma + \left(\int_{\Omega} \mu \varepsilon_{il} \varepsilon_{ik} \Phi_{M,lj} \Phi_{N,kj} \, d\Omega \right) \psi_M$$

$$+ \left(\int_{\Omega} \mu \varepsilon_{jl} \varepsilon_{ik} \Phi_{M,li} \Phi_{N,kj} \, d\Omega \right) \psi_M - \int_{\Omega} \rho \varepsilon_{ik} F_i \Phi_{N,k} \, d\Omega = 0$$

which is put in the form

$$B_{NRM} \psi_R \psi_M + C_{NM} \psi_M = E_N \tag{5-104}$$

where

$$B_{NRM} = - \int_{\Omega} \frac{1}{2} \rho \varepsilon_{ik} \varepsilon_{jm} \varepsilon_{jl} \Phi_{N,ki} \Phi_{R,l} \Phi_{M,m} \, d\Omega$$

$$C_{NM} = \int_{\Omega} \mu \varepsilon_{ik} \varepsilon_{il} \Phi_{N,kj} \Phi_{M,lj} \, d\Omega + \int_{\Omega} \mu \varepsilon_{ik} \varepsilon_{jl} \Phi_{N,kj} \Phi_{M,li} \, d\Omega$$

$$E_N = \int_{\Omega} \rho \varepsilon_{ik} F_i \Phi_{N,k} \, d\Omega + \int_{\Gamma} \varepsilon_{ik} \sigma_{ij} n_j \overset{*}{\Phi}_{N,k} \, d\Gamma - \int_{\Gamma} \frac{1}{2} \varepsilon_{ik} V_j V_j n_i \overset{*}{\Phi}_{N,k} \, d\Gamma$$

To solve the global equations corresponding to (5-104), once again we resort to any one of the iterative schemes given in Sec. 3-3-2, or as described earlier in this section.

Thermal flow problems In motions where temperature differences bring about differences in density, it is necessary to include buoyancy forces in momentum equations and to treat them as impressed body forces. Changes in volume due to temperature differences contribute to these buoyancy forces. Thus the governing equations for steady state free convection flow are

$$\rho V_{i,j} V_j + P_{,i} - \mu V_{i,jj} - \rho \beta T g_i = 0 \qquad (5\text{-}105a)$$

$$V_{i,i} = 0 \qquad (5\text{-}105b)$$

$$\rho c_v T_{,i} V_i - q_{i,i} - \rho h = 0 \qquad (5\text{-}105c)$$

where β is the coefficient of expansion; T is the temperature change, $T = \theta - T_0$ with θ and T_0 being the absolute and reference temperature, respectively; g_i is the gravitation acceleration; $\rho \beta T g_i$ is regarded as the lift force due to thermal expansion; and q_i is the heat flux. In the energy equation (5-105c), the effect of dissipation ($\sigma_{ij} d_{ij}$) is assumed to be negligible.

Typical thermal boundary conditions which may be considered are

$$T = \text{constant} \qquad \text{on } \Gamma_1 \qquad (5\text{-}106a)$$

$$q_i n_i = -q - \bar{\alpha}(T - \hat{T}) \qquad \text{on } \Gamma_2 \qquad (5\text{-}106b)$$

with $\bar{\alpha}$ the heat transfer coefficient and $\hat{T}$ the ambient temperature. The last boundary condition (known as Cauchy boundary condition) occurs naturally in the derivation of finite element equations.

The finite element representation of the variables including velocity, pressure, and temperature is given by

$$V_i = \Phi_N V_{Ni} \qquad P_N = \Psi_N P_N \qquad T = \Theta_N T_N \qquad (5\text{-}107a,b,c)$$

where Φ_N, Ψ_N, and Θ_N represent interpolation functions for the velocity, pressure, and temperature, respectively. It is obvious that the weighting function for the energy is the temperature space. Thus the finite element equation for (5-105c) becomes

$$\int_\Omega (\rho c_v T_{,i} V_i - q_{i,i} - \rho h) \Theta_N \, d\Omega = 0$$

Integrating by parts yields

$$\int_\Omega \rho c_v T_{,i} V_i \Theta_N \, d\Omega - \int_\Gamma q_i n_i \overset{*}{\Theta}_N \, d\Gamma + \int_\Omega q_i \Theta_{N,i} \, d\Omega - \int_\Omega \rho h \Theta_N \, d\Omega = 0$$

Substituting the Fourier law of heat conduction (4-67) and the Cauchy boundary conditions (5-106b) into the above leads to

$$H_{NMiR} V_{Ri} T_M + J_{NM} T_M + K_{NM} T_M = Q_N \qquad (5\text{-}108)$$

where

$$H_{NMiR} = \int_\Omega \rho c_v \Theta_N \Theta_{M,i} \Phi_R \, d\Omega$$

$$J_{NM} = \int_\Omega \kappa \Theta_{N,i} \Theta_{M,i} \, d\Omega$$

$$K_{NM} = \int_{\Gamma} \bar{a} \overset{*}{\Theta}_N \overset{*}{\Theta}_M \, d\Gamma$$

$$Q_N = - \int_{\Gamma} q \overset{*}{\Theta}_N \, d\Gamma + \int_{\Gamma} \bar{a} \hat{T} \overset{*}{\Theta}_N \, d\Gamma + \int_{\Omega} \rho h \Theta_N \, d\Omega$$

The heat flux input q applied normal to the boundary takes a negative sign and thus makes the first term of Q_N a positive quantity unless this input is intended for reduction of the interior temperature.

The finite element equation for the momentum (5-84) is revised to the following form:

$$B_{NiRM} V_{Rj} V_{Mj} + C_{NiMj} V_{Mj} + D_{NMi} P_M + L_{NMi} T_M = E_{Ni}^{(s)} \tag{5-109}$$

where L_{NMi} is the buoyancy matrix given by

$$L_{NMi} = \int_{\Omega} \rho \beta \Phi_N \Theta_M g_i \, d\Omega$$

The continuity matrix, of course, remains the same as (5-86).

Examples

Example 5-5 Solution of Burger's equation. Consider a model problem given by (5-96) for which the analytical solution takes the form [Cole, 1951],

$$u(x, t) = \frac{2\pi v(A_1 \exp(-v\pi^2 t) \sin \pi x + 2A_2 \exp(-4v\pi^2 t) \sin 2\pi x)}{A_0 + A_1 \exp(-v\pi^2 t) \cos \pi x + A_2 \exp(-4v\pi^2 t) \cos 2\pi x}$$

Let the constants A_0, A_1, A_2, and v be given as follows:

$$A_0 = v = 1 \qquad A_1 = \tfrac{1}{4} \qquad A_2 = \tfrac{1}{2}$$

Then the initial condition is of the form

$$u(x, 0) = \frac{2\pi(\tfrac{1}{4} \sin \pi x + \sin 2\pi x)}{1 + \tfrac{1}{4} \cos \pi x + \tfrac{1}{2} \cos 2\pi x}$$

The formula given by (5-98) alone suggests three options with $\theta = 0$, $\theta = \tfrac{1}{2}$, and $\theta = 1$. The explicit scheme $\theta = 1$ requires less computing time for a given time increment Δt and mesh size h but results in less unstable solutions.

The results obtained for $\Delta t = 0.005$ and $h = \tfrac{1}{20}$ with $\theta = \tfrac{1}{2}$ are compared with the analytical solution in Fig. 5-21. The average error is approximately 0.7 percent. Average errors for various combinations of Δt and h with $\theta = 0$ are plotted in Fig. 5-22. It is clear that for a given mesh size, there exists an optimum Δt. Errors tend to decrease with smaller h and Δt to a certain limit, but smaller Δt for a given h causes larger errors on passing this limit.

Example 5-6 Couette flow (velocity–pressure). The finite element solution of equations of the type (5-88) are carried out for Couette flow using the iso-

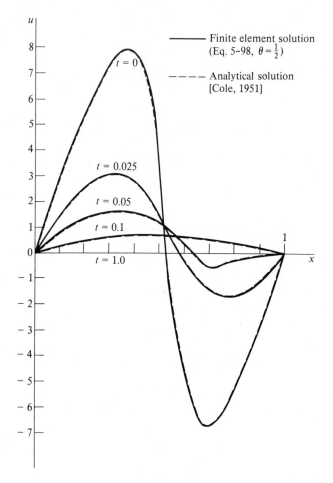

Figure 5-21 Solution of Burger's equation ($\Delta t = 0.005$, $h = 1/20$).

parametric elements (Fig. 5-23a). The results are compared with those of Oden and Wellford [1972] who used the quadratic triangular elements (Fig. 5-23b) and a 4th order Runge–Kutta method for time integration. In both studies, the Newton–Raphson method is used to solve the nonlinear equations. The input data are $V_x = 0.1$ in/s and $V_y = 0$ at $y = 0.2$ in, and $V_x = V_y = 0$ at $y = 0$. The stress on the boundaries $x = 0$ and $x = 2$ in is set equal to zero. Also, the condition $\partial P/\partial y = 0$ is enforced along the boundary $y = 0$. We use the mass density and viscosity of 0.00242 lbf-s^2/in^4 and 0.00362 lbf-s/in, respectively. The time increment for the temporal operator in (5-93) is taken as $\Delta t = 10^{-4}$ s.

The results (Figs. 5-24, 5-25, and 5-26) agree well with those of the analytical solution of Schlichting [1968] and the finite element solution of Oden and Wellford [1972]. Some discrepancies at nodes 7 and 8 for initial

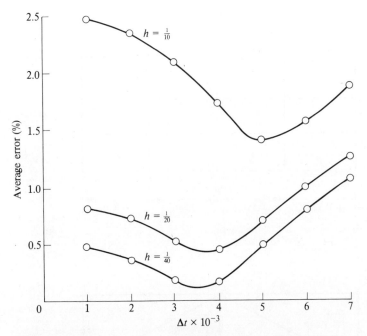

Figure 5-22 Average error for various combinations of Δt and h, $\theta = 1/2$.

periods exist in the triangular element solution, but they disappear in the isoparametric element. This may have been attributed to the time integration schemes, as well as different element configurations.

Example 5-7 Channel flow (velocity–pressure). (a) This example gives a comparison between the Newton–Raphson and perturbation methods for a channel flow with irregular geometries. The steady state equations (5-89) are solved. No-slip conditions on solid walls and $S_i = \sigma_{ij}n_j = 0$ at downstream face are assumed. The results obtained by Kawahara, et al. [1976], are presented in Fig. 5-27. Deviations of the results between the Newton–Raphson and perturbation methods are small for the Reynolds numbers 10, 50, 100, and 150. In these examples, quadratic functions for velocity and linear functions for pressure are used.

(b) The purpose of the next example is to compare the results of the unsteady solution (5-93) reaching the steady state with those of the steady solution (5-89). The results obtained by Kawahara, et al. [1976], are presented in Fig. 5-28. Attainment of steady state at nodes 7, 21, and 23 is rather rapid whereas node 9 appears to reach a steady state over a prolonged time. Interpolation functions used here are the same as the part (a) above.

REMARKS Solutions of Navier–Stokes equation by finite elements have been obtained by numerous other investigators. For example, see Tong [1971],

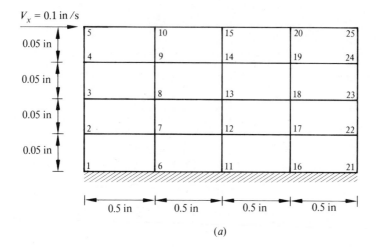

(a)

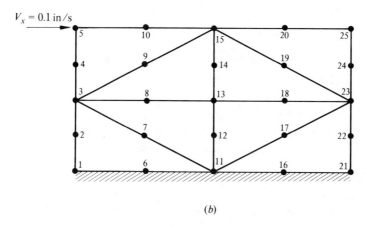

(b)

Figure 5-23 Geometries for Couette flow. (a) Linear isoparametric elements (25 nodes); (b) quadratic triangular elements (25 nodes).

Cheng [1972b], Argyris and Mareczek [1972], and Taylor and Hood [1973], among others.

Example 5-8 Stream function formulation. Steady state solutions by stream function formulation (5-104) are demonstrated for a cavity domain (Fig. 5-29). The boundary conditions are

$$\psi = 0 \qquad \text{for all boundaries}$$

$$\varepsilon_{ij}\psi_{,j}n_i = 0 \qquad \text{for } A - B - C - D$$

$$\psi_{,y} = -u_0 \qquad \text{for } D - A$$

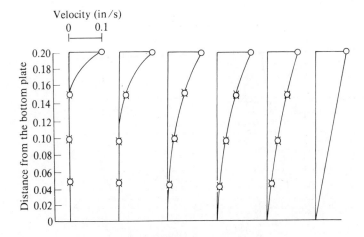

X Linear isoparametric element

O Quadratic triangular element (Oden and Wellford)

—— Exact solution (Shlichting)

Figure 5-24 Velocity profiles for transient Couette flow at $t = 0.001$ s, $t = 0.00042$ s, $t = 0.00376$ s, $t = 0.0669$ s, and $t = \infty$ s.

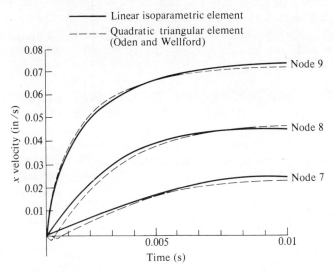

Figure 5-25 Time history of the x velocity at nodes, 7, 8, and 9 for transient Couette flow.

The results reported by Kawahara and Okamoto [1976] using the equations (5-104) and those by Bozeman and Dalton [1973] using the finite difference method are compared for $Re = 100$ in Fig. 5-29b. The finite element solution for $Re = 1000$ is presented in Fig. 5-29c. Bozeman and Dalton used 50×50

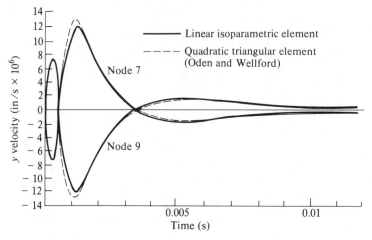

Figure 5-26 Time history of the *y* velocity at nodes 7 and 9 for transient Couette flow.

finite difference mesh whereas 52 nodes were used in the finite element method. Cubic functions with seven nodes, including one at the center, are used in this example. To solve for ten constants, two derivatives as well as the nodal stream function are specified at each of the corner nodes and only the nodal stream function at the center node. This is a simplified scheme of Fig. 2-26. Integration is carried out using the formula of the type (Eq. 2-20). The center node can be eliminated by the reduction scheme described in Sec. 3-3-4.

Example 5-9 Free convection thermal flow. This is the free convection flow heated at the bottom of a container (Fig. 5-30). With standard boundary conditions, the solution can be obtained by (5-105). The results reported by Kawahara, et al. [1976], are shown in Fig. 5-30. The bottom is heated at 5°C with ambient temperature at 0°C. Effective conduction coefficients (κ/c_v) on the wall are taken as 0.002, and heat transfer coefficients for the wall and free surface are assumed to be 0.002 and 0.0001, respectively. In this solution, the interpolation functions identical to those in Example 5-7 are used.

5-6-3 Error Analysis

In the previous section, we dealt with the finite element solution of equations

$$\dot{V}_i + V_{i,j}V_j + P_{,i} - \nu V_{i,jj} = F_i \qquad \text{in } \Omega \qquad (5\text{-}110a)$$

$$V_{i,i} = 0 \qquad \text{in } \Omega \qquad (5\text{-}110b)$$

subject to boundary conditions

$$V_i n_i = 0, g \qquad \text{on } \Gamma_1 \qquad (5\text{-}111a)$$

$$V_{i,j}n_j = 0 \qquad \text{on } \Gamma_2 \qquad\qquad (5\text{-}111b)$$

$$P_{,i}n_i = 0 \qquad \text{on } \Gamma_3 \qquad\qquad (5\text{-}111c)$$

The solution of these equations involve errors which consist of: 1. finite element approximation errors, 2. temporal approximation errors, and 3. Newton–Raphson iteration errors (or errors due to any other iterative nonlinear equation solver).

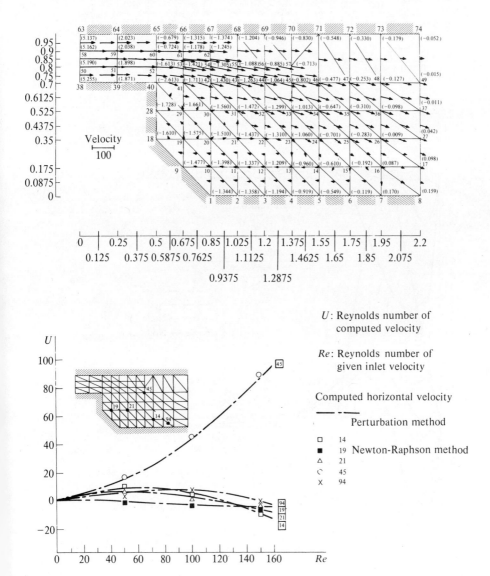

U: Reynolds number of computed velocity

Re: Reynolds number of given inlet velocity

Computed horizontal velocity

——— · — Perturbation method

□   14
■   19   Newton–Raphson method
△   21
○   45
×   94

Figure 5-27 Steady viscous flow through a channel ($Re = 150$, pressure is indicated by parenthesis). After Kawahara [1976]. (*a*) Velocity and pressure distribution (pressure in parenthesis); (*b*) comparison between Newton–Raphson and perturbation methods.

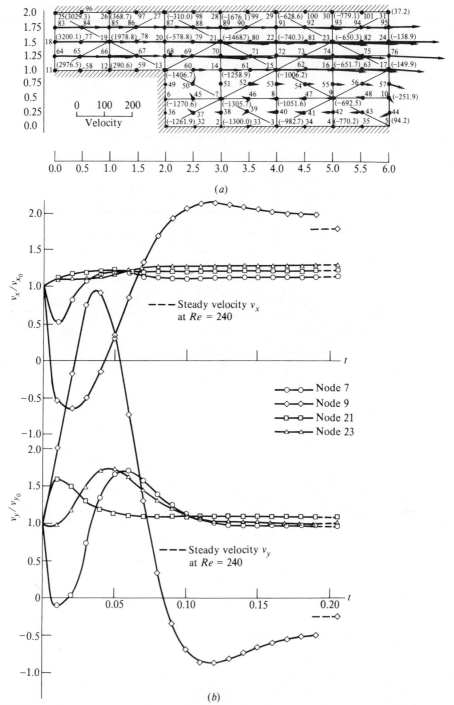

Figure 5-28 Comparison of steady and unsteady solutions ($Re = 240$ at steady state). After Kawahara [1976]. (*a*) Steady state solution ($Re = 240$); (*b*) unsteady solution ($200 \leq Re \leq 240$).

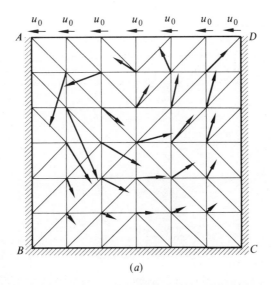

$$(a)$$

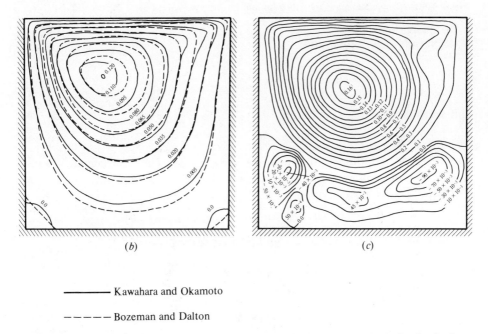

$$(b) \qquad\qquad (c)$$

——————— Kawahara and Okamoto

— — — — — Bozeman and Dalton

Figure 5-29 Stream function formulation for a cavity domain. (*a*) Geometry and velocity distribution; (*b*) streamline contours, $Re = 100$; (*c*) streamline contours, $Re = 1000$.

At present, no theoretical finite element error estimates are available for the unsteady nonlinear two-variable equations with boundary conditions as considered here. Difficulty lies in determining the bound for the nonlinear term $V_{i,j}V_j$. Jamet and Raviart [1974] considered the following problems of steady

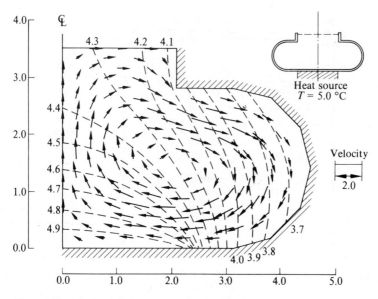

Figure 5-30 Velocity and temperature distributions for free convective flow.

state Navier–Stokes equations:

$$V_{i,j}V_j + P_{,i} - \nu V_{i,jj} = F_i \qquad \text{in } \Omega$$

$$V_{i,i} = 0 \qquad \text{in } \Omega$$

subject to the boundary condition

$$V_i = 0 \qquad \text{on } \Gamma$$

The boundary condition for pressure is unspecified. They considered a weak formulation as follows: given a function $\mathbf{F} \in V'$, find functions $\mathbf{V} \in V$ and $P \in L_2(\Omega)/\mathbf{R}$ such that

$$a(\mathbf{V}, \mathbf{V}, \mathbf{W}) + \nu b(\mathbf{V}, \mathbf{W}) - (P, \operatorname{div} \boldsymbol{\Phi}) = (\mathbf{F}, \boldsymbol{\Phi})$$

for all $\mathbf{V}, \mathbf{W} \in (H^1(\Omega))^N$. Here, $a(\cdot)$ and $b(\cdot)$ are defined as

$$a(\mathbf{V}, \mathbf{W}) = \int_\Omega \mathbf{V}_{,i}\mathbf{W}_{,i} \, d\Omega$$

$$b(\mathbf{V}, \mathbf{V}, \mathbf{W}) = \frac{1}{2} \int_\Omega (\mathbf{V}, \mathbf{V}_{,i}\mathbf{W} - \mathbf{V}, \mathbf{W}_{,i}\mathbf{V}) \, d\Omega$$

Note also that N denotes the N dimensions and $L_2(\Omega)/\mathbf{R}$ is a subset of $L_2(\Omega)$ and denotes the quotient space, implying that boundary pressures are unspecified. Based on this setting, Jamet and Raviart arrive at the error estimates for the velocity and pressure as follows:

$$\| \mathbf{V} - \hat{\mathbf{V}} \|_{(H_0^1(\Omega))^N} = Ch^k (| \mathbf{V} |_{H^{k+1}(\Omega)} + | P |_{H^k(\Omega)}) \qquad (5\text{-}112a)$$

$$\| P - \hat{P} \|_{L_2(\Omega)/\mathbf{R}} = Ch^k(\| \mathbf{V} \|_{H^{k+1}(\Omega)} + |P|_{H^k(\Omega)}) \tag{5-112b}$$

where C is a constant independent of h, k denotes the order of polynomial in finite element approximation, and $\hat{\mathbf{V}}$ and $\hat{P}$ are the finite element solutions. The additional boundary conditions $V_i n_i = g$ and $P_{,i} n_i = 0$ as well as the unsteady conditions, however, may provide an error estimate other than $O(h^k)$ as predicted by (5-112). Nevertheless, the computations in most cases indicate an optimal rate of convergence of $O(h^k)$ in $\hat{V}_i$ in $H_0^1(\Omega)$.

Errors due to temporal operators, together with Newton–Raphson iterations, may still be evaluated as discussed in Sec. 3-3-3. An exact amplification factor cannot be derived for the expression (5-93) with nonlinear terms in the left-hand side matrix. However, if we hold the nonlinear term constant during the iterative cycle, it is then possible to invert the matrix and obtain an approximate amplification property. Note that as the nonlinear terms are updated, the amplification changes, thus altering the stability criteria as calculations progress.

If nonlinear terms are dropped, we have, of course, an explicit error estimate for the velocity as predicted by (3-189),

$$\| \varepsilon(t) \| \le \exp(\gamma t) \| \varepsilon(0) \| + ch^{k+1} \bigg[\| u(t) \|_{H^{k+1}(\Omega)}$$

$$+ \exp(-\gamma t) \| u(0) \|_{H^{k+1}(\Omega)} + \int_0^t \exp(\gamma(s-t)) \left\| \frac{\partial u(s)}{\partial t} \right\|_{H^{k+1}(\Omega)} ds \bigg]$$

This provides $O(h^{k+1})$. In the presence of all boundary conditions such as given by (5-111a,b,c), however, precise error estimates have not been derived.

5-7 WAVE MOTION OF A SHALLOW BASIN

5-7-1 Governing Equations

The wave motion in a shallow basin is of practical interest to many engineering problems, such as the design of a harbor. For simplicity, let us assume an ideal fluid and neglect dissipative effects of boundary friction. The equation of motion neglecting the convective term assumes the form

$$\dot{V}_i - F_i + \frac{1}{\rho} P_{,i} - W_i = 0 \tag{5-113}$$

where F_i denotes the small disturbing forces acting on water in a horizontal canal of small depth H as shown in Fig. 5-31. Note that the vertical force $F_3 - F_z$ may be considered to be constant as z varies from 0 to H with little change in the value of gravitational acceleration g. W_i is due to the wind force acting on the surface

$$W_i = \frac{\tau_i}{H + \zeta} \tag{5-114}$$

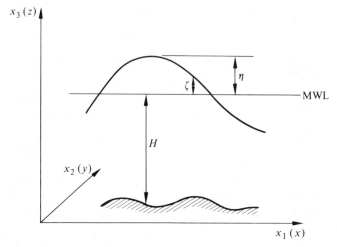

Figure 5-31 Wave motion of water on a basin.

with τ_i = square of the wind velocity (ft^2/s^2). The pressure due to disturbances is given by

$$P = \rho g(H + \zeta) \tag{5-115}$$

where ζ is the variable wave amplitude above the mean water level (MWL). Substituting (5-115) into (5-113) and neglecting F_i yields

$$\dot{V}_i - W_i + g\zeta_{,i} = 0 \tag{5-116}$$

Let $\xi_i(i = 1, 2)$ be the components of displacement. Then the velocity is given by

$$V_i = \dot{\xi}_i \tag{5-117}$$

Since the masses in the disturbed and undisturbed states must be equal, we have

$$\rho(H + \zeta)\left(dx_1 + \frac{\partial \xi_1}{\partial x_1} dx_1\right)\left(dx_2 + \frac{\partial \xi_2}{\partial x_2} dx_2\right) = \rho H\, dx_1\, dx_2$$

Expanding the above and neglecting the second order terms yield

$$\rho\zeta = -\rho H \xi_{i,i} \tag{5-118}$$

Differentiating (5-118) with respect to time and together with (5-103), we arrive at

$$\dot{\zeta} + HV_{i,i} = 0 \tag{5-119}$$

If the height H is variable, (5-119) assumes the form

$$\dot{\zeta} + (HV_i)_{,i} = 0 \tag{5-120}$$

In the finite element analysis, H may be taken as constant (5-119) within a small element and allowed to vary from element to element if variable depths exist.

Finally, by differentiating (5-119) with respect to time and using (5-116), we obtain

$$\ddot{\zeta} - c^2\zeta_{,ii} = -HW_{i,i} \tag{5-121}$$

where c is the speed of propagation or wave velocity given by $c = \sqrt{gH}$. If the external disturbing forces W_i are absent, we have

$$\ddot{\zeta} - c^2\zeta_{,ii} = 0 \tag{5-122}$$

This is the standard type of wave equation in hyperbolic form. Since the predicted motion is harmonic, we introduce a relationship

$$\zeta(x_i, t) = \eta(x_i) \cos \omega t \tag{5-123}$$

where $\eta(x_i)$ is the maximum amplitude of $\zeta(x_i, t)$ and ω is the frequency of wave motion

$$\omega = 2\pi/T$$

with T = harmonic period. Substituting (5-123) into (5-122) leads to the Helmholtz equation of the form

$$c^2\eta_{,ii} + \omega^2\eta = 0 \tag{5-124}$$

The finite element analysis may be performed on (5-121), (5-122), or (5-124), subject to appropriate boundary conditions. Detailed discussions are presented in the following section.

5-7-2 Finite Element Formulations

In engineering applications, we may solve (5-121) to determine the wave amplitudes corresponding to the prescribed disturbing forces. Alternatively, it is possible to formulate the eigenvalue problem based on (5-122) or (5-124) to obtain the frequencies and mode shapes of wave motion through solutions of eigenvalues and eigenvectors.

Let us first examine the equation of motion (5-121). To obtain the variational principle for this equation, we may proceed in standard manner as discussed in Chap. 1. However, once again the Galerkin's weighted residual method is convenient and will be shown here. First we approximate the wave amplitude $\eta(x_i)$ as usual in the form

$$\zeta(x_i) = \Phi_N(x_i)\zeta_N$$

and construct the Galerkin integral

$$\int_\Omega (\ddot{\zeta} - c^2\zeta_{,ii} + HW_{i,i})\Phi_N \, d\Omega = 0 \tag{5-125}$$

Integrating by parts or using the Green–Gauss theorem, we have

$$\left[\int_\Omega \Phi_N\Phi_M \, d\Omega\right]\ddot{\zeta}_M - \int_\Gamma c^2\zeta_{,i}n_i\overset{*}{\Phi}_{N,i} \, d\Gamma + \left[\int_\Omega c^2\Phi_{N,i}\Phi_{M,i} \, d\Omega\right]\zeta_M$$

$$+ \int_\Gamma HW_i n_i\overset{*}{\Phi}_N \, d\Gamma - \int_\Omega HW_i\Phi_{N,i} \, d\Omega = 0$$

Writing in a compact form yields

$$A_{NM}\ddot{\zeta}_M + B_{NM}\zeta_M = F_N \tag{5-126}$$

where A_{NM}, B_{NM}, and F_N are commonly known as the mass matrix, stiffness matrix, and force vector, respectively,

$$A_{NM} = \int_{\Omega} \Phi_N \Phi_M \, d\Omega$$

$$B_{NM} = \int_{\Omega} c^2 \Phi_{N,i} \Phi_{M,i} \, d\Omega$$

$$F_N = F_N^{(1)} + F_N^{(2)} + F_N^{(3)}$$

with

$$F_N^{(1)} = -\int_{\Omega} H W_i \Phi_{N,i} \, d\Omega \qquad F_N^{(2)} = \int_{\Gamma} H W_i n_i \overset{*}{\Phi}_N \, d\Gamma \qquad F_N^{(3)} = \int_{\Gamma} c^2 \zeta_{,i} n_i \overset{*}{\Phi}_N \, d\Gamma$$

If the disturbing forces W_i are absent in the domain but applied only at the boundaries, then $F_N^{(1)}$ is zero. Note that $\overset{*}{\Phi}_N$ is the interpolation function for the variation of $W_i n_i$ and $\zeta_{,i} n_i$ along the boundaries. It should also be mentioned that $F_N^{(2)}$ acts as the forcing function whereas $F_N^{(3)}$ is the Neumann boundary condition. Along the boundaries we must assure the condition $\zeta_{,i} n_i = \partial \zeta / \partial n = 0$.

To solve (5-126), we may use the temporal operators for hyperbolic type discussed in Sec. 3-3-3. The procedure is the same as demonstrated in Sec. 5-6. It should be noted that W_i is the function of $(H + \zeta)^{-1}$ in (5-114) but it can be held constant within each time step and updated with advance of time increments. The solution of (5-126) provides the vertical nodal wave amplitudes, and it is possible from this information to calculate the horizontal displacements in each element. In view of (5-116) and (5-117), the components of acceleration can be found as

$$\ddot{\xi}_r = W_r - g\zeta_{,r} \qquad (r = 1, 2) \tag{5-127}$$

Once again this expression can be calculated via suitable temporal operators to determine horizontal displacements ξ_r in terms of the vertical displacements ζ_r which we determine from (5-126).

An alternate formulation leads to a simultaneous solution of (5-113), together with continuity equation. For general applications (nonshallow basin), it is necessary to include the convective (advective) term. Thus the governing equations are

$$\dot{V}_i + V_{i,j} V_j + g\zeta_{,i} - W_i = 0 \tag{5-128a}$$

$$V_{i,i} = 0 \tag{5-128b}$$

The finite element solution of these equations is identical to that of (5-93). Solutions of the equations of the type (5-128) associated with estuaries have been studied by Taylor and Davis [1975] and Berkhoff [1975], among others.

5-7-3 Eigenvalue Solutions

Although the solution procedure outlined in the previous section is straightforward, the input disturbing forces $F_N^{(1)}$ and $F_N^{(2)}$ may not be known in practical situations. It is therefore often advantageous in solving eigenvalue problems to find frequencies and relative values of maximum wave amplitudes. Toward this end, we make a substitution of the type (5-123) in the homogeneous part of (5-126)

$$(-\omega^2 A_{NM} + B_{NM})\eta_M \cos \omega t = 0$$

or

$$(B_{NM} - \omega^2 A_{NM})\eta_M = 0 \tag{5-129}$$

Upon assembly of the element matrices, we have the global form

$$(B_{ij} - \omega^2 A_{ij})\eta_j = 0 \tag{5-130}$$

The expression (5-129) is identical to the finite element equation that may be derived from (5-122) or (5-124) except for the boundary condition $F_N^{(3)}$, which is the consequence of the integration by parts of the spatial derivative terms of (5-122) or (5-124). To illustrate, we have from (5-124)

$$\int_\Omega (c^2\eta_{,ii} + \omega^2\eta)\Phi_N \, d\Omega = 0 \tag{5-131}$$

with $\eta(x_i) = \Phi_N(x_i)\eta_N$. Proceeding in the usual manner, we get

$$(B_{NM} - \omega^2 A_{NM})\eta_M = F_N^{(3)}$$

This result can also be obtained from (5-126) with $F_N^{(3)}$ retained. The global form of (5-131) is

$$(B_{ij} - \omega^2 A_{ij})\eta_j = F_i^{(3)} \tag{5-132}$$

It should be noted that $\eta_{,i}n_i = 0$ along the boundaries and thus $F_i^{(3)} = 0$. This leads to (5-130). For a nontrivial solution of η_j in (5-130), we must have the determinant of the terms in the parentheses equal to zero. Thus,

$$|B_{ij} - \omega^2 A_{ij}| = 0 \tag{5-133a}$$

or in matrix form

$$|\mathbf{B} - \omega^2 \mathbf{A}| = 0 \tag{5-133b}$$

This is the standard eigenvalue problem. Here the matrix A_{ij} is nonsingular whereas the matrix B_{ij} is singular. Therefore, we multiply (5-133b) by A_{ik}^{-1} and write

$$|A_{ik}^{-1} B_{kj} - \omega^2\delta_{ij}| = 0 \tag{5-134a}$$

or

$$|\mathbf{A}^{-1}\mathbf{B} - \omega^2\mathbf{I}| = 0 \tag{5-134b}$$

where $\mathbf{I}$ is the identity matrix. It should be noted that $\mathbf{A}^{-1}\mathbf{B}$ is unsymmetric although $\mathbf{A}$ and $\mathbf{B}$ are both symmetric. It is preferable to have the product $\mathbf{A}^{-1}\mathbf{B}$ converted to a symmetric form before going into eigenvalue solutions.

This can be achieved as follows: First, let $\mathbf{A}$ be written in the form

$$\mathbf{A} = \mathbf{L}\mathbf{L}^T$$

where $\mathbf{L}$ is the lower triangular matrix having zeros above the diagonal and T denotes a transpose. Then

$$\mathbf{A}^{-1} = (\mathbf{L}^T)^{-1}\mathbf{L}^{-1} \tag{5-135}$$

Writing (5-130) in matrix form premultiplied by $\mathbf{A}^{-1}$ yields

$$(\mathbf{A}^{-1}\mathbf{B} - \omega^2\mathbf{I})\boldsymbol{\eta} = 0 \tag{5-136}$$

Substituting (5-135) into (5-136) and premultiplying by $\mathbf{L}^T$ gives

$$(\mathbf{L}^{-1}\mathbf{B} - \omega^2\mathbf{L}^T)\boldsymbol{\eta} = 0$$

Let

$$\mathbf{L}^T\boldsymbol{\eta} = \mathbf{Y}$$

$$\boldsymbol{\eta} = (\mathbf{L}^T)^{-1}\mathbf{Y} \tag{5-137}$$

Then we obtain

$$(\mathbf{Z} - \omega^2\mathbf{I})\mathbf{Y} = 0$$

where

$$\mathbf{Z} = \mathbf{L}^{-1}\mathbf{B}(\mathbf{L}^{-1})^T$$

Thus the eigenvalue problem takes the form

$$|\mathbf{Z} - \omega^2\mathbf{I}| = 0 \tag{5-138}$$

It is now seen that the matrix $\mathbf{Z}$ is symmetric. The eigenvalue solution consists of determining the eigenvalues ω^2 and the eigenvectors $\mathbf{Y}$. From these we obtain the frequencies ω and the actual eigenvectors $\boldsymbol{\eta}$ from the relation (5-137). It should be noted that in general, we are interested in only a few low frequency modes for design purpose.

The procedures for eigenvalue solutions are well documented in the literature [Ralston and Wilf, 1967]. Exhaustive discussion of this subject is beyond the scope of this book.

Example 5-9 Consider a square surface geometry as shown in Fig. 5-32. The surface is divided into 25 elements with 36 nodes. Linear isoparametric elements are used in this solution. Assuming that the mean water level is constant and $gH = 1000 \text{ s}^{-2}$ with $g = 32.174 \text{ ft/s}^2$ and $H = 31.08 \text{ ft}$, the solution of the eigenvalue problem (5-133) was carried out by the Givens–Householder method [Ralston and Wilf, 1967].

The frequencies in Hertz given by square roots of eigenvalues are summarized in Table 5-4. Because the geometry is square, only 6 of the 36 modes have independent, distinct eigenvalues. They are the modes 1, 4, 9, 16, 33, and 36. Other modes appear as pairs. If the geometry does not have a pair of symmetric axes, then every mode is expected to be independently distinct and no pairs of the identical eigenvalues would occur.

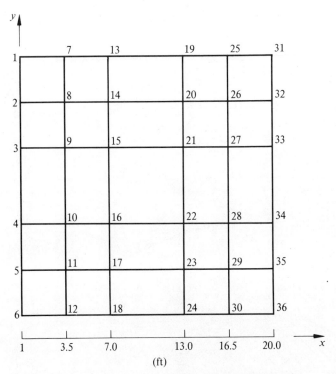

Figure 5-32 Mean water level surface finite element discretization.

Table 5-4 Frequences for the wave motion problem

Mode no.	Frequency (Hz)	Mode no.	Frequency (Hz)
1	0	19	4.31652
2	0.80049	20	4.31652
3	0.80049	21	4.58771
4	1.11321	22	4.58771
5	1.74800	23	4.59109
6	1.74800	24	4.59109
7	1.92258	25	4.66035
8	1.92258	26	4.66035
9	2.47205	27	4.91259
10	2.62165	28	4.91259
11	2.62165	29	4.98644
12	2.74114	30	4.98644
13	2.74114	31	5.28688
14	3.15097	32	5.28688
15	3.15097	33	5.99859
16	3.70758	34	6.25057
17	4.24164	35	6.25057
18	4.24164	36	6.49278

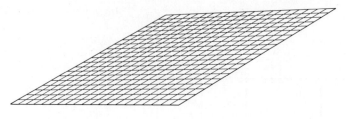

(*a*) ISOPAR. ELMT/MODE 3 OMEGA = 0.800496 + 00 CYCLES PER SEC

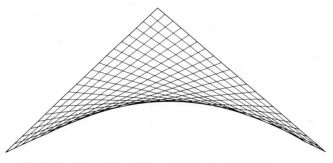

(*b*) ISOPAR. ELMT/MODE 4 OMEGA = 0.113207 + 01 CYCLES PER SEC

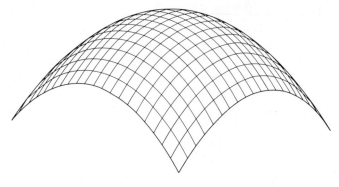

(*c*) ISOPAR. ELMT/MODE 5 OMEGA = 0.174800 + 01 CYCLES PER SEC

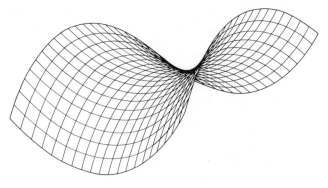

(*d*) ISOPAR. ELMT/MODE 6 OMEGA = 0.174800 + 01 CYCLES PER SEC

Figure 5-33 Mode shapes.

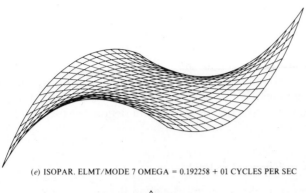

(e) ISOPAR. ELMT/MODE 7 OMEGA = 0.192258 + 01 CYCLES PER SEC

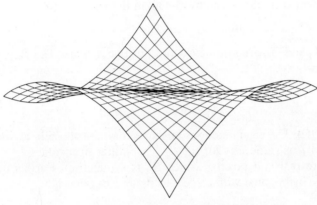

(f) ISOPAR. ELMT/MODE 8 OMEGA = 0.192258 + 01 CYCLES PER SEC

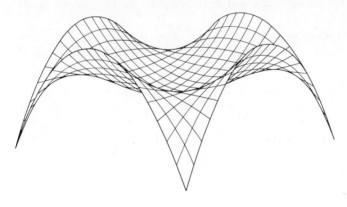

(g) ISOPAR. ELMT/MODE 9 OMEGA = 0.247205 + 01 CYCLES PER SEC

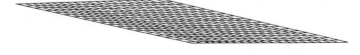

(h) ISOPAR. ELMT/MODE 10 OMEGA = 0.262166 + 01 CYCLES PER SEC

Figure 5-33 Mode shapes—*continued.*

The mode shapes or eigenvectors corresponding to some of the eigenvalues are shown in Fig. 5-33. The first mode represents the rigid body motion with the entire plane floating up by one unit. Figure 5-33 shows the computer plots between the third and tenth modes. These mode shapes are viewed at the origin 220° counterclockwise from the x axis and 20° above the plane. The eigenvectors are extrapolated by the second degree least square fit with 20 divisions across the x and y axes.

5-7-4 Error Analysis in Eigenvalue Problems

The eigenvalue problem is characterized by (5-130) in the form

$$B_{ij}u_j = \lambda A_{ij}u_j \tag{5-139}$$

where $\lambda = \omega^2$ and B_{ij} and A_{ij} are symmetric and positive definite. The Rayleigh quotient is defined as

$$R = \frac{b(v^h, v^h)}{(v^h, v^h)} = \frac{b(\Phi_i, \Phi_j)u_iu_j}{(\Phi_k, \Phi_l)u_ku_l} = \frac{B_{ij}u_iu_j}{A_{kl}u_ku_l} \tag{5-140}$$

where $v^h = \Phi_iu_i \in H^m(\Omega)$. The eigenfunctions (eigenvectors) are those η_i at which the Rayleigh quotient R is stationary for the corresponding eigenvalues $\lambda_n = R_n$. These eigenvalues are real and positive (since B_{ij} is symmetric). Furthermore, the eigenfunctions are orthogonal with respect to A_{ij} and B_{ij}; that is

$$A_{(ij)}u_iu_j = B_{(ij)}u_iu_j = \delta_{ij}$$

where δ_{ij} is the Kronecker delta and indices inside the parenthesis are not to be summed.

An important observation in the eigenvalue problem is the maximum–minimum principle [Courant and Hilbert, 1953], stated as follows. If $R(v)$ is maximized over an l-dimensional subspace S_l, then the minimum possible value for this maximum is λ_n

$$\lambda_l = \min_{S_n} \max_{v \in S_l} R(v) \tag{5-141}$$

Note that the fundamental frequency in which the stationary point is actually minimum is given by

$$\lambda_1 = \min_{v \in H^1(\Omega)} R(v) \tag{5-142}$$

Let λ_l^h be the finite element solution and Π the Rayleigh–Ritz projection. Then it follows from the maximum–minimum principle that

$$\lambda_l^h \leq \min_{v^h \in S_l} R(v^h) = \max \frac{b(\Pi u, \Pi u)}{(\Pi u, \Pi u)} \tag{5-143}$$

After some algebra [see Strang and Fix, 1973], we arrive at the error estimate for eigenvalues:

$$\lambda_l \leq \lambda_l^h \leq \lambda_l + 2ch^{2(k+1-m)}\lambda_l^{(k+1)/m}$$

or
$$\lambda_l^h - \lambda_l \le Ch^{2(k+1-m)} \lambda_l^{(k+1)/m} \tag{5-144}$$

The corresponding eigenfunction error estimates are governed by [Strang and Fix, 1973]

$$\| u_l - \hat{\Pi} u_l^h \|_{H^0(\Omega)} \le \hat{C} [h^{k+1} + h^{2(k+1-m)}] \lambda_l^{(k+1)/2m} \tag{5-145}$$

Following these error estimates, we now return to the eigenvalue problem discussed in Sec. 5-7-3. We conclude that for linear interpolation function $(k = 1)$ and $2m$th order equation $(m = 1)$, the error estimates are

$$\lambda_h^l - \lambda_h \le Ch^2 \lambda_l^2$$

and

$$\| u_l - \hat{\Pi} u_l^h \|_{H^0(\Omega)} \le \hat{C} [h^2 + h^2] \lambda_l = 2\hat{C} h^2 \lambda_l \tag{5-146}$$

Thus we have $0(h^2)$ for both eigenvalues and eigenfunctions in $H^0(\Omega)$.

5-8 ROTATIONAL FLOW

5-8-1 General

The assumption of ideal flow is based on its irrotationality. In more general flow fields, however, we often encounter local rotational fluid motions known as vortex flow.

The vorticity ω is related by the rotation $\mathbf{w}$ in the form

$$\omega = 2\mathbf{w} = \nabla \times \mathbf{V}$$

For two-dimensional flow, we have

$$\omega_3 = 2w_3 = V_{2,1} - V_{1,2}$$

The circulation around any closed curve containing a given set of fluid particles remains constant with time as long as the external force field remains conservative. This phenomenon is known as the Kelvin circulation theorem. It follows from the Kelvin theorem that a vortex line can neither begin nor end in the fluid; hence, it appears as a closed loop or ends on a boundary.

Vortex motions cannot be formed in the ideal fluid, but regions of localized vorticity are often produced in portions of real fluids, which affects the fluid motion. It is well known that the action of viscosity is in general a prerequisite to the appearance of vorticity. The regions in which rotational motion and circulation exist being relatively small, the effect of vortexes on fluid motion is felt through the velocities which they "induce" throughout the rest of the flow field.

In certain real fluid flows, we encounter flow patterns which may be idealized as perfect fluids containing infinite or semi-infinite rows of vortexes. The occurrence of vortex streets is usually associated with the formation of wakes behind bodies, which would lead to an instability of possible arrangements in the kinematics of vortex.

5-8-2 Galerkin Approach

Assume that the fluid is incompressible and the fluid motion is described by the Navier–Stokes equation (5-77) along with (5-76). Taking the curl and the divergence of (5-77) yields

$$\mathbf{V} \times \left[\frac{\partial \mathbf{V}}{\partial t} + (\mathbf{V} \cdot \mathbf{V})\mathbf{V} - \mathbf{F} + \frac{1}{\rho} \nabla P - \nu \nabla^2 \mathbf{V} \right] = 0 \qquad (5\text{-}147a)$$

$$\mathbf{V} \cdot \left[\frac{\partial \mathbf{V}}{\partial t} + (\mathbf{V} \cdot \mathbf{V})\mathbf{V} - \mathbf{F} + \frac{1}{\rho} \nabla P - \nu \nabla^2 \mathbf{V} \right] = 0 \qquad (5\text{-}147b)$$

It should be noted that $(\mathbf{V} \cdot \mathbf{V})\mathbf{V}$ in the above expressions may be replaced by

$$(\mathbf{V} \cdot \mathbf{V})\mathbf{V} = \mathbf{V}\left(\frac{V^2}{2}\right) - \mathbf{V} \times \boldsymbol{\omega}$$

Neglecting the body force and substituting (5-75) and (5-76) into the above expressions give the vorticity transport equation and pressure distribution, respectively

$$\frac{\partial \boldsymbol{\omega}}{\partial t} + (\mathbf{V} \cdot \mathbf{V})\boldsymbol{\omega} - \nu \nabla^2 \boldsymbol{\omega} = 0 \qquad (5\text{-}148a)$$

$$\nabla^2 P + \rho \mathbf{V} \cdot (\mathbf{V} \cdot \mathbf{V})\mathbf{V} = 0 \qquad (5\text{-}149)$$

It is seen that we may write (5-148a) in an alternative form as

$$\frac{\partial \boldsymbol{\omega}}{\partial t} + \mathbf{V} \times (\boldsymbol{\omega} \times \mathbf{V}) - \nu \mathbf{V} \times (\nabla^2 \mathbf{V}) = 0 \qquad (5\text{-}148b)$$

In index notations, we may write (5-148a) and (5-149) in the form

$$\dot{\omega}_i + \omega_{i,j} V_j - \nu \omega_{i,jj} = 0 \qquad (5\text{-}148c)$$

$$P_{,ii} + \rho(V_{i,j} V_j)_{,i} = 0 \qquad (5\text{-}150)$$

Equivalently from (5-148b) follows

$$\dot{\omega}_i + \varepsilon_{ijk}\varepsilon_{kmn}(\omega_m V_n)_{,j} - \nu \omega_{i,jj} = 0 \qquad (5\text{-}151)$$

To these expressions we add the vorticity

$$\mathbf{V} \times \mathbf{V} = \varepsilon_{ijk} V_{k,j}\mathbf{i}_i = \varepsilon_{ijk}\varepsilon_{km}\psi_{,mj}\mathbf{i}_i = \omega_i \mathbf{i}_i$$

or for two-dimensional problems, we have $\omega_3 = \omega$ and

$$-\nabla^2 \psi = -\psi_{,ji} = \omega \qquad (5\text{-}152)$$

Substituting the relation $V_i = \varepsilon_{ij}\psi_{,j}$ in (5-148c) or (5-151), we obtain

$$\dot{\omega}_i + \varepsilon_{jk}\psi_{,k}\omega_{i,j} - \nu \omega_{i,jj} = 0 \qquad (5\text{-}153)$$

To obtain the finite element equations, we construct interpolation spaces for vorticity, stream function, and pressure such that

$$\omega = \alpha_N \omega_N \qquad \psi = \Phi_N \psi_N \qquad P = \Psi_N P_N \qquad (5\text{-}154)$$

Linear interpolation functions may be considered reasonable unless in high speed flows, in which case higher order approximations are required particularly for the stream function. For two-dimensional problems, the vorticity transport equation becomes

$$\dot{\omega} + \varepsilon_{ij}\psi_{,j}\omega_{,i} - v\omega_{,ii} = 0 \tag{5-155}$$

The required finite element equations may be obtained from orthogonal projection of the residual spaces of (5-155), (5-152), and (5-150) to the interpolation spaces of (5-154) so that

$$\int_\Omega (\dot{\omega} + \varepsilon_{ij}\psi_{,j}\omega_{,i} - v\omega_{,ii})\alpha_N \, d\Omega = 0$$

$$\int_\Omega (\psi_{,ii} + \omega)\Phi_N \, d\Omega = 0$$

$$\int_\Omega [P_{,ii} + \rho(V_{i,j}V_j)_{,i}]\Psi_N \, d\Omega = 0$$

or

$$\int_\Omega (\dot{\omega}\alpha_N - \varepsilon_{ij}\psi_{,j}\omega\alpha_{N,i} + v\omega_{,i}\alpha_{N,i}) \, d\Omega = \int_\Gamma v\omega_{,i}n_i\overset{*}{\alpha}_N \, d\Gamma - \int_\Gamma \varepsilon_{ij}\psi_{,j}\omega n_i\overset{*}{\alpha}_N \, d\Gamma$$

$$\int_\Omega (\psi_{,i}\Phi_{N,i} - \omega\Phi_N) \, d\Omega = \int_\Gamma \psi_{,i}n_i\overset{*}{\Phi}_N \, d\Gamma$$

$$\int_\Omega P_{,i}\Psi_{N,i} \, d\Omega = -\int_\Omega \rho Q_i\Psi_{N,i} \, d\Omega + \int_\Gamma P_{,i}n_i\overset{*}{\Psi}_N \, d\Gamma + \int_\Gamma \rho Q_i n_i\overset{*}{\Psi}_N \, d\Gamma$$

with $Q = V_{i,j}V_j$. These equations are written in compact forms

$$A_{NM}\dot{\omega}_M + B_{NMP}\psi_M\omega_P + C_{NM}\omega_M = D_N \tag{5-156a}$$

$$E_{NM}\psi_M - F_{NM}\omega_M = G_N \tag{5-156b}$$

$$H_{NM}P_M = L_N \tag{5-156c}$$

where

$$A_{NM} = \int_\Omega \alpha_N\alpha_M \, d\Omega$$

$$B_{NMP} = -\int_\Omega \varepsilon_{ij}\alpha_{N,i}\Phi_{M,j}\alpha_P \, d\Omega$$

$$C_{NM} = \int_\Omega v\alpha_{N,i}\alpha_{M,i} \, d\Omega$$

$$D_N = \int_\Gamma v\omega_{,i}n_i\overset{*}{\alpha}_N \, d\Gamma - \int_\Gamma \varepsilon_{ij}\psi_{,j}\omega n_i\overset{*}{\alpha}_N \, d\Gamma$$

$$E_{NM} = \int_{\Omega} \Phi_{N,i} \Phi_{M,i} \, d\Omega$$

$$F_{NM} = \int_{\Omega} \Phi_{N} \alpha_{M} \, d\Omega$$

$$G_{N} = \int_{\Gamma} \psi_{,i} n_{i} \overset{*}{\Phi}_{N} \, d\Gamma$$

$$H_{NM} = \int_{\Omega} \Psi_{N,i} \Psi_{M,i} \, d\Omega$$

$$L_{N} = -\int_{\Omega} \rho Q_{i} \Psi_{N,i} \, d\Omega + \int_{\Gamma} P_{,i} n_{i} \overset{*}{\Psi}_{N} \, d\Gamma + \int_{\Gamma} \rho Q_{i} n_{i} \overset{*}{\Psi}_{N} \, d\Gamma$$

Instead of using (5-153), one may start from (5-151), which is known as conservation form. For two-dimensional problems, we replace (5-155) by

$$\dot{\omega} + (V_i \omega)_{,i} - v\omega_{,jj} = 0 \qquad (5\text{-}157a)$$

In terms of Reynolds number, the nondimensionalized equation takes the form

$$\dot{\omega} + (V_i \omega)_{,i} - \frac{1}{Re} \omega_{,jj} = 0 \qquad (5\text{-}157b)$$

Integrating this over the domain yields

$$\frac{\partial}{\partial t} \int_{\Omega} \omega \, d\Omega + \int_{\Omega} (V_i \omega)_{,i} \, d\Omega - \frac{1}{Re} \int_{\Omega} \omega_{,ii} \, d\Omega$$

Using the Green–Gauss theorem leads to

$$\frac{\partial}{\partial t} \int_{\Omega} \omega \, d\Omega = -\int_{\Gamma} V_i \omega n_i \, d\Gamma + \frac{1}{Re} \int_{\Gamma} \omega_{,i} n_i \, d\Gamma$$

This indicates that the time rate of accumulation of vorticity in Ω is equal to the net convective (or advective) flux rate of vorticity across the boundary plus the net diffusion flux rate of vorticity entering normal to the boundary. These boundary terms appear in the finite element equations rederived from (5-157) as

$$\int_{\Omega} (\dot{\omega} + (V_i \omega)_{,i} - v\omega_{,jj}) \alpha_N \, d\Omega = 0$$

$$A_{NM} \dot{\omega}_M + B_{NM} \omega_M + C_{NM} \omega_M = D_N + Q_N \qquad (5\text{-}158)$$

where

$$B_{NM} = -\int_{\Omega} V_i \alpha_{N,i} \alpha_M \, d\Omega$$

$$Q_N = \int_{\Gamma} V_i \omega n_i \overset{*}{\alpha}_N \, d\Gamma$$

If the assembled equations corresponding to (5-158) and (5-156b) are used, we can linearize the B_{NM} matrix by holding V_i constant at each time increment and updated at the following time steps. If (5-156a) is used instead, the nonlinear term may be linearized at each time step by holding ψ_M constant and updated at the following time steps.

With initial and boundary conditions (see Sec. 5-8-3), the temporal operators can be applied to the time derivatives of vorticity to obtain solutions marching in time. First, vorticity transport equations are solved; then from (5-156b) we iterate for new stream functions to calculate the velocities via $V_i = \varepsilon_{ij}\psi_{,j}$. Calculate new boundary values of vorticity and repeat the process until sufficient time has been advanced.

5-8-3 Boundary Conditions

The boundary conditions involving both stream functions and vorticities have been studied extensively in finite element calculations [Roache, 1972, pp. 139–173]. Let us consider a system of equations

$$\dot{\omega} + \varepsilon_{ij}\psi_{,j}\omega_i - v\omega_{,ii} = 0 \qquad \text{in } \Omega \qquad (5\text{-}159a)$$

$$\psi_{,ii} = -\omega \qquad \text{in } \Omega \qquad (5\text{-}159b)$$

$$\psi = 0, \qquad \psi_{,i}n_i = 0 \qquad \text{on } \Gamma \qquad (5\text{-}159c)$$

Figure 5-34 depicts typical boundaries. The vorticity transport equation describes the advection and diffusion of vorticity dictated by $\dot{\omega}$. The total vorticity is conserved in the interior

$$\int_{\Omega} \omega \, d\Omega = 0$$

with

$$\int_{\Gamma} \omega_{,i}n_i \, d\Gamma = 0$$

at the solid boundary, which automatically appears in the finite element equations (5-156). At no-slip boundaries, the vorticity ω is produced. The vorticity production at the solid walls Γ is the dominant physical mechanism which drives the problem.

Consider a uniform inflow along a–b. Then, initially at $t = 0$, the vorticity vanishes everywhere in Ω, including Γ along a–b. The stream functions along a–b are determined as a linear function of y from the input data. As observed in inviscid flow, we set the stream functions equal to zero along the solid wall a–f–e–d. Likewise, along b–c, stream functions are constant and calculated from the input velocity along a–b.

The vorticity along no-slip boundaries (b–c and a–f–e–d) is determined by the interior stream functions. Since stream function is constant along the wall

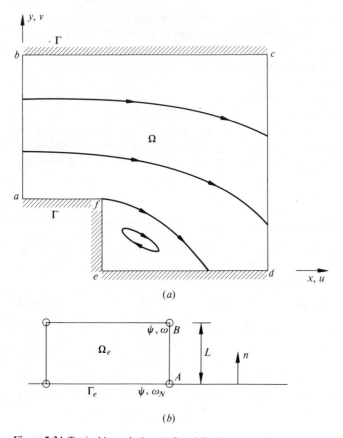

Figure 5-34 Typical boundaries. (a) Ω and Γ; (b) Ω_e and Γ_e.

($\partial\psi/\partial\sigma = \partial\psi/\partial x = 0$), (5-159$b$) takes the form

$$\frac{\partial^2\psi}{\partial y^2} = -\omega$$

or

$$\frac{\partial^2\psi}{\partial n^2} = -\omega \tag{5-160}$$

Assuming a linear distribution of vorticity between the two points A and B in Fig. 5-34b and integrating (5-160) between these two points yields

$$\int_{\psi_A}^{\psi_B} \frac{\partial^2\psi}{\partial n^2}\, dn = -\int_0^L \left[\left(1 - \frac{n}{L}\right)\omega_A + \frac{n}{L}\,\omega_B\right] dn$$

$$\omega_A = -\left[\frac{3(\psi_B - \psi_A)}{L^2} + \frac{\omega_B}{2}\right] \tag{5-161}$$

which is well known in the finite difference literature. In using the vorticity boundary condition (5-161), we must install nodes A and B normal to the boundary surface. Furthermore, the distance between these two points must be small as the Reynolds number becomes large.

For the case of a corner such as at f in Fig. 5-34, the wall vorticity may be taken as an average of two values obtained from (5-161) corresponding to the normals constructed along a–f and e–f.

If uniform flow prevails at the outflow boundary along c–d, it is reasonable to specify $u = \partial\psi/\partial y = \text{constant}$ and $v = -\partial\psi/\partial x = 0$. In general flow conditions, the normal derivatives of both vorticity and stream function are set equal to zero.

5-8-4 Variational Approach

It is well known that exact variational principles do not exist for some nonlinear problems in fluid mechanics. The works of Vainberg [1964] followed by Tonti [1969], Finlayson [1972], and Oden and Reddy [1975], among others, established the general guidelines of searching for the variational principles. A brief account is given below. We start with the nonlinear differential equation of the form

$$N(u) = 0 \qquad (5\text{-}162)$$

We define the derivative of the nonlinear operator N as the Fréchet derivative $\mathscr{D}_u N$. We take the Fréchet derivatives of (5-162) in the directions ϕ and ψ such that

$$\mathscr{D}_u N(u)\phi = \lim_{\varepsilon \to 0} \frac{N(u + \varepsilon\phi) - N(u)}{\varepsilon} = \left[\frac{\partial}{\partial \varepsilon} N(u + \varepsilon\phi) \right]_{\varepsilon = 0} \qquad (5\text{-}163a)$$

$$\mathscr{D}_u N(u)\psi = \lim_{\varepsilon \to 0} \frac{N(u + \theta\psi) - N(u)}{\theta} = \left[\frac{\partial}{\partial \theta} N(u + \theta\psi) \right]_{\theta = 0} \qquad (5\text{-}163b)$$

Integrating by parts yields

$$\int_\Omega \mathscr{D}_u N(u)\phi \, d\Omega = \int_\Omega \phi N(u) \, d\Omega + \text{boundary terms}$$

$$\int_\Omega \mathscr{D}_u N(u)\psi \, d\Omega = \int_\Omega \psi N(u) \, d\Omega + \text{boundary terms}$$

If the path integral is independent of the path taken, then it follows that

$$\int_\Omega \frac{N(u + \varepsilon\phi) - N(u)}{\varepsilon} \psi \, d\Omega = \int_\Omega \frac{N(u + \theta\psi) - N(u)}{\theta} \phi \, d\Omega$$

With $\varepsilon \to 0$ and $\theta \to 0$ in the limits, we have

$$\int_\Omega \psi \mathscr{D}_u N(u)\phi \, d\Omega = \int_\Omega \phi \mathscr{D}_u N(u)\psi \, d\Omega \qquad (5\text{-}164)$$

This is the well-known symmetry condition for existence of a variational principle due to Vainberg [1964]. Unfortunately, such a symmetry condition does not hold true for the full Navier–Stokes equation [Finlayson, 1972].

In solving the engineering problems in an approximate manner, one may resort to the approximate variational principles. Consider the vorticity transport equation (5-155) for steady state,

$$v\omega_{,ii} - \varepsilon_{ij}\psi_{,j}\omega_{,i} = 0 \qquad (5\text{-}165a)$$

or

$$v\nabla^4\psi + \frac{\partial\psi}{\partial x}\frac{\partial\omega}{\partial y} - \frac{\partial\psi}{\partial y}\frac{\partial\omega}{\partial x} = 0 \qquad (5\text{-}165b)$$

For an axisymmetric geometry, commas implying partial derivatives should be replaced by strokes denoting covariant derivatives. This results in

$$\frac{v}{r}\left[\frac{\partial^4\psi}{\partial z^4} + 2\frac{\partial^4\psi}{\partial z^2\,\partial r^2} + \frac{\partial^4\psi}{\partial r^4} - \frac{2}{r}\frac{\partial^3\psi}{\partial r^3} + \frac{3}{r^2}\frac{\partial^2\psi}{\partial r^2} - \frac{3}{r^3}\frac{\partial\psi}{\partial r} - \frac{2}{r}\frac{\partial^4\psi}{\partial z^2\,\partial r}\right]$$
$$+ \frac{1}{r}\left[\frac{\partial\psi}{\partial z}\left(\frac{\partial\omega}{\partial z} - \frac{\omega}{r}\right) - \frac{\partial\psi}{\partial r}\frac{\partial\omega}{\partial z}\right] = 0 \qquad (5\text{-}166)$$

Here the vorticity is defined for cartesian coordinates and axisymmetric curvilinear coordinates, respectively, as

$$-\omega = \frac{\partial^2\psi}{\partial x^2} + \frac{\partial^2\psi}{\partial y^2} \qquad (5\text{-}167a)$$

and

$$-\omega = \frac{1}{r}\left(\frac{\partial^2\psi}{\partial z^2} + \frac{\partial^2\psi}{\partial r^2} - \frac{1}{r}\frac{\partial\psi}{\partial r}\right) \qquad (5\text{-}167b)$$

It follows from (5-165a) that the variational statement takes the form

$$\delta I(\psi) = \int_R (v\psi_{,iijj} - \varepsilon_{ij}\psi_{,j}\omega_{,i})\delta\psi \; d\Omega = 0$$

Integrating by parts, we obtain[†]

$$\delta I(\psi) = \int_\Gamma v\psi_{,iij}n_j\delta\psi \; d\Gamma - \int_R v\psi_{,iij}\delta\psi_{,j} \; dR$$
$$- \int_\Gamma \varepsilon_{ij}\psi_{,j}\omega n_i\delta\psi \; d\Gamma + \int_\Omega \varepsilon_{ij}\psi_{,j}\omega\delta\psi_{,i} \; d\Omega$$

[†] In this integration, $\psi_{,j}$ in the second term is held constant but is to be updated in the iterative solution process [Olson, 1975].

$$
= \int_{\Gamma} v\psi_{,iij} n_j \delta\psi \, d\Gamma - \int_{\Gamma} v\psi_{,ii} n_j \delta\psi_{,j} \, d\Gamma + \int_{\Omega} \psi_{,ii} \delta\psi_{,jj} \, d\Omega
$$

$$
- \int_{\Gamma} \varepsilon_{ij}\psi_{,j}\omega n_i \delta\psi \, d\Gamma + \int_{\Omega} \varepsilon_{ij}\psi_{,j}\omega \delta\psi_{,i} \, d\Omega
$$

$$
= \delta \left[\int_{\Gamma} v\omega_{,j} n_j \psi \, d\Gamma - \int_{\Gamma} v\omega n_j \psi_{,j} \, d\Gamma - \int_{\Gamma} \varepsilon_{ij}\psi_{,j}\omega n_i \psi \, d\Gamma \right.
$$

$$
\left. - \int_{\Omega} \tfrac{1}{2}\psi_{,ii}\psi_{,jj} \, d\Omega + \int_{\Omega} \varepsilon_{ij}\psi_{,j}\omega\psi_{,i} \, d\Omega \right] = 0 \tag{5-168}
$$

Thus the variational principle is written as

$$
I(\psi) = \int_{\Omega} \left[\tfrac{1}{2}(\nabla^2\psi)^2 - \varepsilon_{ij}\psi_{,j}\omega\psi_{,i} \right] d\Omega - \int_{\Gamma} \left[v\omega_{,j} n_j \psi - v\omega n_j \psi_{,j} - \varepsilon_{ij}\psi_{,j}\omega n_i \psi \right] d\Gamma
$$

$$
\tag{5-169}
$$

Similar derivations may be carried out for the axisymmetric case. Setting $\omega = -\psi_{,ii}$, with the finite element interpolations applied in (5-169) we obtain

$$
I(\psi) = \tfrac{1}{2}Q_{NM}\psi_N\psi_M + R_{NMP}\psi_M\psi_P\psi_N - W_N\psi_N \tag{5-170}
$$

Thus the finite element equation is of the form

$$
Q_{NM}\psi_M + R_{NMP}\psi_M\psi_P = W_N \tag{5-171}
$$

The global form of (5-171) results in typical nonlinear equations which may then be solved by standard methods.

In the method described here, the equation given by (5-156b) is not required and the pressure field may be determined by (5-156c) once the stream functions and subsequently the velocities are calculated. Napolitano [1976] reports hybrid variational principles in which multitensorial fields are considered. This is in line with the hybrid method developed by Pian [see Pian and Tong, 1969]. For additional discussions of variational methods see Sec. 6-4-2.

5-8-5 Example Problems

A simple example problem that may be considered is a rectangular duct flow in steady state. The time-dependent term in (5-156a) is dropped from the solution. Let us consider the geometry discretized into 60 linear isoparametric elements as shown in Fig. 5-35. The results obtained for $Re = 200$ agree well with those of Baker [1975] using 264 linear triangular elements. The fully-developed longitudinal velocity distributions as shown in Fig. 5-36 also agree excellently with those determined analytically [Schlichting, 1968]. Figure 5-37 shows the computed stream function and vorticity compared with the analytical solution. Experiments with refined mesh systems reveal no changes in the results, confirming the accuracy of the 30-element solution.

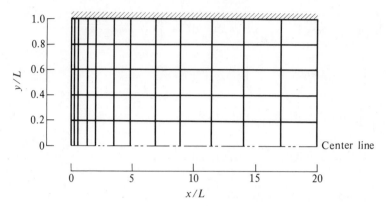

Figure 5-35 Discretization of rectangular duct into 60 isoparametric elements.

Olson [1975] presents the finite element solution using the high-precision triangular element [Cowper, et al., 1969] with quintic polynomial. In his work, the approximate variational functional as given by (5-169) is utilized to determine the flow field over a circular cylinder. We note that the functional $I(\psi)$ in (5-170) contains derivatives of ψ up to the second order. Thus the continuity of the first derivatives of ψ (known as C^1 continuity) as well as the continuity of ψ itself must be assured for convergent solution. The high precision triangular

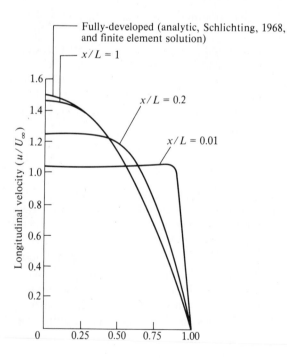

Figure 5-36 Longitudinal velocity distributions for rectangular duct flow, $Re = 200$, 60 isoparametric elements.

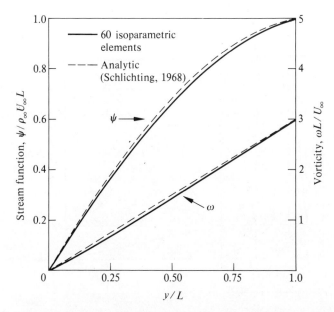

Figure 5-37 Fully-developed stream function and vorticity for rectangular duct flow $Re = 200$.

element of Cowper, et al., satisfying this requirement can be provided with ψ, $\partial\psi/\partial x$, $\partial\psi/\partial y$, $\partial^2\psi/\partial x^2$, $\partial^2\psi/\partial y^2$, and $\partial^2\psi/\partial x\,\partial y$ at each corner node (Fig. 5-38). This yields a total of 18 degrees of freedom. We begin with a full fifth degree polynomial having 21 parameters and introduce three constraints for cubic variations of $\partial\psi/\partial n$ along the edges of the triangle. This operation leads to the 18×18 coefficient matrices in (5-149b). The integrals over the triangular area are calculated from the formula (2-20).

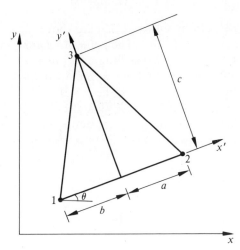

Figure 5-38 High precision triangular element with first and second derivatives of ψ specified at corner nodes.

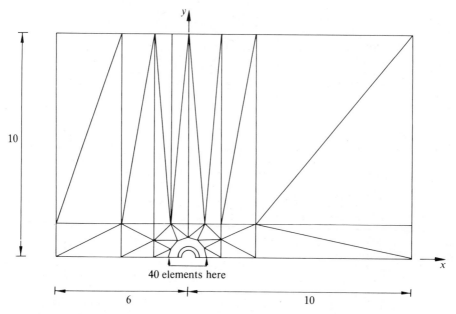

Figure 5-39 Finite element discretization (75 elements) for flow over a circular cylinder ($r = 0.5$).

Consider the geometry as shown in Fig. 5-39 and the boundary conditions:

1. $\psi = y$ $\partial\psi/\partial x = 0$ on the upstream edge
2. $\psi = 0$ on the x-axis
3. $\partial\psi/\partial y = 1$ along the upper edge
4. $\partial\phi/\partial x = 0$ on the downstream edge
5. To satisfy $\psi = 0$ and $\partial\psi/\partial n$ along the cylinder, the nodal variables are transformed from the global x, y coordinates to local coordinates (tangential, normal to surface) and all variables except $\partial^2\psi/\partial n^2$ are zeroed and eliminated.

The streamline patterns obtained for $Re = 20$ by Olsen [1975] are shown in Fig. 5-40 compared with those of Dennis and Chang [1970]. The point of separation in both solutions agrees quite well. Olson reports that drag and pressure coefficients do not compare well due to the boundary condition approximations on the cylinder surface. Circulation problems in lakes have also been studied by Cheng [1972a], Gallagher [1975], Gallagher, et al. [1973a, 1973b], and Leonard and Melfi [1972], among others. On the other hand, Bratanow, Ecer, and Kobiske [1973] report finite element analysis of unsteady flow around oscillating obstacles.

5-9 BOUNDARY LAYER FLOW

Viscosity plays a major role also in boundary layer flow. There is a marked velocity change from zero at the wall of a body or a channel to a velocity greater

Finite elements

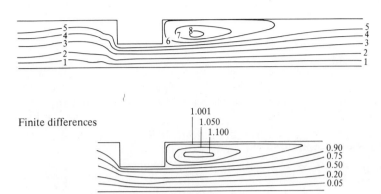

Finite differences

Figure 5-40 Streamlines over circular cylinder, $Re = 20$, finite element solution after Olson (1975) and finite difference solution after Dennis and Chang (1970).

than zero at some distance from the wall. In the flow outside the boundary layer, viscosity or internal friction has little influence.

In the study of boundary layer, we consider a very thin layer in the immediate neighborhood of the body where the velocity gradient normal to the wall is very large, contributing to high shear stresses. Away from the surface of the body or channel, no such large velocity gradients occur, and the flow becomes frictionless and potential. The boundary layer thickness is proportional to the square root of kinematic viscosity. More specifically, we have the relationship for boundary layer on a flat plate

$$\mu \frac{U}{\delta^2} \sim \frac{\rho U^2}{L} \tag{5-172}$$

where U is the free stream velocity, μ is the dynamic viscosity, and L is the plate length. Thus

$$\delta \sim \sqrt{\frac{\mu L}{\rho U}} = \sqrt{\frac{\nu L}{U}} \tag{5-173}$$

Blasius showed that for laminar flow

$$\delta = 5 \sqrt{\frac{\nu L}{U}} \tag{5-174}$$

or

$$\frac{\delta}{L} = \frac{5}{\sqrt{R_L}} \tag{5-175}$$

where $R = UL/\nu$ is the Reynolds number. It is seen that as the Reynolds number becomes infinity, the boundary layer thickness vanishes.

In the following sections, we discuss finite element applications of the vortex flow and boundary layers. Although these subjects are involved in both incompressible and compressible fluids, we limit our discussions to the incompressible flow for simplicity of presentations.

5-9-1 Laminar Boundary Layer

Although the governing equations for viscous fluids presented in the earlier chapters may be used in the boundary layer problems, various forms of simplified equations are used in analytical solutions. Prandtl showed how the Navier–Stokes equations could be simplified to yield approximate solutions and clarified the essential influence of viscosity in flows at high Reynolds numbers.

The Prandtl's boundary layer equations for two-dimensional flow on a flat plate are written

$$\frac{\partial u}{\partial t} + u\frac{\partial u}{\partial x} + v\frac{\partial u}{\partial y} = -\frac{1}{\rho}\frac{\partial P}{\partial x} + v\frac{\partial^2 u}{\partial y^2} \tag{5-176}$$

$$\frac{\partial u}{\partial x} + \frac{\partial v}{\partial y} = 0 \tag{5-177}$$

with the boundary conditions

$$y = 0, \quad u = v = 0; \quad y = \infty, \quad u = U_{(x)}$$

Note that the term $v(\partial^2 u/\partial x^2)$ is neglected due to its smallness in comparison with $v(\partial^2 u/\partial y^2)$. In the case of steady flow, the pressure depends only on x. Therefore

$$U\frac{dU}{dx} = -\frac{1}{\rho}\frac{dP}{dx} \tag{5-178}$$

and using this relation in (5-176) for steady flow, we obtain

$$u\frac{\partial u}{\partial x} + v\frac{\partial u}{\partial y} = U\frac{dU}{dx} + v\frac{\partial^2 u}{\partial y^2} \tag{5-179}$$

The finite element equations based on (5-176) and (5-179) together with (5-177) can be easily formulated as outlined in earlier chapters. It should be noted, however, that these simplifications are meaningful in the analytical solution or conventional methods whereas in the finite element procedure, the solution with all terms included may be obtained without difficulty. Chung and Chiou [1974] solved via finite elements the thermal boundary layer problem of nonsteady compressible flow between two parallel plates behind the normal shock. Results of their studies are given in Sec. 6-3-2. See Baker [1972] for three-dimensional boundary layer flows.

We are reminded that there are classical boundary layer equations such as Falkner–Skan,

$$f''' + \alpha f f'' + \beta(1 - f'^2) = 0$$

or its special case given by Blasius

$$f''' + ff'' = 0$$

The solution of these equations by the finite element method is possible but its usefulness is limited because of the inherent limitations of these boundary layer equations themselves. Furthermore, implementation of boundary conditions encounters difficulties in these equations unlike that in the general governing equations.

5-9-2 Turbulent Boundary Layer

It is well known that the motion of fluids may occur in irregular fluctuations, mixing, or eddying. Such motions are called the turbulent flows. When the Reynolds number is increased, internal flows and boundary layers around solid bodies change from laminar to turbulent. Such transition is influenced by geometries, pressure gradients, suction compressibility, and heat transfer. The exact mathematical treatment of turbulence is hopelessly complex. In turbulence, the motion is such as though the viscosity were increased tremendously. At large Reynolds numbers, there exists a continuous transport of energy from the main flow into the large eddies. The velocity and pressure at a fixed point in space do not remain constant with time but undergo very irregular fluctuations of high frequency. The motion is then assumed to consist of a mean motion and eddying motion. For compressible flow, this leads to

$$V_i = \bar{V}_i + V_i'$$

$$P = \bar{P} + P'$$

$$\rho = \bar{\rho} + \rho'$$

$$T = \bar{T} + T'$$

where the bars and primes denote "mean" and "eddying," respectively. The pressure of fluctuation contributes to increased apparent viscosity of the mean stream. Thus, in the case of a two-dimensional incompressible flow, we have

$$\bar{u}\frac{\partial \bar{u}}{\partial x} + \bar{v}\frac{\partial \bar{u}}{\partial y} = -\frac{1}{\rho}\frac{d\bar{P}}{dx} + \frac{\partial}{\partial y}\left[(v + \overset{*}{v})\frac{\partial \bar{u}}{\partial y}\right] \tag{5-180}$$

$$\frac{\partial \bar{u}}{\partial x} + \frac{\partial \bar{v}}{\partial y} = 0 \tag{5-181}$$

in which $\overset{*}{v}$ is called kinematic eddy viscosity

$$\overset{*}{v} = A_\tau/\rho \tag{5-182}$$

with A_τ being dynamic eddy viscosity. The apparent shear stress τ is defined as

$$\tau = -\overline{\rho u' v'} = A_\tau \frac{d\bar{u}}{dy} = \rho\overset{*}{v}\frac{d\bar{u}}{dy} \tag{5-183}$$

or
$$\tau = \rho l^2 \left| \frac{d\bar{u}}{dy} \right| \frac{d\bar{u}}{dy} \tag{5-184}$$

where l is known as Prandtl's mixing length and

$$\overset{*}{v} = l^2 \left| \frac{d\bar{u}}{dy} \right| \tag{5-185}$$

von Kármán showed that the mixing length satisfies the equation

$$l = k \left| \frac{d\bar{u}/dy}{d^2 \bar{u}/dy^2} \right| \tag{5-186}$$

where k is empirical dimensionless constant. It is seen that the mixing length is not a property of a fluid but rather a function of velocity distribution. In the finite element formulation, we simply substitute (5-185) into (5-180) and the standard procedure will be followed. Thus the finite element analog for (5-180) and (5-181) are

$$\int_{\Omega} \left(\bar{u} \frac{\partial \bar{u}}{\partial x} + \bar{v} \frac{\partial \bar{u}}{\partial y} + \frac{1}{\rho} \frac{\partial \bar{P}}{\partial x} - \frac{\partial}{\partial y} \left\{ (v + \overset{*}{v}) \frac{\partial \bar{u}}{\partial y} \right\} \right) \Phi_N \, d\Omega = 0 \tag{5-187a}$$

$$\int_{\Omega} \left(\frac{\partial \bar{u}}{\partial x} + \frac{\partial \bar{v}}{\partial y} \right) \Psi_N \, d\Omega = 0 \tag{5-187b}$$

with $\bar{u} = \Phi_N \bar{u}_N$, $\bar{v} = \Phi_N \bar{v}_N$, and $P = \Psi_N P_N$, resulting in

$$A_{NMR} u_M u_R + B_{NMR} v_M u_R + D_{NM} u_M = E_N + J_N \tag{5-188a}$$

$$F_{NM} u_M = G_N \tag{5-188b}$$

where

$$A_{NMR} = \int_{\Omega} \Phi_N \Phi_M \frac{\partial \Phi_R}{\partial x} \, d\Omega \qquad\qquad B_{NMR} = \int_{\Omega} \Phi_N \Phi_M \frac{\partial \Phi_R}{\partial y} \, d\Omega$$

$$D_{NM} = \int_{\Omega} (v + \overset{*}{v}) \frac{\partial \Phi_N}{\partial y} \frac{\partial \Phi_M}{\partial y} \, d\Omega$$

$$J_N = \left[\int_{\Omega} \frac{1}{\rho} \frac{\partial \Psi_M}{\partial x} \, d\Omega \right] P_M \qquad\qquad E_N = \int_0^L (v + \overset{*}{v}) \frac{\partial \bar{u}}{\partial y} \sin \theta \, ds$$

$$F_{NM} = \int_{\Omega} \left(\frac{\partial \Psi_N}{\partial x} \Phi_M + \frac{\partial \Psi_N}{\partial y} \Phi_M \right) d\Omega \qquad\qquad G_N = \int_0^L (\bar{u} \cos \theta + \bar{v} \sin \theta) \overset{*}{\Psi}_N \, ds$$

Note that θ is the angle between the x-axis and the direction normal to the boundary line along ds, and the pressure is regarded as a given impressed force.

The finite element equations (5-188a) are similar to (5-88a) except for the additional nonlinear terms brought in by the kinematic eddy viscosity $\overset{*}{v}$. In the Newton–Raphson procedure, $\overset{*}{v}$ is calculated from nodal velocities of the previous iterative cycle. If isoparametric coordinates are used, $\overset{*}{v}$ can be most conveniently determined at the center of an element where the coordinate values are zero.

Note that the second derivative in the denominator of the mixing length (5-186) becomes zero for a linear velocity interpolation function. This dictates that in turbulence, the velocity must be second order or higher in finite element approximations. Pressure may still be considered a linear function of spatial domain. For Newton–Raphson solution procedure see Sec. 5-6-2.

5-10 THREE-DIMENSIONAL ANALYSIS

A duct flow with noncircular variable cross sections or a flow over a missile with angle of attack cannot be idealized to a two-dimensional flow. In the analysis of such flow problems, we require three-dimensional elements as discussed in Sec. 2-4. Either a tetrahedral element or a hexahedral element can be used. An hexahedral element may be divided into five tetrahedral elements as shown in Fig. 5-41. However, the most convenient approach is clearly the use of an hexahedral element itself since this element provides ease in integration via the Gaussian quadrature as well as less cumbersome work in keeping up with element geometry information. Typical hexahedral elements in a three-dimensional duct with rectangular cross sections are shown in Fig. 5-42. Note that the symmetry of geometry should be exploited as much as possible to reduce the domain of study to a minimum. For example, if the cross sections of Fig. 5-41 are squares, then only a quadrant need be analyzed.

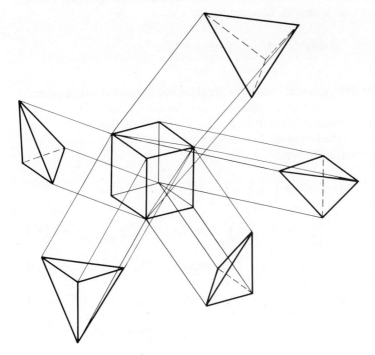

Figure 5-41 Five tetrahedral elements from a hexahedron.

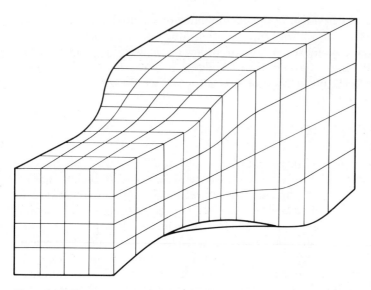

Figure 5-42 Discretization of three-dimensional nozzle with hexahedral elements.

In the three-dimensional analysis, the stream function formulation is impractical. If the fluid is inviscid, the potential function approach is feasible. For viscous flow, it is best to solve the Navier–Stokes equations directly. For a steady state incompressible viscous flow, we have

$$V_{i,j}V_j - F_i + \frac{1}{\rho}P_{,i} + \nu V_{i,jj} = 0$$

$$V_{i,i} = 0$$

with $i, j = 1, 2, 3$. The finite element equations are the same as in a two-dimensional case except that we now have an additional equation plus extra terms for the third dimension. The solution procedure is also identical to that in the two-dimensional analysis. If the hexahedral element is used, the Gaussian quadrature integration for any function becomes

$$\int_V g(x, y, z) \, dx \, dy \, dz = \int_{-1}^{1} \int_{-1}^{1} \int_{-1}^{1} f(\xi, \eta, \zeta) \, d\xi \, d\eta \, d\zeta$$

$$= \sum_{i=1}^{n} \sum_{j=1}^{n} \sum_{k=1}^{n} w_i w_j w_k f(\xi_i, \eta_j, \zeta_k) \qquad (5\text{-}189)$$

Although no special difficulties are encountered in the three-dimensional analysis, the computer time required will be considerable.

An example of a three-dimensional duct flow is shown in Fig. 5-43. Axial velocity development at duct center line for creeping flow is compared with the results obtained by the finite difference technique [Carlson and Hornbeck, 1973; Zienkiewicz and Godbole, 1975].

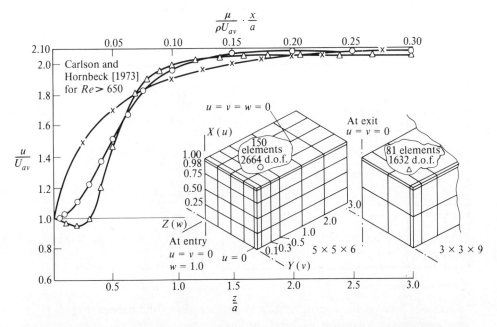

Figure 5-43 Three-dimensional creeping duct flow, axial velocity at center line, after Zienkiewicz and Godbole, 1975.

5-11 BOUNDARY SINGULARITIES

The well-defined error estimates given by (3-177), (3-189), and (3-197) are applicable only to smooth boundaries. If the boundary contains a corner, then the solution u will have derivatives which are unbounded at the corner, resulting in deterioration of the solution. Use of refined mesh around the corner is an effective remedy (Fig. 5-44). However, an implementation of singular functions to deal with corners has been studied extensively [Lehman, 1959] and it is only logical to take advantage of such development [see Kellogg, 1971; Barnhill and Whiteman, 1973; Whiteman, 1975].

Consider the homogeneous Dirichlet problem for Poisson's equation ($m = 1$) of the form

$$Au = -\nabla^2 u = f \qquad \text{in } \Omega \qquad (5\text{-}190a)$$

$$u = 0 \qquad \text{on } \Gamma \qquad (5\text{-}190b)$$

together with

$$(Au, v) = \int_0^{r_0} r\,dr \int_0^{\alpha\pi} \left[\frac{\partial u}{\partial r}\frac{\partial v}{\partial r} + \frac{1}{r^2}\frac{\partial u}{\partial \theta}\frac{\partial v}{\partial \theta} - fv \right] d\theta \qquad v \in H^1(\Omega) \qquad (5\text{-}191)$$

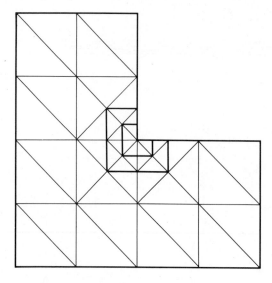

Figure 5-44 Refined mesh at a corner.

In terms of local polar coordinates (r, θ) with origin at the corner (Fig. 5-45), a local asymptotic expansion of u takes the form

$$u(r, \theta) = \sum_{i=1}^{N} a_i \phi_i(r, \theta) \tag{5-192}$$

Let the internal angle of the corner be $\phi = \alpha\pi$. Then the asymptotic expansions [Lehman, 1959] are, for example

$\alpha = 2$:

$$u(r, \theta) = a_1 r^{1/2} \sin \frac{\theta}{2} + a_2 r \sin \theta + a_3 r^{3/2} \sin \frac{3\theta}{2} + \cdots \tag{5-193a}$$

$\alpha = \frac{3}{2}$:

$$u(r, \theta) = a_1 r^{2/3} \sin \frac{2\theta}{3} + r^{4/3} \left[a_2 \sin \frac{4\theta}{3} + a_3 \left(1 - \cos \frac{4\theta}{3} \right) \right]$$

$$+ r^{5/3} \left[a_4 \left(\cos \frac{5\theta}{3} - \cos \frac{\theta}{3} \right) \right] + \cdots \tag{5-193b}$$

Note that $\partial u / \partial r$ is unbounded at $r = 0$ and that for $\alpha > 1$ (re-entrant), u is in H^1 but not in H^2. It is now possible to write the exact solution as a sum of a smooth function w plus singular functions:

$$u = w + \sum_{i=1}^{N} a_i \phi_i \tag{5-194}$$

The purpose of (5-194) is to approximate u by the singular functions ϕ_i near the corner. For example, we may choose

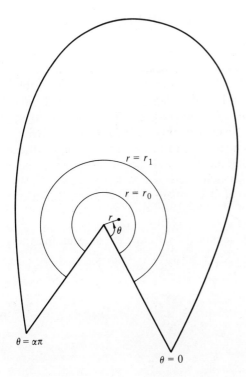

Figure 5-45 A domain with singularity.

$$
\psi_i(r,\theta) = \begin{cases} \psi_i(r,\theta) & 0 \le r \le r_0 \\ F_i(r)G_i(\theta) & 0 \le r \le r \\ 0 & r_1 < r \end{cases} \tag{5-195}
$$

Here the $F_i(r)$ are Hermite polynomials with $\phi_i(r,\theta) \in H^m(\Omega)$ and $F_i(r) \in S^{2m-1}$, S being the space of interpolating polynomials. The $G_i(\theta)$ refer to appropriate functions satisfying the homogeneous boundary conditions on the arms of the corner; and $i = 1, 2, \ldots, N$ with N chosen so that $w \in H^n(\Omega)$, $n \ge 2$. For example, let $\phi = 2\pi$ and $w \in H^2(\Omega)$. Then from (5-193a), the minimal N is 1, resulting in $w_1(r,\theta) = r^{1/2} \sin \theta/2$. Note that the constants a_i must be determined so as to make $w \in H^n(\Omega)$. The error estimate is now in the form

$$
\left\| u - \sum_{i=1}^{N} a_i w_i - \tilde{u} \right\|_{H^m} \le K h^{k+1-m} \| w \|_{H^{k+1}} \tag{5-196}
$$

Suppose that triangular elements are used in the boundary singularity problem. Then with one singular function

$$
\tilde{u} = a + bx + cy + a_1 w_1(r,\theta) \tag{5-197}
$$

Integrations involving both cartesian and polar coordinates can be performed analytically or by using higher order numerical quadratures.

5-12 FINITE ELEMENTS VERSUS FINITE DIFFERENCES

The reader should be aware that the method of finite differences has long been a driving force in the field of computational fluid dynamics since its beginning early in this century [Richardson, 1910; see also Roache, 1972]. The fact is that at this writing, the finite difference method is still dominating the computational fluid dynamics and has been quite successful in many respects. The question then naturally arises: "Where does the finite element method (FEM) stand relative to the finite difference method (FDM)?" The answer to this question is of a rather philosophical nature. The opinions among the scientists and engineers are somewhat divided. However, it can be shown that there is a striking analogy, and sometimes an identity, between the two, and yet there exist certain advantages and disadvantages of one over the other.

Elliptic equations To show that FEM is identical to FDM, let us examine the Laplace equation:

$$\frac{\partial^2 \psi}{\partial x^2} + \frac{\partial^2 \psi}{\partial y^2} = 0$$

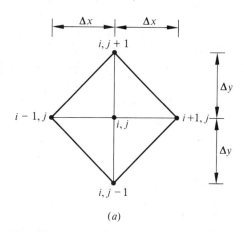

(a)

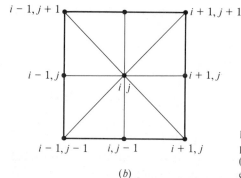

(b)

Figure 5-46 FDM and FEM meshes. (a) Five-point FDM and linear FEM with four elements; (b) nine-point FDM and linear FEM with eight elements.

The finite difference analog takes the form (see Fig. 5-46a)

$$\frac{\psi_{i+1,j} - 2\psi_{i,j} + \psi_{i-1,j}}{\Delta x^2} + \frac{\psi_{i,j+1} - 2\psi_{i,j} + \psi_{i,j-1}}{\Delta y^2} = 0 \qquad (5\text{-}198)$$

Now return to the global finite element equation (5-34). Upon assembly of four triangular elements (Fig. 5-46a), expanding the equation corresponding to the node i,j results in the expression identical to (5-198). If the discretization is as shown in Fig. 5-46b, however, the finite element equation becomes identical to the nine-point finite difference formula [see Arakawa, 1966, and Jespersen, 1974].

The FEM and FDM analogs to the Laplace equation are no longer identical if the Neumann boundary conditions ($\partial \psi / \partial n = 0$ or $\partial \psi / \partial n = g$) are considered. In the FEM, the Neumann conditions are automatically satisfied by simply calculating F_α in (5-34). This is not true of FDM [see Crandall, 1956, p. 263]. Irregular boundaries are troublesome in FDM but the use of a coordinate transformation makes it possible to solve the transformed equations on a regular mesh [Thompson, Thames, and Mastin, 1974].

The differences between FEM and FDM can be illustrated by a simple one-dimensional problem. Let us return to Example 1-7 for the solution of differential equation

$$\frac{d^2 u}{dx^2} + u + x = 0 \qquad 0 < x < 1 \qquad (5\text{-}199)$$

with $u(0) = 0$ and $u(1) = 0$. Typical equations for FDM and FEM written at point i are

FDM:

$$\frac{u_{i+1} - 2u_i + u_{i-1}}{h^2} + u_i + x_i = 0 \qquad (5\text{-}200)$$

FEM, linear:

$$\frac{u_{i+1} - 2u_i + u_{i-1}}{h^2} + \frac{u_{i+1} + 4u_i + u_{i-1}}{6} + \frac{x_{i+1} + 4x_i + x_{i-1}}{6} = 0 \qquad (5\text{-}201a)$$

FEM, quadratic:

$$\frac{-u_{i+1} + 8u_{i+1/2} - 14u_i + 8u_{i-1/2} - u_{i-1}}{h^2}$$

$$+ \frac{-u_{i+1} + 2u_{i+1/2} + 8u_i + 2u_{i-1/2} - u_{i-1}}{10}$$

$$+ \frac{-x_{i+1} + 2x_{i+1/2} + 8x_i + 2x_{i-1/2} - x_{i-1}}{10} = 0 \qquad (5\text{-}201b)$$

The comparison between FEM and FDM for the solution of (5-199) subject to boundary conditions $u(0) = u(1) = 0$ is summarized in Table 5-5 and Fig. 5-47.

It is seen that the convergence of FEM is lower bound whereas the FDM tends to converge upper bound. This trend is influenced by the different approximations for u and x in the differential equation. Note that the results of FDM are comparable to those of FEM for the linear element, the pointwise error being $0(h^{2k}) = 0(h^2)$. For the quadratic element in FEM, the results indicate that the pointwise error is of $0(h^4)$. The errors in L_2 and H^1 norms can also be confirmed to be of $0(h^{k+1})$ and $0(h^k)$, respectively. These results are the same as predicted by the formulas in Sec. 3-4-4. If the boundary conditions were $u(0) = 1$ and $du(1)/dx = 0$, then the FEM results would be considerably more accurate than the FDM results.

Upwind FDM and FEM Spalding [1972] considered the steady state heat transfer given by the differential equation of the form

$$P\frac{\partial T}{\partial x} - \frac{\partial^2 T}{\partial x^2} = 0 \qquad 0 < x < 1 \tag{5-202}$$

Table 5-5 Comparison between FDM and FEM for the solutions of (5-199). Boundary conditions $u(0) = u(1) = 0$ satisfied for all cases

		$x = 0.25$	$x = 0.5$	$x = 0.75$	Pointwise error average %	$\dfrac{\ln(|e_1|/|e_2|)}{\ln(h_1/h_2)}$
FDM	2 elem.	—	0.0714000	—	2.36900	
	4 elem.	0.0442740	0.0701559	0.0604030	0.58510	$2.017 \approx 2$
	8 elem.	0.0440783	0.0698485	0.0601423	0.14520	$2.011 \approx 2$
	16 elem.	0.0440298	0.0697723	0.0600776	0.03545	$2.034 \approx 2$
FEM linear	2 elem.	—	0.0682000	—	-2.22217	
	4 elem.	0.0437579	0.0693453	0.0597154	-0.57476	$1.951 \approx 2$
	8 elem.	0.0439493	0.0696459	0.0599704	-0.14463	$1.991 \approx 2$
	16 elem.	0.0439975	0.0697216	0.0600347	-0.03619	$1.999 \approx 2$
FEM quadratic	2 elem.	0.0440006	0.0697403	0.0600262	-0.02967	$\left.\begin{array}{c}5.62\\[6pt]3.03\end{array}\right\} \approx 4$
	4 elem.	0.0440110	0.0697445	0.0600510	-0.00060	
	8 elem.	0.0440135	0.0697468	0.0600561	-0.00007	
Exact sol. $u = \dfrac{\sin x}{\sin 1} - x$		0.0440136	0.0697469	0.0600562		

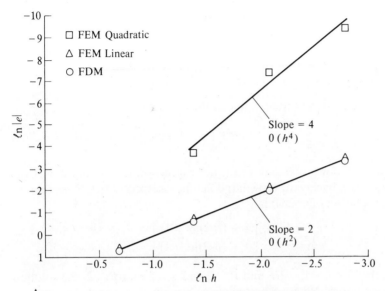

Figure 5-47 Pointwise error (average) for FDM and FEM (Table 5-5).

where P is the Peclet number

$$P = \frac{cu}{\kappa}$$

with c the specific heat, u the mass flow rate, and κ the thermal conductivity coefficient. With the boundary conditions $T(0) = 1$, $T(1) = 1$, he showed that the central difference for $\partial T/\partial x$ was unstable whereas the *upwind* difference (backward difference) was stable.

Subsequently, Heinrich, Huyakorn, Zienkiewicz, and Mitchell [1977] studied the finite element analog for (5-202). They showed that particular choices of weighting functions with orders higher than the trial functions make it possible to generate the upwind scheme. Their choices are

$$T = \Phi_N T_N \qquad (5\text{-}203a)$$

$$W_N = \Phi_N + \Psi_N \qquad (5\text{-}203b)$$

with

$$\Phi_N = \left[1 - \frac{x}{h}, \frac{x}{h} \right] \qquad (5\text{-}203c)$$

$$\Psi_N = \left[\frac{3\alpha}{h^2}(x^2 - hx), \quad \frac{-3\alpha}{h^2}(x^2 - hx) \right] \qquad (5\text{-}203d)$$

where α is a constant with its sign negative if u is negative. Thus the Galerkin-finite element equation is obtained as

$$\int \left(P \frac{\partial T}{\partial x} - \frac{\partial^2 T}{\partial x^2} \right) W_N \, dx = 0$$

or
$$(A_{NM} + B_{NM})u_M = 0$$

with

$$A_{NM} = \frac{P}{2}\begin{bmatrix} \alpha - 1 & 1 - \alpha \\ -1 - \alpha & 1 + \alpha \end{bmatrix} \qquad B_{NM} = \frac{1}{h}\begin{bmatrix} 1 & -1 \\ -1 & 1 \end{bmatrix}$$

The finite element equation written at node i then becomes

$$\left(1 + \frac{\hat{P}}{2}(\alpha - 1)\right)T_{i+1} - (2 + \hat{P}\alpha)T_i + \left(1 + \frac{\hat{P}}{2}(\alpha + 1)\right)T_{i-1} = 0 \qquad (5\text{-}204)$$

with $\hat{P} = Ph$. Clearly, for $\alpha = 0$ and $\alpha = 1$, the expression (5-204) represents the central difference and backward (upwind) difference, respectively.† Note that an exact difference solution of (5-204) is

$$T_i = A + B\left[\frac{1 + (\hat{P}/2)(\alpha + 1)}{1 + (\hat{P}/2)(\alpha - 1)}\right]^i \qquad (5\text{-}205)$$

Thus it is seen that for $\alpha \geq 1 - 2/\hat{P}$ and $\hat{P} \geq 2$ and $\alpha = 0$ and $\hat{P} < 2$, the solution (5-202) is not oscillatory. No error occurs when

$$\alpha = \left(\coth \frac{\hat{P}}{2}\right) - \frac{2}{\hat{P}}$$

These observations are indicative of a close relationship between FDM and FEM.

In Fig. 5-48, we demonstrate the behavior of the various schemes applied to computations. It is clear that the performance of the upwind FDM is superior to the centered FDM which diverges for high Peclet numbers. In contrast, the regular FEM (weighting functions equal to the trial functions) produces results very close to the exact solution, but the solution oscillates with a different number of elements. The upwind FEM, however, gives the results, clearly tending to convergence as the mesh is refined.

Time-dependent problems To deal with time-dependent problems in FEM, we have shown in Sec. 3-3-3 that various temporal operators can be used. These temporal operators are nothing other than the finite difference operators. The Taylor series expansion permits various order accuracies, depending on appropriate truncations of the series. An alternative approach is to treat the time coordinate like a spatial dimension. For example, consider the Burger's equation in one dimension:

$$\frac{\partial u}{\partial t} + u\frac{\partial u}{\partial x} - v\frac{\partial^2 u}{\partial x^2} = 0 \qquad (5\text{-}206)$$

Choosing the interpolation function $\Phi_i(x, t)$ such that

$$u(x, t) = \Phi_i(x, t)u_i$$

---

† If u turns negative and the sign of α is not changed, the resulting equation represents the forward difference.

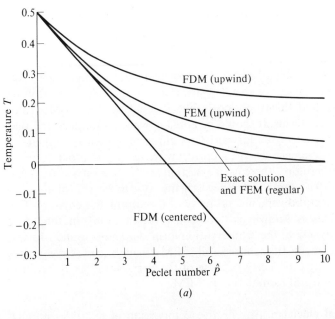

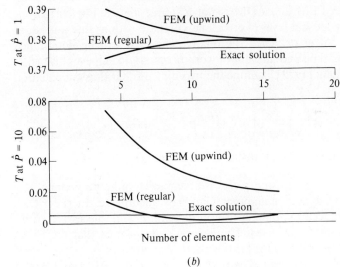

Figure 5-48 Temperature at $x = 0.5$. (a) Temperature versus Peclet number with four elements; (b) convergence of FEM.

the finite element equation can be written in the form

$$\int_t \int_x \left(\frac{\partial u}{\partial t} + u \frac{\partial u}{\partial x} - v \frac{\partial^2 u}{\partial x^2} \right) \Phi_i(x, t) \, dx \, dt = 0 \qquad (5\text{-}207)$$

which now constitutes the two-dimensional space.

The FDM analog for Burger's equation may be written as

$$\frac{u_i^{(n+1)} - u_i^{(n)}}{\Delta t} + \frac{u_i^{(n)}(u_{i+1}^{(n)} - u_{i-1}^{(n)})}{2h} - v\frac{u_{i+1}^{(n)} - 2u_i^{(n)} + u_{i-1}^{(n)}}{h^2} = 0 \qquad (5\text{-}208)$$

This is called the forward-time and centered-space differences. Comparing (5-208) with the finite element equation expanded at an arbitrary node and arbitrary time step (5-102a), we find that they look alike except that the FEM analog carries considerably more terms. It should be noted however, if the centered-time differences are used instead of the forward-time differences, then the scheme is unstable, although the difference scheme is correct with the second order accuracy [Roache, 1972]. Furthermore, there is a question of "conservative property" which must prevail in the finite difference scheme. It can be shown that the preservation of the integral divergence property of the continuum equation fails if $v = 0$ in (5-206). This is known as "non-conservation form" in the FDM literature. For a remedy, use of the "conservation form" [Richtmyer and Morton, 1967] has been advocated. In contrast, all FEM analogs can be generally assured of the conservative property because of the very nature of the control volume built into the integral formulation in FEM.

Relative computational efficiency An effective comparison of relative computational aspects between FEM and FDM is not a simple matter. For a rigorous comparison, we may choose a level of error and compute with as fine a mesh as necessary to obtain that error. Then the relative efficiency can be determined by required computing time and amount of work necessary to achieve this task.

Eisenstat and Schultz [1972] examine the two-dimensional Poisson equation with variable coefficients:

$$-\frac{\partial}{\partial x}\left(p(x,y)\frac{\partial u}{\partial x}\right) - \frac{\partial}{\partial y}\left(q(x,y)\frac{\partial u}{\partial y}\right) + c(x,y)u = f(x,y)$$

$$(x,y)\in \Omega \equiv \{(x,y): 0 < x,y < 1\}; \quad u(x,y) = 0$$

The FEM solution is obtained using piecewise bicubic Hermite polynomials as basis functions integrated by a nine-point Gaussian quadrature. In the FDM solution, the five-point central difference approximation and successive over-relaxation [Southwell, 1956] are used. The general conclusion is that the FEM is much more efficient than the FDM for this particular problem.

On the other hand, Swartz and Wendroff [1974] compare FEM and FDM in a hyperbolic problem defined by

$$\frac{\partial u}{\partial t} = c\frac{\partial u}{\partial x}$$

$$u(x,0) = \exp(2\pi i \omega x) \qquad 0 \le x \le 1$$

In the FEM solution, various order polynomials are used to generate differential-difference equations of the type discussed in Sec. 3-3-3. The basic point is that a fair comparison of schemes can be made only if each scheme is so optimized for a given error. Then the computational work to attain this level of error is a reliable scale for the comparison. With this in mind, the conclusion derived from the data reported by Swartz and Wendroff appears to be noncommittal.

We have cited only two model problems. Relative efficiency can also be problem-dependent. This makes the overall comparison more complicated. However, based on the current state of the art, we may provide some concluding remarks below.

Concluding remarks

1. The standard FDM is analogous to FEM if linear approximations are used in FEM for Dirichlet problems.
2. FEM satisfies the Neumann boundary conditions via an integral form exactly as derived from the variational approach. This is not true of FDM.
3. Mathematical error estimates are more rigorously handled in FEM than in FDM.
4. Some finite difference schemes can be obtained by the FEM analog for irregular domains and boundaries. If this is done, FDM is brought closer to FEM.
5. The computational work required to obtain the same level of error by FEM and FDM varies, depending on problems and various schemes employed in FEM and FDM.

SIX

COMPRESSIBLE FLOW

6-1 GENERAL

The physics and governing equations of the most general fluid flows are highly complicated. The velocity, density, pressure, and temperature are functions of spatial coordinates and time. Dependence of any of these quantities on another renders the resulting governing equations extremely nonlinear. For example, viscosity and density depend on temperature in compressible flow, which in turn affects the velocity and pressure. Therefore, the equations of momentum, continuity, energy, and state must be solved simultaneously to calculate velocity, density, temperature, and pressure.

In what follows, we demonstrate the finite element formulation for compressible, viscous, and unsteady flow problems. For simplicity, we choose the constitutive relationship for the case of a Stokesian flow of a perfect gas. The formulation as given here applies in general to all speed levels as long as there is no shock in the flow field. The problems of shock waves in transonic and supersonic compressible flows are discussed in Sec. 6-5.

6-2 INTERPOLATION FUNCTIONS

To derive the finite element analogs for equations of momentum, continuity, energy, and state, it is necessary to introduce interpolation functions for the velocity V_i, density ρ, temperature T, and pressure P in the form

$$V_i(\mathbf{x}, t) = \Phi_N(\mathbf{x}) V_{Ni}(t) \tag{6-1a}$$

$$\rho(\mathbf{x}, t) = \Lambda_N(\mathbf{x})\rho_N(t) \tag{6-1b}$$

$$T(\mathbf{x}, t) = \Theta_N(\mathbf{x})T_N(t) \tag{6-1c}$$

$$P(\mathbf{x}, t) = \Psi_N(\mathbf{x})P_N(t) \tag{6-1d}$$

where V_N, ρ_N, T_N, and P_N are the nodal values of velocity, density, temperature, and pressure which are functions of time only. Note that the interpolation functions Φ_N, Λ_N, Θ_N, and Ψ_N are dependent on the spatial coordinates $\mathbf{x}$ but free from dependence on time.

The most general approach to the compressible fluid is to construct as usual the Galerkin weighted integrals for each of the four governing equations. In each case, once again suitable weighting functions are needed. The general guidelines for choosing the weighting functions are the same as in incompressible fluids. For completeness, however, we shall repeat here together with additional information for compressible fluids.

For the momentum equation, the obvious variable to be used as the weighting function is the velocity since the inner product of the residual of the momentum equation and the velocity yields an invariant.† Thus we write

$$I_1 = (\varepsilon_i^{(1)}, V_i) = (\varepsilon_i^{(1)}, \Phi_N(\mathbf{x})V_{Ni}(t)) \tag{6-2a}$$

Therefore, the weighting function is the velocity interpolation function $\Phi_N(\mathbf{x})$. Note that the density ρ, pressure P, and temperature T are disqualified because they are scalars and do not possess a free index i required to yield an invariant as a result of an inner product with the residual $\varepsilon_i^{(1)}$ of the momentum equation.

Likewise for the continuity equation, we form an invariant as an inner product of the residual $\varepsilon^{(2)}$ and the density. Thus

$$I_2 = (\varepsilon^{(2)}, \rho) = (\varepsilon^{(2)}, \Lambda_N(\mathbf{x})\rho_N(t)) \tag{6-2b}$$

Obviously the density interpolation function $\Lambda_N(\mathbf{x})$ is the appropriate weighting function, although it would have been the pressure function for incompressible fluids.

For the energy equation, we find that temperature, pressure, density, and velocity are all involved as variables. The residual is denoted $\varepsilon^{(3)}$ and all variables except the velocity appear to provide an invariant from the inner product. In the absence of dissipative energy due to velocity gradients, the spatial distribution of energy is contributed by the gradient of heat flux or $\nabla^2 T$. Thus it is apparent that an invariant is constructed as

$$I_3 = (\varepsilon^{(3)}, T) = (\varepsilon^{(3)}, \Theta_N(\mathbf{x})T_N) \tag{6-2c}$$

and the temperature interpolation $\Theta_N(\mathbf{x})$ is the weighting function.

Lastly, the equation of state for a perfect gas assumes an invariant from the

---

† Contrary to this physical interpretation, we recall that one may choose test functions of order higher than that of trial functions. For certain combinations of physical and geometric parameters, the methods such as employed in upwind finite elements (Sec. 5-12) may improve the accuracy and stability of solutions. Applications of this idea to compressible flows must await further studies.

inner product with the pressure. Therefore the pressure interpolation function $\Psi_N(\mathbf{x})$ is taken as the weighting function

$$I_4 = (\varepsilon^{(4)}, P) = \left(\varepsilon^{(4)}, \Psi_N(\mathbf{x})P_N\right) \tag{6-2d}$$

With these inner product spaces, we are now ready to discuss finite element formulations.

6-3 VISCOUS COMPRESSIBLE FLOW

6-3-1 Finite Element Formulations

The procedures for obtaining the finite element equations in compressible flow are identical to those discussed in Chap. 5, except for additional variables to be taken into account. However, for the sake of completeness and in an effort to keep this chapter for compressible flow reasonably self-contained, some repetitions will occur in the following discussions.

Momentum equation The inner product (6-2a) leads to

$$\int_\Omega \left[\rho(\dot{V}_i + V_j V_{i,j}) - \rho F_i - \sigma_{ij,j}\right]\Phi_N \, d\Omega = 0 \tag{6-3}$$

For Stokesian fluid, the stress tensor assumes the form

$$\sigma_{ij} = -P\delta_{ij} + \lambda d_{rr}\delta_{ij} + 2\mu d_{ij} \tag{6-4a}$$

where, for Stokes hypothesis, the viscosity constants μ and λ are related by $\lambda = -\frac{2}{3}\mu$, and the deformation rate tensor d_{ij} is

$$d_{ij} = \frac{1}{2}(V_{i,j} + V_{j,i}) \tag{6-4b}$$

The viscosity constants are temperature dependent in general. We may thus set $\mu = aT$ and $\lambda = bT$ where a and b are constants.

The partial integration of (6-3) can be carried out similarly as in Sec. 5-6-2 for incompressible flow. Thus the local Galerkin-finite element equation for momentum is written as[†]

$$A_{NMQ}\rho_Q \dot{V}_{Mi} + B_{NiRMQ}\rho_Q V_{Mj}V_{Rj} + C_{NiM}P_M + D_{NMQik}T_Q V_{Mk} = E_{Ni}^{(b)} + E_{Ni}^{(s)} \tag{6-5}$$

where

$$A_{NMQ} = \int_Q \Phi_N \Phi_M \Lambda_Q \, d\Omega$$

$$B_{NiRMQ} = -\int_\Omega \frac{1}{2}\Phi_{N,i}\Phi_R \Phi_M \Lambda_Q \, d\Omega$$

[†] We assume here that the flow is irrotational. If rotational, $V_{i,j}V_j = \frac{1}{2}(V_j V_j)_{,i} - \varepsilon_{kji}\varepsilon_{mnj}V_k V_{n,m}$, and the finite element equation for the convective term will assume a non-conservative form.

$$C_{NiM} = -\int_\Omega \Phi_{N,i}\Psi_M \, d\Omega$$

$$D_{NMQik} = \int_\Omega [b\Phi_{N,i}\Phi_{M,k} + a(\Phi_{N,j}\Phi_{M,j}\delta_i^k + \Phi_{N,j}\Phi_{M,i}\delta_j^k)]\Theta_Q \, d\Omega$$

$$E_{Ni}^{(s)} = \int_\Omega F_i \Phi_N \Lambda_M \, d\Omega \rho_M$$

$$E_{Ni}^{(s)} = \int_\Gamma S_i \overset{*}{\Phi}_N \, d\Gamma - \int_\Gamma \tfrac{1}{2}\rho V_j V_j n_i \overset{*}{\Phi}_N \, d\Gamma$$

The nonlinearity present in the Navier–Stokes equation for incompressible flow is now coupled with additional variables of density and temperature for the compressible flow.

Equation of continuity According to the inner product (6-2b), the finite element equation takes the form

$$\int_\Omega [\dot\rho + (\rho V_i)_{,i}]\Lambda_N \, d\Omega = 0$$

or

$$W_{NM}\dot\rho_M + F_{NiMQ}\rho_M V_{Qi} = G_N \tag{6-6}$$

where

$$W_{NM} = \int_\Omega \Lambda_N \Lambda_M \, d\Omega$$

$$F_{NiMQ} = -\int_\Omega \Lambda_{N,i}\Lambda_M \Phi_Q \, d\Omega$$

$$G_N = -\int_\Gamma \rho V_i n_i \overset{*}{\Lambda}_N \, d\Gamma$$

Here G_N is the boundary normal velocity with $\overset{*}{\Lambda}$ being the interpolation function describing the normal velocity distribution along the boundaries.

Energy equation For the linear momentum to be conserved, the first law of thermodynamics requires the balance of energy given by

$$\rho \frac{D\varepsilon}{Dt} - \sigma_{ij}d_{ij} - q_{i,i} - \rho h = 0$$

where ε is the internal energy per unit mass, q_i is the heat flux per unit surface area, h is the heat supply per unit mass. The residual function of the energy equation is

$$\varepsilon^{(3)} = \rho \frac{D\varepsilon}{Dt} - \sigma_{ij}d_{ij} - q_{i,i} - \rho h = \rho\dot\varepsilon + \rho V_i \varepsilon_{,i} - \sigma_{ij}d_{ij} - q_{i,i} - \rho h$$

Once again using the Galerkin's approximation with a subspace Θ_N as represented by (6-2a), we get

$$\int_\Omega (\rho\dot\varepsilon + \rho V_i\varepsilon_{,i} - \sigma_{ij}d_{ij} - q_{i,i} - \rho h)\Theta_N \, d\Omega = 0 \tag{6-7}$$

Here for a perfect gas

$$\varepsilon = c_p T - P/\rho \tag{6-8}$$

where c_p is the specific heat at constant pressure and q_i is defined by $q_i = \kappa T_{,i}$. In view of (5-106b), it follows that

$$\int_\Omega q_{i,i}\Theta_N \, d\Omega = \int_\Gamma q_i n_i \overset{*}{\Theta}_N \, d\Gamma - \int_\Omega q_i\Theta_{N,i} \, d\Omega$$

$$= -\int_\Gamma [q + \bar\alpha(T - T')]\overset{*}{\Theta}_N \, d\Gamma - \int_\Omega \kappa\Theta_{N,i}\Theta_{M,i} \, d\Omega T_M \tag{6-9}$$

Noting also that

$$\int_\Omega \sigma_{ij}d_{ij}\Theta_N \, d\Omega = \int_\Omega \tfrac{1}{2}\sigma_{ij}(V_{i,j} + V_{j,i})\Theta_N \, d\Omega$$

$$= \int_\Omega [-PV_{i,i} + aT(V_{i,j}V_{i,j} + V_{i,j}V_{j,i} - \tfrac{2}{3}V_{i,i}V_{j,j})]\Theta_N \, d\Omega \tag{6-10}$$

and substituting (6-8), (6-9), and (6-10) into (6-7) yields

$$L_{NMQ\rho Q}\dot T_M + H_{NMiQR\rho R}V_{Qi}T_M + R_{NM}\dot P_M + S_{NMQi}V_{Mi}P_Q + C_{NMQi}V_{Qi}P_M$$
$$+ D^{(1)}_{NMQjRj}T_M V_{Qi}V_{Ri} + D^{(2)}_{NMQjRi}T_M V_{Qi}V_{Rj}$$
$$+ D^{(3)}_{NMQiRj}T_M V_{Qi}V_{Rj} + (J_{NM} + K_{NM})T_M$$
$$= Q^{(b)}_N + Q^{(q)}_N \tag{6-11}$$

where

$$L_{NMQ} = \int_\Omega c_p\Theta_N\Theta_M\Lambda_Q \, d\Omega \qquad H_{NMiQR} = \int_\Omega c_p\Theta_N\Theta_{M,i}\Phi_Q\Lambda_R \, d\Omega$$

$$R_{NM} = -\int_\Omega \Theta_N\Psi_M \, d\Omega \qquad S_{NMQi} = -\int_\Omega \Theta_N\Phi_M\Psi_{Q,i} \, d\Omega$$

$$C_{NMQi} = \int_\Omega \Theta_N\Psi_M\Phi_{Q,i} \, d\Omega \qquad D^{(1)}_{NMQjRj} = -\int_\Omega a\Theta_N\Theta_M\Phi_{Q,j}\Phi_{R,j} \, d\Omega$$

$$D^{(2)}_{NMQjRi} = -\int_\Omega a\Theta_N\Theta_M\Phi_{Q,j}\Phi_{R,i} \, d\Omega \qquad D^{(3)}_{NMQiRj} = \int_\Omega \frac{2}{3} a\Theta_N\Theta_M\Phi_{Q,i}\Phi_{R,j} \, d\Omega$$

$$J_{NM} = \int_\Omega \kappa\Theta_{N,i}\Theta_{M,i} \, d\Omega \qquad K_{NM} = \int_\Gamma \bar\alpha\overset{*}{\Theta}_N\overset{*}{\Theta}_M \, d\Gamma$$

$$Q_N^{(h)} = \int_\Omega h\Theta_N \Lambda_M \rho_M \, d\Omega \qquad\qquad Q_N^{(q)} = -\int_\Gamma q\overset{*}{\Theta}_N \, d\Gamma + \int_\Gamma \bar{\alpha}\hat{T}\overset{*}{\Theta}_N \, d\Gamma$$

If internal energy is defined as

$$\varepsilon = c_v T \tag{6-12}$$

then the terms with R_{NM} and S_{NMQi} drop out from (6-11) and c_p is replaced by c_v in the matrices L_{NMQ} and H_{NMiQR}.

Equation of state The equation of state which will now complete the description of compressible thermal flow is given by

$$P = \rho R T \tag{6-13}$$

where R is the gas content. Proceeding as before but this time with a subspace Ψ_N, we obtain

$$\int_\Omega (P - \rho R T)\Psi_N \, d\Omega = 0$$

or

$$E_{NM}P_M = F_N \tag{6-14}$$

where

$$E_{NM} = \int_\Omega \Psi_N \Psi_M \, d\Omega$$

$$F_N = R \int_\Omega \Lambda_L \Psi_N \Theta_M \, d\Omega \rho_L T_M$$

6-3-2 Solution Procedure and Examples

The simultaneous solution of the global equations corresponding to (6-5), (6-6), (6-11), and (6-14) follows the same procedures outlined in Sec. 5-6-2 for incompressible flow. We begin with the global equations of the form

$$A_{\alpha\beta\gamma}\rho_\gamma \dot{V}_{\beta i} + B_{\alpha i\beta\gamma\delta}\rho_\delta V_{\gamma j}V_{\beta j} + C_{\alpha i\beta}P_\beta + D_{\alpha\beta\gamma}T_\gamma V_{\beta i} = E_{\alpha i} \tag{6-15a}$$

$$W_{\alpha\beta}\dot{\rho}_\beta + F_{\alpha i\beta\gamma}\rho_\beta V_{\gamma i} = G_\alpha \tag{6-15b}$$

$$L_{\alpha\beta\gamma}\rho_\gamma \dot{T}_\beta + H_{\alpha\beta i\gamma\eta}\rho_\eta V_{\gamma i}T_\beta + C_{\alpha i\beta\gamma}V_{\gamma i}P_\beta + D^{(1)}_{\alpha j\beta j\gamma\eta}T_\eta V_{\beta i}V_{\gamma i} + D^{(2)}_{\alpha j\beta i\gamma\eta}T_\eta V_{\beta j}V_{\gamma i}$$

$$+ D^{(3)}_{\alpha j\beta i\gamma\eta}T_\eta V_{\beta i}V_{\gamma j} + (J_{\alpha\beta} + K_{\alpha\beta})T_\beta = Q_\alpha \tag{6-15c}$$

$$E_{\alpha\beta}P_\beta = F_\alpha \tag{6-15d}$$

Here, as usual, the indices α, β, γ, η, and δ represent the global node and i and j denote the spatial dimensions. The energy equation is based on the internal energy given by (6-12). Let the time-dependent nodal variables be written in terms of a temporal operator. If forward differences are used, this will lead to a recursive time integration form similar to (5-96)

$$
\begin{bmatrix}
\hat{A}_{\alpha\beta ij} & 0 & 0 & C_{\alpha i\beta} \\
0 & \hat{B}_{\alpha\beta} & 0 & 0 \\
\hat{D}_{\alpha\beta j} & 0 & \hat{L}_{\alpha\beta} & \hat{C}_{\alpha\beta} \\
0 & 0 & 0 & E_{\alpha\beta}
\end{bmatrix}
\begin{bmatrix}
V_{\beta j} \\
\rho_\beta \\
T_\beta \\
P_\beta
\end{bmatrix}^{(n+1)}
=
\begin{bmatrix}
\Delta t E_{\alpha i} \\
\Delta t G_\alpha \\
\Delta t Q_\alpha \\
F_\alpha
\end{bmatrix}^{(n+1)}
+
\begin{bmatrix}
\hat{E}_{\alpha i} \\
\hat{G}_\alpha \\
\hat{Q}_\alpha \\
0
\end{bmatrix}^{(n)}
\tag{6-16}
$$

where

$$
\hat{A}_{\alpha\beta ij} = A_{\alpha\beta\gamma}\rho_\gamma^{(n+1)}\delta_{ij} + \Delta t(1-\theta)(B_{\alpha i\beta\gamma\eta}\rho_\eta^{(n+1)})V_{\gamma j}^{(n+1)} + D_{\alpha\beta\gamma}T_\gamma^{(n+1)}\delta_{ij})
$$

$$
\hat{B}_{\alpha\beta} = W_{\alpha\beta} + \Delta t(1-\theta)F_{\alpha i\beta\gamma}V_{\gamma i}^{(n+1)}
$$

$$
\hat{D}_{\alpha\beta j} = \Delta t(1-\theta)D_{\alpha j\beta i\gamma\eta}^{(1)}T_\eta^{(n+1)}V_{\gamma i}^{(n+1)} + D_{\alpha j\beta i\gamma\eta}^{(2)}T_\eta^{(n+1)}V_{\gamma i}^{(n+1)}
$$

$$
+ D_{\alpha j\gamma i\beta\eta}^{(3)}T_\eta^{(n+1)}V_{\gamma i}^{(n+1)}
$$

$$
\hat{L}_{\alpha\beta} = L_{\alpha\beta\gamma}\rho_\gamma^{(n+1)} + \Delta t(1-\theta)(H_{\alpha\beta i\gamma\eta}\rho_\eta^{(n+1)}V_{\gamma i}^{(n+1)} + J_{\alpha\beta} + K_{\alpha\beta})
$$

$$
\hat{C}_{\alpha\beta} = C_{\alpha i\beta\gamma}V_{\gamma i}^{(n+1)}
$$

$$
\hat{E}_{\alpha i} = [A_{\alpha\beta ij} - \Delta t\theta(B_{\alpha i\beta\gamma\delta}\rho_\delta V_{\gamma j}^{(n)} + D_{\alpha\beta\gamma}T_\gamma^{(n)}\delta_{ij})]V_{\beta j}^{(n)}
$$

$$
\hat{G}_\alpha = (W_{\alpha\beta} - \Delta t\theta F_{\alpha i\beta\gamma}V_{\gamma i}^{(n)})\rho_\beta^{(n)}
$$

$$
\hat{Q}_\alpha = [L_{\alpha\beta\gamma}\rho_\gamma^{(n)} - \Delta t\theta(H_{\alpha\beta i\gamma\delta}\rho_\delta^{(n)}V_{\gamma i}^{(n)} + J_{\alpha\beta} + K_{\alpha\beta})]T_\beta^{(n)}
$$

$$
- \Delta t\theta[(D_{\alpha j\beta j\gamma\delta}^{(1)}T_\delta^{(n)}V_{\gamma i}^{(n)} + D_{\alpha i\beta j\gamma\delta}^{(2)}T_\delta^{(n)}V_{\gamma j}^{(n)}
$$

$$
+ D_{\alpha j\beta i\gamma\delta}^{(3)}T_\delta^{(n)}V_{\gamma j}^{(n)})V_{\beta i}^{(n)} - C_{\alpha i\beta\gamma}V_{\gamma i}^{(n)}P_\beta^{(n)}]
$$

Details of the solution procedure for equations of the type (6-16) are given in Sec. 5-6-2. Thus only some minor comments are presented here. It follows from (6-16) that

$$
A_{mnp}X_n^{(n+1)}X_p^{(n+1)} = Y_m^{(n)}
\tag{6-17}
$$

$$
X_m^{(n+1),(r+1)} = X_m^{(n+1),(r)} - [J_{mn}^{(n+1),(r)}]^{-1}R_n^{(n+1),(r);(n)(0)}
\tag{6-18}
$$

$$
J_{mn}^{(n+1),(r)} = \frac{\partial R_m^{(n+1),(r);(n),(0)}}{\partial X_n}
$$

$$
R_n^{(n+1),(r);(n),(0)} = A_{npq}X_p^{(n+1),(r)}X_q^{(n+1),(r)} - Y_n^{(n),(0)}
$$

We begin with initial and boundary conditions and the Newton–Raphson iterations processed within each time increment until convergence. As shown in incompressible flow, we may resort to a simplified approach in which the Jacobian matrix is set equal to

$$
J_{mn}^{(n+1),(r)} = A_{mnp}X_p^{(n+1),(r)}
$$

$$
J_{ij}^{(n)} \approx A_{ijk}X_k^{(n)}
\tag{6-19}
$$

which is equivalent to the one-dimensional analog given by (5-99), resulting in a scheme similar to (5-100).

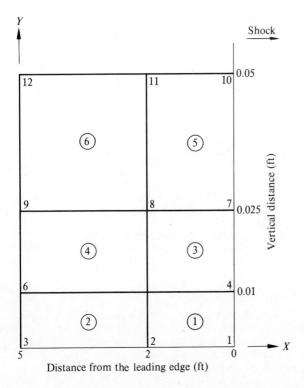

Figure 6-1 Finite element model for the boundary layer behind a normal shock wave.

To demonstrate the computational procedures described above, let us choose a compressible thermal boundary layer problem. In the finite element method, no simplified equations characteristic of boundary layers such as those of Blasius or Falkner–Skan are necessary. The global form of completely general equations (6-15) will be used here. In the solution obtained by Chung and Chiou [1976], nonsteady thermal boundary layer behind the normal shock was considered (Fig. 6-1). The fluid properties, initial and boundary conditions are summarized in Table 6-1.

The analytical formulas for the boundary layer velocity and temperature as used in Mirels [1956] are

$$u = U_\infty f'(\eta) \tag{6-20}$$

and

$$\frac{T - T_\infty}{T_\infty} = \frac{\gamma - 1}{2} M_\infty^2 r(\eta) + \left\{ \frac{\gamma - 1}{2} M_\infty^2 r(0) + 1 - \frac{T_w}{T_\infty} \right\} s(\eta) \tag{6-21}$$

Here U_∞ is the velocity outside the boundary layer, $f(\eta)$ is the dimensionless streamline function, T_∞ is the temperature outside the boundary layer, $\gamma = c_p/c_v$, M_∞ is the Mach number outside the boundary layer, T_w and T_a are wall and recovery temperatures, respectively, and the functions $r(\eta)$ and $s(\eta)$ are solutions satisfying the equations for linear momentum, continuity, energy, and state.

Table 6-1 Input Data

I Fluid properties

External flow velocity	$u_e = 347.5$ ft/s
External flow temperature	$T_e = 200°$R
External flow pressure	$P_e = 10$ lbf/ft^2
External fluid density	$\rho_e = 3.38 \times 10^{-5}$ slug/ft^3
External fluid viscosity	$\mu_e = 1.424 \times 10^{-7}$ lbf/s/ft^2
External fluid thermal conductivity	$\kappa_e = 1.536 \times 10^{-6}$ lbf/s/ft^2
External flow Reynolds number	$Re = 4.125 \times 10^5$ at $L = 5$ ft
Prandtl number	$Pr = 0.72$
Wall temperature	$T_w = 180°$R
Constant pressure specific heat	$c_p = 13.73$ Btu/Slug-°R
Constant volume specific heat	$c_v = 5.46$ Btu/Slug-°R
Free stream Mach no.	$M_\infty = 2.65$
Constants	$a = 0.712 \times 10^{-9}$ lbf/ft^2 °R, $b = -(\frac{2}{3})a$

II Initial conditions

$u = -695$ ft/s at $y = 0$
$u = -347.5$ ft/s elsewhere
$v = 0$ everywhere
$T = 180°$R at $y = 0$
$T = 200°$R elsewhere
$P = 10$ lb/ft^2 everywhere
$\rho = 3.76 \times 10^{-5}$ slug/ft^3 at $y = 0$
$\rho = 3.38 \times 10^{-5}$ slug/ft^3 elsewhere

III Boundary conditions

$u = -695$ ft/s at $y = 0$
$u = -347.5$ ft/s at $y = 0.05$ ft and $x = 0$
$v = 0$ at $y = 0$
$T = 180°$R at $y = 0$
$T = 200°$R at $y = 0.05$ ft and $x = 0$
$P = 10$ lb/ft^2 at $y = 0.05$ ft and $x = 0$
$\rho' = 3.38 \times 10^{-5}$ slug/ft^3 at $y = 0.05$ ft
and $x = 0$

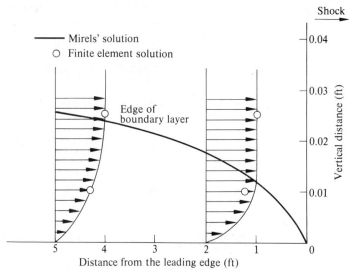

Figure 6-2 Velocity distribution in the laminar boundary layer behind a normal shock wave.

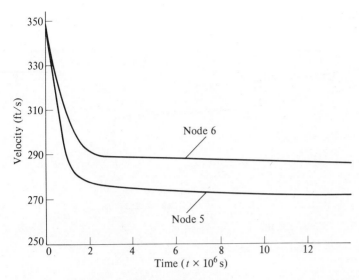

Figure 6-3 Velocity versus time at nodes 5 and 6.

The finite element computer solution of the transient boundary layer problem behind the normal shock wave as shown in Fig. 6-1 is carried out. Four-node linear isoparametric elements, six-point Gaussian quadrature integration, and Newton–Raphson method (6-18) with approximate Jacobian (6-19) are used in this analysis. Figure 6-2 shows the velocity versus time plots for nodes 5 and 6. A comparison of the results with those of Mirels [1956] indicates a good agreement. The nonsteady velocity profiles at distances of 2 ft and 5 ft behind the shock are shown in Fig. 6-3. Steady state velocities are reached at approximately

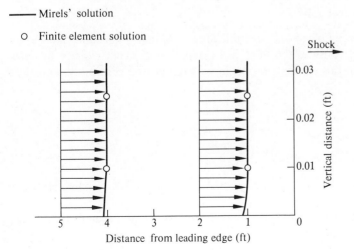

Figure 6-4 Temperature distribution in the laminar boundary layer behind a normal shock wave.

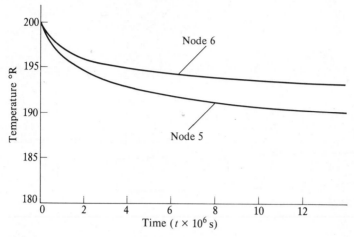

Figure 6-5 Temperature versus time at nodes 5 and 6.

3×10^{-6} s. It should be noted that these plots represent velocities in the direction of the shock relative to the wall.

The temperature versus time plots as shown in Fig. 6-4 indicate that the steady state is reached more slowly than in the case of velocity. The comparison between the finite element solution and analytical solution is again favorable. The temperature distribution which has reached a steady state at approximately 12×10^{-6} s is presented in Fig. 6-5.

6-4 INVISCID COMPRESSIBLE FLOW

6-4-1 Galerkin Approach

When viscosity is negligible in compressible fluids, the analysis becomes simplified. Physically, we may consider the fluid to be a perfect gas, particularly in aerodynamics design. In this context, we begin with Euler and continuity equations of the form

$$V_{i,j}V_j + \frac{1}{\rho} P_{,i} = 0 \tag{6-22}$$

$$(\rho V_i)_{,i} + \frac{e\rho V_2}{x_2} = 0 \tag{6-23}$$

where $e = 0$ and $e = 1$ for two-dimensional flow and axially symmetric flow, respectively.

The total enthalphy $\hat{H}$ and entropy η are constant along each streamline. Therefore, we must have

$$d\hat{H} = \hat{H}_{,i}V_i \, dt = 0$$

or

$$\hat{H}_{,i} V_i = 0 \tag{6-24}$$

and

$$d\eta = \eta_{,i} V_i \, dt = 0$$

or

$$\eta_{,i} V_i = 0 \tag{6-25}$$

Here it should be noted that (6-25) is valid everywhere except for passage through a shock wave. Along the streamline, the velocity of sound is

$$a = \left(\sqrt{\frac{\partial P}{\partial \rho}} \right)_{\eta = \text{const}} = \sqrt{\frac{\gamma P}{\rho}} = \sqrt{\gamma R T} \tag{6-26}$$

The relation between pressure and density in an ideal gas with c_p and c_v constant is given by

$$\frac{P}{\rho^\gamma} = \frac{P_0}{\rho_0^\gamma} \exp\left(\frac{\eta - \eta_0}{c_v}\right) = C \exp\left(\frac{\eta}{c_v}\right) \tag{6-27}$$

where P_0, ρ_0, and η_0 are the initial conditions and $\gamma = c_p/c_v$, and the function C remains constant along each streamline. In view of (6-26) and (6-23), we obtain

$$P_{,i} = a^2 \rho_{,i} - \frac{\rho a^2}{c_p} \eta_{,i} - \frac{\rho a^2}{\gamma} (\ln C)_{,i} \tag{6-28}$$

Rewriting (6-22) in the form

$$\tfrac{1}{2}(V_j V_j)_{,i} - \varepsilon_{ijk} \varepsilon_{kmn} V_j V_{n,m} + \frac{1}{\rho} P_{,i} = 0 \tag{6-29}$$

and from the definitions of enthalpy and entropy

$$H = c_p T = \frac{\gamma}{\gamma - 1} \frac{P}{\rho} \tag{6-30}$$

$$\eta = c_v \ln \frac{P}{\rho^\gamma} \tag{6-31}$$

we obtain

$$T\eta_{,i} + \varepsilon_{ijk} \varepsilon_{kmn} V_j V_{n,m} - (H + \tfrac{1}{2} V_j V_j)_{,i} = 0 \tag{6-32}$$

For the direction normal to the streamline, (6-32) becomes

$$T \frac{\partial \eta}{\partial n} - \frac{\partial \hat{H}}{\partial n} + \varepsilon_{ijk} \varepsilon_{kmn} V_j V_{n,m} n_i = 0$$

where

$$\hat{H} = H + \tfrac{1}{2} V_j V_j \tag{6-33}$$

For two-dimensional or axially symmetric flow

$$\varepsilon_{ijk}\varepsilon_{kmn}V_jV_{n,m}n_i = -\overset{*}{V}(V_{2,1} - V_{1,2}) \tag{6-34}$$

where

$$\overset{*}{V} = V_2n_1 - V_1n_2 \tag{6-35}$$

Hence,

$$V_{2,1} - V_{1,2} = -\frac{1}{\overset{*}{V}}\left(\frac{\partial\hat{H}}{\partial n} - \frac{a^2}{\gamma R}\frac{\partial\eta}{\partial n}\right) \tag{6-36}$$

Multiplying (6-22) by V_i and substituting (6-26) and (6-23) into (6-22) yields

$$V_iV_jV_{i,j} + \frac{a^2}{\rho}V_i\rho_{,i} = 0$$

$$V_iV_jV_{i,j} - a^2\left(V_{k,k} + \frac{eV_2}{x_2}\right) = 0 \tag{6-37}$$

Expanding (6-37) gives

$$\left[1 - \left(\frac{V_1}{a}\right)^2\right]V_{1,1} + \left[1 - \left(\frac{V_2}{a}\right)^2\right]V_{2,2} - \frac{V_1V_2}{a^2}(V_{1,2} + V_{2,1}) + \frac{eV_2}{x_2} = 0 \tag{6-38}$$

Adding and subtracting $(V_1V_2/a^2)V_{1,2}$ in (6-38) and subtracting from (6-36) yields

$$\left[1 - \left(\frac{V_1}{a}\right)^2\right]V_{1,1} + \left[1 - \left(\frac{V_2}{a}\right)^2\right]V_{2,2} - \frac{2V_1V_2}{a^2}V_{1,2} + \frac{eV_2}{x_2} = E \tag{6-39}$$

in which

$$E = \frac{V_1V_2}{\overset{*}{V}}\left(\frac{1}{\gamma R}\frac{\partial\eta}{\partial n} - \frac{1}{a^2}\frac{\partial\hat{H}}{\partial n}\right) \tag{6-40}$$

This is the most general form of the governing equation for two-dimensional or axially symmetric inviscid compressible flows for subsonic, transonic, and super-sonic speeds.

The solution of (6-39) for a two-dimensional case may be obtained by the method of characteristics by transforming into hodograph plane. This is not possible, however, for axially symmetric flow because of the presence of the term eV_2/x_2. If the entropy η and enthalpy H are constants in the direction normal to the surface, then the quantity E in (6-39) vanishes. In this case, (6-39) written in terms of the velocity potential takes the form for two-dimensional flow

$$\left[1 - \frac{1}{a^2}\left(\frac{\partial\phi}{\partial x}\right)^2\right]\frac{\partial^2\phi}{\partial x^2} + \left[1 - \frac{1}{a^2}\left(\frac{\partial\phi}{\partial y}\right)^2\right]\frac{\partial^2\phi}{\partial x^2} - \frac{2}{a^2}\frac{\partial\phi}{\partial x}\frac{\partial\phi}{\partial y}\frac{\partial^2\phi}{\partial x\,\partial y} = 0 \tag{6-41}$$

where

$$a^2 = A + Bq^2$$

with $A = a_0^2 = a_\infty^2 - Bq_\infty^2$, subscripts 0 and ∞ indicating the stagnation point and undisturbed state, $B = (1 - \gamma)/2$, and $q^2 = V_1^2 + V_2^2$.

To proceed with finite element formulations from (6-41), it is convenient to write (6-41) in index notation as follows.

$$\phi_{,ii} - \frac{1}{a^2}\phi_{,i}\phi_{,j}\phi_{,ij} = 0 \tag{6-42}$$

Setting $\phi = \Phi_N\phi_N$, we obtain the local Galerkin-finite element equation

$$\int_\Omega \left(\phi_{,ii} - \frac{1}{a^2}\phi_{,i}\phi_{,j}\phi_{,ij}\right)\Phi_N \, d\Omega = 0 \tag{6-43}$$

Integrating by parts with first derivatives of ϕ held constant, we get

$$A_{NM}\phi_M = G_N + F_N + S_N \tag{6-44}$$

where

$$A_{NM} = \int_\Omega \Phi_{N,i}\Phi_{M,i} \, d\Omega \tag{6-45a}$$

$$G_N = \int_\Omega \frac{1}{a^2}\Phi_{M,i}\Phi_{P,j}\Phi_{Q,i}\Phi_{N,j} \, d\Omega \, \phi_M\phi_P\phi_Q \tag{6-45b}$$

$$F_N = \int_\Gamma \phi_{,i}n_i\overset{*}{\Phi}_N \, d\Gamma \tag{6-45c}$$

$$S_N = -\int_\Gamma \frac{1}{a^2}\phi_{,i}\phi_{,j}\phi_{,i}n_j\overset{*}{\Phi}_N \, d\Gamma \tag{6-45d}$$

Here, F_N represents the Neumann boundary conditions on solid surfaces or at free stream, and S_N is the nonlinear boundary condition on the surface of pressure discontinuity, which will be elaborated in Sec. 6-5. For a shockless domain we set $S_N = 0$. It should be remembered that the partial integration of (6-43) for the nonlinear term is not exact† but the momentum is conserved more exactly than the case in which this term is not integrated at all, commonly known as *nonconservation form*.

For $G_N = S_N = 0$, the equation is reduced to an ideal incompressible flow. The assembled form of (6-44) can be solved using standard techniques for nonlinear equations:

$$A_{ij}\phi_j = G_i + F_i$$

subject to proper boundary conditions. Instead of using techniques such as the Newton–Raphson method, some authors prefer to use a simple iterative procedure by first setting $G_i = 0$ and calculating G_i from the initial solution of ϕ_j. Then

---

† Two of the nonlinear terms can be integrated exactly by setting

$$\left(\frac{\partial\phi}{\partial x}\right)^2\frac{\partial^2\phi}{\partial x^2} = \frac{\partial}{\partial x}\left[\frac{1}{3}\left(\frac{\partial\phi}{\partial x}\right)^3\right] \quad \text{and} \quad \left(\frac{\partial\phi}{\partial y}\right)^2\frac{\partial^2\phi}{\partial y} = \frac{\partial}{\partial y}\left[\frac{1}{3}\left(\frac{\partial\phi}{\partial y}\right)^3\right]$$

In this case, G_N and S_N should be reduced by $\frac{2}{3}$ of the integrated values corresponding to these terms. If so, the approximation ($\phi_{,i}$ held constant) affects only the cross derivative term in (6-41).

G_i is updated for the rest of the iterative cycles until convergence:

$$A_{ij}\phi_j^{(n+1)} = G_i^{(n)} + F_i^{(n)} \tag{6-46}$$

where $(n + 1)$ and (n) denote the current and previous iterative cycles.

For simplicity, a less rigorous form based on the small perturbation theory may be used:

$$(1 - M_\infty^2)\frac{\partial V_1}{\partial x_1} + \frac{\partial V_2}{\partial x_1} + \frac{eV_2}{x_2} = M_\infty^2(1 + \gamma)\frac{V_1}{U}\frac{\partial V_1}{\partial x_1}$$

where U is the free stream velocity and $M_\infty = U/a$ is the free stream Mach number. In terms of potential function, we may also write

$$(1 - M_\infty^2)\frac{\partial^2 \phi}{\partial x_1^2} + \frac{\partial^2 \phi}{\partial x_2^2} + \frac{e}{x_2}\frac{\partial \phi}{\partial x_2} = M_\infty^2\frac{(1 + \gamma)}{U}\frac{\partial \phi}{\partial x_1}\frac{\partial^2 \phi}{\partial x_1^2} \tag{6-47}$$

This equation is valid for transonic flow. If the right-hand side of (6-46) vanishes, then it is valid for subsonic and supersonic flows.

The pressure coefficients are given approximately (higher order terms neglected) by

$$C_p = -\frac{2V_1}{U} - \frac{eV_2^2}{U^2} \tag{6-48a}$$

or

$$C_p = -\frac{2}{U}\frac{\partial \phi}{\partial x_1} - \frac{e}{U^2}\left(\frac{\partial \phi}{\partial x_2}\right)^2 \tag{6-48b}$$

The finite element equation corresponding to (6-44) is now given by the matrices for two-dimensional flow

$$A_{NM} = \int_\Omega \left(\frac{\partial \Phi_N}{\partial x}\frac{\partial \Phi_M}{\partial x} + \frac{\partial \Phi_N}{\partial y}\frac{\partial \Phi_M}{\partial y}\right) d\Omega \tag{6-49a}$$

$$G_N = \left\{\int_\Omega \left[M_\infty^2 + \tfrac{1}{2}M_\infty^2\left(\frac{1 + \gamma}{U}\right)\frac{\partial \Phi_P}{\partial x}\phi_P\right]\frac{\partial \Phi_N}{\partial x}\frac{\partial \Phi_M}{\partial x} d\Omega\right\}\phi_M \tag{6-49b}$$

$$F_N = \int_\Gamma \left(\frac{\partial \phi}{\partial x}n_1 + \frac{\partial \phi}{\partial y}n_2\right)\overset{*}{\Phi}_N d\Gamma \tag{6-49c}$$

$$S_N = -\int_\Gamma \left[M_\infty^2\frac{\partial \phi}{\partial x} + \tfrac{1}{2}M_\infty^2\left(\frac{1 + \gamma}{U}\right)\left(\frac{\partial \phi}{\partial x}\right)^2\right]n_1\overset{*}{\Phi}_N d\Gamma \tag{6-49d}$$

where $d\Omega = dx\,dy$. Note that, in contrast to (6-45), G_N and S_N are exactly integrated here. The solution procedure in the small perturbation theory is the same as in (6-44), but with a considerably simpler nonlinear term.

In general, for transonic and supersonic flows around an airfoil, shock waves are expected to develop, thus leading to pressure discontinuities across the shock. Because of the presence of shock waves, solution of the equations presented here will suffer a breakdown. A remedy of this situation as well as general discussions of shock waves will be given in Sec. 6-5.

6-4-2 Variational Formulations

The variational principles for compressible flow have been studied by Bateman [1929, 1930], Herrivel [1955], and Serrin [1959], among others. The basic idea of deriving the variational principle in nonlinear equations is to determine the existence of symmetric Fréchet derivative (5-164). If the Fréchet derivative is not symmetric, no variational principle exists in general. However, if the flow is irrotational we may use the so-called Bateman principle [Bateman, 1929, 1930] to derive the variational principle for compressible flow,

$$I = \int_\Omega P \, d\Omega + \int_\Gamma f\phi \, d\Gamma$$

where $P = A + \beta\rho^\gamma$ with A, B, and γ being the constants. The first variation of the integral $\int_\Omega P \, d\Omega$ attaining the maximum is equivalent to the conservation of mass

$$(\rho V_i)_{,i} = 0$$

To prove this we begin with the relation $a^2 = \partial P/\partial\rho$, or

$$a^2 = B\gamma\rho^{\gamma-1} = a_\infty^2 + \frac{\gamma-1}{2}(q_\infty^2 - q^2)$$

Denoting $\hat{q}^2 = 2a_\infty^2/(\gamma - 1) + q_\infty^2$ and neglecting the constants, the variational integral takes the form

$$\int_\Omega (\hat{q}^2 - q^2)^{\gamma/(\gamma-1)} \, d\Omega = \text{maximum}$$

The first variation of the above becomes

$$\delta I = \delta \int_\Omega (\hat{q}^2 - q^2)^{\gamma/(\gamma-1)} \, d\Omega = 0$$

With some algebra it can easily be shown that the first variation can be simplified to

$$\delta I = \int_\Omega (\hat{q}^2 - \phi_{,j}\phi_{,j})^{1/(\gamma-1)} \phi_{,i} \delta\phi_{,i} \, d\Omega = 0 \qquad (6\text{-}50)$$

or

$$\delta I = \int_\Omega \rho\phi_{,i}\delta\phi_{,i} \, d\Omega = \int_\Omega \left[(\rho\delta\phi\phi_{,i})_{,i} - \delta\phi(\rho\phi_{,i})_{,i} \right] d\Omega = 0$$

Finally,

$$\delta I = -\int_\Gamma \rho\delta\phi\phi_{,i}n_i \, d\Gamma - \int_\Omega \delta\phi(\rho\phi_{,i})_{,i} \, d\Omega = 0$$

Since $\phi_{,i} n_i = 0$ or $\delta\phi = 0$ on the boundaries, we now have

$$(\rho\phi_{,i})_{,i} = (\rho V_i)_{,i} = 0$$

Thus the validity of (6-50a) has been confirmed.

At this point, we wish to examine the relationship between the variational principle (6-50) and the full potential equation (6-42),

$$\frac{\partial^2 \phi}{\partial x^2} + \frac{\partial^2 \phi}{\partial y^2} - G(\phi) = 0 \qquad \text{in } \Omega \tag{6-51}$$

with

$$G(\phi) = \frac{1}{a^2}\left[\left(\frac{\partial\phi}{\partial x}\right)^2 \frac{\partial^2 \phi}{\partial x^2} + 2\frac{\partial\phi}{\partial x}\frac{\partial\phi}{\partial y}\frac{\partial^2 \phi}{\partial x\,\partial y} + \left(\frac{\partial\phi}{\partial y}\right)^2 \frac{\partial^2 \phi}{\partial x\,\partial y}\right]$$

subject to the boundary conditions

$$\phi = g(x, y) \qquad \text{on } \Gamma_1$$
$$\phi_{,i} n_i = 0 \qquad \text{on } \Gamma_2$$

If we choose $\gamma = 2$, $a^2 = q^2$ or $\hat{q}^2 = a^2$, and keep the first derivatives $\phi_{,i}\phi_{,j}$ held constant, then the expression (6-50) can be integrated by parts as follows:

$$\delta I(\phi) = \int_\Omega \left(1 - \frac{1}{a^2}\,\phi_{,j}\phi_{,j}\right)\phi_{,i}\delta\phi_{,i}\,d\Omega = 0 \tag{6.52a}$$

or

$$\delta I(\phi) = \int_\Gamma \left(\phi_{,i} n_i - \frac{1}{a^2}\,\phi_{,j}\phi_{,i}\phi_{,i} n_j\right)\delta\phi\,d\Gamma$$

$$- \int_\Omega \left(\phi_{,ii} - \frac{1}{a^2}\,\phi_{,i}\phi_{,j}\phi_{,ij}\right)\delta\phi\,d\Omega \tag{6-52b}$$

We notice that the integrand of the last term of (6-52b) becomes zero as it satisfies (6-51). Setting (6-52a) equal to (6-52b), we get

$$\delta I(\phi) = \int_\Omega \left(1 - \frac{1}{a^2}\,\phi_{,j}\phi_{,j}\right)\phi_{,i}\delta\phi_{,i}\,d\Omega + \int_\Gamma f\delta\phi\,d\Gamma \tag{6-52c}$$

where

$$f = -\phi_{,i} n_i + \frac{1}{a^2}\,\phi_{,j}\phi_{,i}\phi_{,i} n_j$$

The derivation in this context implies that the Bateman principle (6-50) does not lead to the full potential equation (6-51) unless the various assumptions as proposed here are valid. Rewriting (6-52c) gives

$$\delta I(\phi) = \delta\left[\int_\Omega \frac{1}{2}\left(\phi_{,i}\phi_{,i} - \frac{2}{a^2}\,\phi_{,i}\phi_{,i}\phi_{,j}\phi_{,j}\right)d\Omega + \int_\Gamma f\phi\,d\Gamma\right] \tag{6-52d}$$

Thus the variational principle valid only under the assumptions made above takes the form

$$I(\phi) = \int_\Omega \left(\frac{1}{2} \phi_{,i}\phi_{,i} - \frac{1}{a^2} \phi_{,i}\phi_{,i}\phi_{,j}\phi_{,j} \right) d\Omega + \int_\Gamma f\phi \, d\Gamma \qquad (6\text{-}52e)$$

There are other means of developing the variational principle in an approximate manner using Fréchet differentials [Finlayson, 1972]. Toward this end, we consider the nonlinear problem

$$N(u) - f = 0$$

and the variational integral

$$I(u, v) = \int_\Omega [vN(u) - vf - ug] \, d\Omega$$

The first variation takes the form

$$\delta I = \int_\Omega \{[N(u) - f]\delta v + [\overset{*}{N}(u, v) - g]\delta u\} \, d\Omega = 0$$

This may be called an *adjoint variational principle*. An alternative approach is to write

$$\delta I = \int_\Omega [N(u^0) - f]\delta u \, d\Omega = 0$$

in which variations are taken with respect to u while holding u^0 constant. Upon the variation we set $u = u^0$ and recover the original governing equation as the "Euler equation." The variational principle of this type is referred to as a *restricted variational principle* [Finlayson, 1972]. In this case, the functional is not stationary unless

$$\overset{*}{N}(u, u) - g = 0$$

where $v = u = u^0$. It can be shown that the restricted variational principle is equivalent to the integral given by (5-168) for the incompressible flow. We demonstrate this idea for the compressible flow below.

The derivation of a restricted variational principle for (6-51) begins with

$$\delta I(\phi) = \int_\Omega \left(\phi_{,ii} - \frac{1}{a^2} \phi_{,i}\phi_{,j}\phi_{,ij} \right) \delta\phi \, d\Omega = 0 \qquad (6\text{-}53)$$

With the first derivatives of ϕ in (6-53) held constant, we integrate by parts:

$$\delta I(\phi) = \int_\Gamma \phi_{,i}n_i\delta\phi \, d\Gamma - \int_\Omega \phi_{,i}\delta\phi_{,i} \, d\Omega - \int_\Gamma \frac{1}{a^2} \phi_{,i}\phi_{,j}\phi_{,i}n_j\delta\phi \, d\Gamma$$

$$+ \int_\Omega \frac{1}{a^2} \phi_{,i}\phi_{,j}\phi_{,i}\delta\phi_{,j} \, d\Omega$$

or

$$\delta I(\phi) = \delta \left[\int_\Omega \frac{1}{2} \left(\phi_{,i}\phi_{,i} - \frac{2}{a^2} \phi_{,i}\phi_{,j}\phi_{,i}\phi_{,j} \right) d\Omega \right.$$
$$\left. - \int_\Gamma \phi_{,i}n_i\phi \, d\Gamma + \int_\Gamma \frac{1}{a^2} \phi_{,i}\phi_{,j}\phi_{,i}n_j\phi \, d\Gamma \right] \qquad (6\text{-}54)$$

The variational functional to be made stationary for the solution of ϕ assumes the form

$$I(\phi) = \int_\Omega \tfrac{1}{2}\phi_{,i}\phi_{,i} \, d\Omega - \int_\Omega \frac{1}{a^2} \phi_{,i}\phi_{,j}\phi_{,i}\phi_{,j} \, d\Omega$$
$$- \int_\Gamma \phi_{,i}n_i\phi \, d\Gamma + \int_\Gamma \frac{1}{a^2} \phi_{,i}\phi_{,j}\phi_{,i}n_j\phi \, d\Gamma \qquad (6\text{-}55)$$

The integral $\int_\Gamma \phi_{,i}n_i\phi \, d\Gamma$ provides the Neumann boundary condition on solid surfaces whereas the last term of (6-55) denotes the pressure discontinuity. Notice that (6-55) is identical to (6-52e) which is a special case derived from the Bateman principle. In terms of the finite element interpolation functions, we may write (6-55) in the form

$$I(\phi) = \tfrac{1}{2}A_{NM}\phi_N\phi_M - B_{NMPQ}\phi_M\phi_P\phi_Q\phi_N - F_N\phi_N + S_N\phi_N \qquad (6\text{-}56)$$

Minimizing this functional with respect to the nodal values of the potential yields

$$A_{NM}\phi_M - B_{NMPQ}\phi_M\phi_P\phi_Q - F_N + S_N = 0 \qquad (6\text{-}57)$$

Setting the second term equal to G_N, the expression (6-57) is seen to be identical to (6-44).

The finite element solution of the flow around a circular cylinder at low Mach numbers formulated from the restricted variational principle of the type (6-55) was first carried out by Norrie and deVries [1972] but without the partial integration of the nonlinear term, resulting in nonconservation form. Their results are shown in Fig. 6-6.

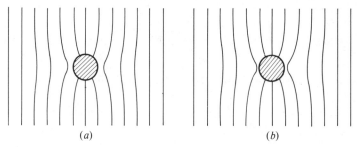

(a) (b)

Figure 6-6 Isopotential lines for isentropic compressible flow past a circular cylinder. (a) For $M = 0.3$; (b) for $M = 0.7$, after Norrie and deVries [1972].

An alternative approach proposed by Carey [1975] deals with the streamline formulation for two-dimensional, compressible, inviscid, irrotational flow. The governing equation is given by

$$\psi_{,ii}\left[1 + M_\infty^2\left(\frac{\gamma - 1}{2}\right)\right] + G(\psi) = 0 \qquad (6\text{-}58)$$

with

$$G(\psi) = \psi_{,ii}\frac{M_\infty^2}{U^2}\left(\frac{\gamma - 1}{2}\right)\varepsilon_{kj}\varepsilon_{kl}\psi_{,j}\psi_{,l} + \psi_{,i}\frac{M_\infty^2}{U^2}\left(\tfrac{1}{2}\varepsilon_{kj}\varepsilon_{kl}\psi_{,j}\psi_{,l}\right)_{,i} = 0 \qquad (6\text{-}59)$$

Although Carey utilizes the perturbation expansion of ψ, it may be simpler to follow the procedure for the variational functional approach as demonstrated for the velocity potential. This takes the form

$$\delta I(\psi) = \int_\Omega\left\{\nabla^2\psi\left[1 + M_\infty^2\left(\frac{\gamma - 1}{2}\right)\right] + G(\psi)\right\}\delta\psi\,d\Omega \qquad (6\text{-}60)$$

Thus the variational principle (holding $\psi_{,j}\psi_{,l}$ constant) is derived as

$$I(\psi) = \int_\Omega\left\{\tfrac{1}{2}\psi_{,i}\psi_{,i}\left[1 + M_\infty^2\left(\frac{\gamma - 1}{2}\right)\right] - \frac{1}{2}\frac{M_\infty^2}{U^2}\left(\frac{\gamma + 1}{2}\right)\varepsilon_{kj}\varepsilon_{kl}\psi_{,i}\psi_{,i}\psi_{,j}\psi_{,l}\right\}d\Omega$$
$$- \int_\Gamma\left\{\psi_{,i}n_i\psi - \frac{M_\infty^2}{U^2}\left(\frac{\gamma + 1}{2}\right)\varepsilon_{kj}\varepsilon_{kl}\psi_{,i}\psi_{,j}\psi_{,l}n_i\psi\right\}d\Gamma \qquad (6\text{-}61)$$

With suitable finite element interpolation functions, we can write (6-62) in the form

$$I(\psi) = \tfrac{1}{2}A_{NM}\psi_N\psi_M - B_{NMPQ}\psi_N\psi_M\psi_P\psi_Q - F_N\psi_N + S_N\psi_N \qquad (6\text{-}62)$$

The corresponding finite element equation becomes

$$A_{NM}\psi_M - B_{NMPQ}\psi_M\psi_P\psi_Q - F_N + S_N = 0 \qquad (6\text{-}63)$$

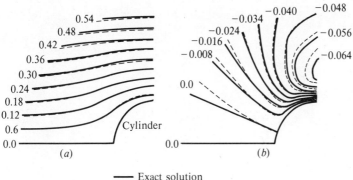

— Exact solution

--- Finite element solution

Figure 6-7 (a) Stream function contours of incompressible flow; (b) first-order contribution of compressibility corrections. After Carey [1975].

with the appearance of this equation identical to that for the velocity potential formulation (6-57).

Carey [1975], using the form (6-58) together with the perturbation expansion of ψ instead of the approach (6-63), presents the finite element solution of the fully-infinite compressible flow past a cylinder. This is shown in Fig. 6-7. Note that in the perturbation solution, the Laplace equation is solved first and the further solutions repeated with effects of compressibility added, the basic idea being similar to the solution of (6-63) in which the first term refers to the incompressible flow whereas the second term represents a compressibility correction.

6-4-3 Linearization in Local Equations

Noting that the solution procedure for (6-57) may suffer nonconvergence with higher Mach numbers, Shen and Habashi [1974] proposed to linearize the

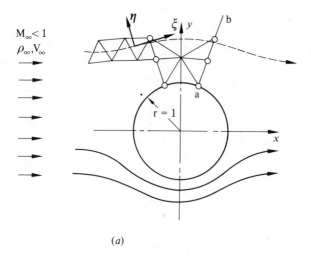

(a)

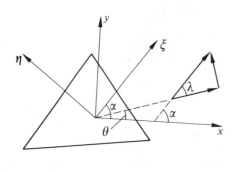

(b)

Figure 6-8 Compressible flow over a cylinder. (*a*) Triangular finite element discretization; (*b*) transformation between flow coordinates (ξ, η) and global coordinates.

operator in each element with respect to the mean flow direction of the previous iterate. Toward this end, we begin with the Bateman variational principles of subsonic flows,

$$I(\phi) = \int_\Omega P(\phi) \, d\Omega + \int_\Gamma \rho\phi \, \frac{\partial\phi}{\partial n} \, d\Gamma \tag{6-64}$$

Now consider a triangular element system as shown in Fig. 6-8 with the local cartesian coordinates ξ and η in the directions tangent and normal to the flow. We assume that the velocity potential can be given by

$$\phi = U_e\xi + \phi'(\xi,\eta) \tag{6-65}$$

where U_e is the average resultant velocity within the element, defined for the $n + 1$ step,

$$U_e^{(n+1)} = [U_e^{(n)} + u' ; v'] \tag{6-66}$$

with $\qquad u'/U_e \ll 1, \qquad v'/U_e \ll 1, \qquad u' = \partial\phi'/\partial\xi, \qquad v' = \partial\phi'/\partial\eta$

Denoting the pressure ratio within each element,

$$\frac{P}{P_e} = \left[1 - \frac{\gamma - 1}{2} M_e^2 \frac{\nabla\phi \cdot \nabla\phi}{V_e^2}\right]^{\frac{\gamma}{\gamma-1}}$$

we write, in view of (6-66),

$$P \simeq P_e - \frac{\rho_e}{2}\left[(1 - M_e^2)u'^2 + v'^2 + 2u'\right] + 0(\varepsilon^3) \tag{6-67}$$

Substituting (6-67) into (6-64), we obtain the global average variational principle,

$$I(\phi) = \sum_e \frac{\rho}{2} \int_\Omega \left[(1 - M_e^2)\left(\frac{\partial\phi}{\partial\xi}\right)^2 + \left(\frac{\partial\phi}{\partial\eta}\right)^2 + 2U_e M_e^2 \frac{\partial\phi}{\partial\xi}\right] d\xi \, d\eta$$

$$- \int_\Gamma \rho_e\phi \, \frac{\partial\phi}{\partial n} \, d\Gamma \tag{6-68}$$

where ρ_0, U_e, and M_e are lagged one step behind in iterations.

To formulate in terms of stream functions, we write

$$\left[\left(\frac{\rho a}{\rho_0}\right)^2 - \left(\frac{\partial\psi}{\partial y}\right)^2\right]\frac{\partial^2\psi}{\partial x^2} + 2\frac{\partial\psi}{\partial x}\frac{\partial\psi}{\partial x}\frac{\partial^2\psi}{\partial x \, \partial y} + \left[\left(\frac{\rho a}{\rho_0}\right)^2 - \left(\frac{\partial\psi}{\partial x}\right)^2\right]\frac{\partial^2\psi}{\partial y^2} = 0 \tag{6-69}$$

$$\frac{\rho_0}{\rho} = \left(\frac{a_0^2}{a^2}\right)^{\frac{1}{\gamma-1}} = \left[1 + \frac{\gamma - 1}{2a^2}\left(\frac{\rho_0}{\rho}\right)^2 (\nabla\psi \cdot \nabla\psi)\right]^{\frac{1}{\gamma-1}} \tag{6-70}$$

with the corresponding variational principle of Bateman–Kelvin,

$$I(\psi) = \int_\Omega (P + \rho v^2) \, d\Omega \tag{6-71}$$

A similar approach as followed in the velocity potential yields

$$I(\psi) = \sum_e \int_\Omega \frac{1}{2\rho_e}\left[(1 - M_e^2)\left(\frac{\partial\psi}{\partial\xi}\right)^2 + \left(\frac{\partial\psi}{\partial\eta}\right)^2\right]d\xi\,d\eta - \int_\Gamma \frac{1}{\rho}\psi\frac{\partial\psi}{\partial n}\,d\Gamma \qquad (6\text{-}72)$$

In an alternative approach to derive the iterative variational principle (6-68), we may substitute (6-65) into (6-50) which gives the governing equation for the perturbation potential ϕ',

$$(1 - M_e^2)\frac{\partial^2\phi'}{\partial\xi^2} + \frac{\partial^2\phi'}{\partial\eta^2} = 0 \qquad (6\text{-}73)$$

By means of the Prandtl–Glaurt coordinate stretching, the perturbation equation is transformed into the Laplace equation, from which the desired iterative variational principle can be obtained.

Habashi [1976] used cylindrical coordinates with linear triangles to derive the influence coefficient matrix. Let us consider a stream function formulation and write

$$\psi = \Phi_N(r, \theta)\psi_N \qquad (6\text{-}74)$$

where $\Phi_N(r, \theta)$ refers to the standard linear triangle interpolation function. Then the variational integral (6-72) takes the form

$$I(\psi) = \tfrac{1}{2}A_{NM}\psi_N\psi_M - F_N\psi_N$$

Minimization of this functional yields

$$A_{NM}\psi_N = F_N$$

where

$$A_{NM} = \int_\Omega \frac{1}{\rho_e}\left[(1 - M_e^2)\frac{\partial\Phi_N}{\partial\xi}\frac{\partial\Phi_M}{\partial\xi} + \frac{\partial\Phi_N}{\partial\eta}\frac{\partial\Phi_M}{\partial\eta}\right]dx\,dy$$

with

$$\begin{bmatrix}\dfrac{\partial\Phi_N}{\partial\xi}\\[2ex]\dfrac{\partial\Phi_N}{\partial\eta}\end{bmatrix} = \begin{bmatrix}\cos\alpha & \sin\alpha\\[2ex]-\sin\alpha & \cos\alpha\end{bmatrix}\begin{bmatrix}\cos\theta & -\dfrac{\sin\theta}{r}\\[2ex]\sin\theta & \dfrac{\cos\theta}{r}\end{bmatrix}\begin{bmatrix}\dfrac{\partial\Phi_N}{\partial r}\\[2ex]\dfrac{\partial\Phi_N}{\partial\theta}\end{bmatrix}$$

$$\alpha = \theta + \lambda \qquad \lambda = \tan^{-1}(V_r/V_\theta)$$

$$V_\theta = \frac{1}{r}\frac{\partial\psi}{\partial\theta} \qquad V_r = -\frac{\partial\psi}{\partial r}$$

Thus the coefficient matrix A_{NM} assumes the form

$$A_{NM} = \int_\Omega \frac{1}{\rho_e}[(1 - M_e^2)B_{NM}(r, \theta) + C_{NM}(r, \theta)]r\,dr\,d\theta \qquad (6\text{-}75)$$

Here integration via Gaussian quadrature is an appropriate choice.

In general airfoil problems, it is often advantageous to use a mapping such as the inverse Joukowsky transformation into a near circle (Fig. 6-9). The finite

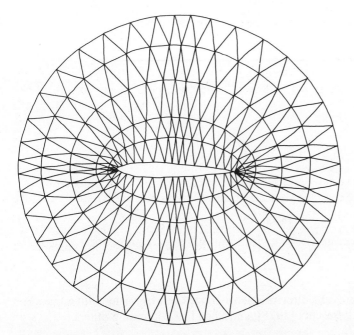

Figure 6-9 Automated mesh generation around NACA 0012, using mapping-remapping procedure. After Habashi [1976].

element mesh can be automatically generated in an r-θ cartesian plane. Then the mesh is transformed back to airfoil plane to carry out the solution. Further details are given in Habashi [1976]. As an illustration, the velocity distribution over a circular cylinder and Mach numbers over a NACA 0012 airfoil obtained by the procedures described above are shown in Fig. 6-10 and Fig. 6-11, respectively.

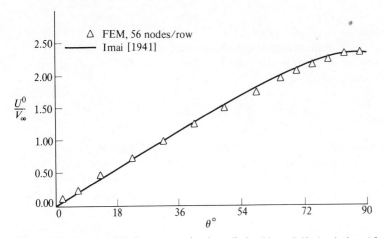

Figure 6-10 Compressible flow over a circular cylinder $M_\infty = 0.42$, ϕ solution. After Habashi [1976].

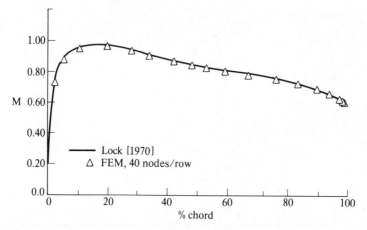

Figure 6-11 Finite element solution for NACA 0012, zero angle of attack, $M_\infty = 0.72$, ϕ formulation. After Habashi [1976].

Research in compressible flow problems carried out by various investigators include Gelder [1971], Periaux [1975], and Shen [1976], among others.

6-5 TRANSONIC AERODYNAMICS

6-5-1 General

Transonic gas flows are governed by partial differential equations of mixed types. The theory was initiated in the historical memoir of Tricomi [1923] in the form

$$y \frac{\partial^2 \phi}{\partial x^2} + \frac{\partial^2 \phi}{\partial y^2} = 0 \qquad (6\text{-}76)$$

which is hyperbolic for $y < 0$ and elliptic for $y > 0$.

The extreme lateral influence produced by an obstacle in a stream near Mach number unity may be studied by the linearized equation for two-dimensional flow with small perturbations

$$(1 - M_\infty^2) \frac{\partial^2 \phi}{\partial y^2} + \frac{\partial^2 \phi}{\partial y^2} = 0 \qquad (6\text{-}77)$$

This shows that when $M_\infty = 1$, we have $\partial^2 \phi / \partial y^2 = 0$, and therefore the perturbation velocity $V_y = \partial \phi / \partial y$ depends only on x. This implies that the disturbance produced by the body is propagated laterally with undiminished strength. Since the curvature of the streamlines remains finite as y increases, it follows that infinite pressure differences can exist between the surface of the body and points at great lateral distances from the body. Thus, the linear theory invalidates the physical conditions and we may resort to the nonlinear theory of small perturbation

$$(1 - M_\infty^2)\frac{\partial^2 \phi}{\partial x^2} + \frac{\partial^2 \phi}{\partial y^2} = M_\infty^2 \frac{(1 + \gamma)\partial \phi}{U} \frac{\partial^2 \phi}{\partial x \partial x^2} \tag{6-78}$$

or the more rigorous and general equation of transonic flow

$$(a^2 - u^2)\frac{\partial u}{\partial x} + (a^2 - v^2)\frac{\partial v}{\partial y} - 2uv\frac{\partial u}{\partial y} = 0 \tag{6-79}$$

Written alternatively,

$$\nabla^2 \phi - G = 0 \tag{6-80}$$

with

$$G = \frac{1}{a^2}\left[\left(\frac{\partial \phi}{\partial x}\right)^2 \frac{\partial^2 \phi}{\partial x^2} + 2\frac{\partial \phi}{\partial x}\frac{\partial \phi}{\partial y}\frac{\partial^2 \phi}{\partial x \partial y} + \left(\frac{\partial \phi}{\partial y}\right)^2 \frac{\partial^2 \phi}{\partial y^2}\right]$$

For unsteady conditions, the small perturbation theory provides

$$\frac{\partial^2 \phi}{\partial x^2} + \frac{\partial^2 \phi}{\partial y^2} - 2M_\infty^2 \frac{\partial}{\partial t}\left(\frac{\partial \phi}{\partial x}\right) - M_\infty^2 \frac{\partial^2 \phi}{\partial t^2} - G = 0 \tag{6-81}$$

with

$$G = \left(M_\infty^2 + M_\infty^2 \frac{(1 + \gamma)}{U}\frac{\partial \phi}{\partial x}\right)\frac{\partial^2 \phi}{\partial x^2}$$

If the amplitude of airfoil oscillation is small in comparison with the airfoil thickness, it is possible to linearize the unsteady problem by taking $\hat{\phi}$ to be a small perturbation on ϕ. This gives the following linearized differential equation for $\hat{\phi}$ in nondimensionalized form:

$$(1 - M_\infty^2)\frac{\partial^2 \hat{\phi}}{\partial x^2} - M_\infty^2(\gamma + 1)\frac{\partial}{\partial x}\left(\frac{\partial \phi}{\partial x}\frac{\partial \hat{\phi}}{\partial x}\right) + \frac{\partial^2 \hat{\phi}}{\partial y^2} + \frac{e}{y}\frac{\partial \hat{\phi}}{\partial y}$$

$$- 2M_\infty^2 \frac{\partial}{\partial t}\left(\frac{\partial \hat{\phi}}{\partial x}\right) - M_\infty^2 \frac{\partial^2 \hat{\phi}}{\partial t^2} = 0 \tag{6-82}$$

with $x = x'/a$, $y = y'/a$, $t = t'U/a$, $\phi = \phi'/Ua$.

For the flow over a harmonically oscillating body, we set

$$\hat{\phi} = Re\,[\overline{\phi}\exp{(ikt)}] \tag{6-83}$$

where $\overline{\phi}$ is a complex function and k is the frequency of oscillation. Substituting (6-83) into (6-82) yields

$$(1 - M_\infty^2)\frac{\partial^2 \overline{\phi}}{\partial x^2} - M_\infty^2(\gamma + 1)\frac{\partial}{\partial x}\left(\frac{\partial \phi}{\partial x}\frac{\partial \overline{\phi}}{\partial x}\right) + \frac{\partial^2 \overline{\phi}}{\partial y^2} + \frac{e}{y}\frac{\partial \overline{\phi}}{\partial y}$$

$$- 2ikM_\infty^2 \frac{\partial \overline{\phi}}{\partial x} + k^2 M_\infty^2 \overline{\phi} = 0 \tag{6-84}$$

with $i = \sqrt{-1}$.

The physics represented by these equations is predicted reasonably well qualitatively. Some significant quantitative results from wind tunnel experiments are also available. Shock waves are an essential feature of transonic flow and their appearance on a moving body leads to a rapid increase in drag coefficient with increasing Mach numbers. The qualitative view of flow past a simple wedge at various free stream Mach numbers is shown in Fig. 6-12 whereas the flow through the converging-diverging nozzle is depicted in Fig. 6-13.

The analytical solution of the nonlinear-mixed type equations is a formidable task. In recent years, significant achievements have been made by workers on computational fluid dynamics using the method of finite difference. The finite difference method in shock wave problems was studied by von Neumann and Richtmyer [1950], among others. They introduced an artificial dissipative mechanism so that the shock transition would become smooth without the

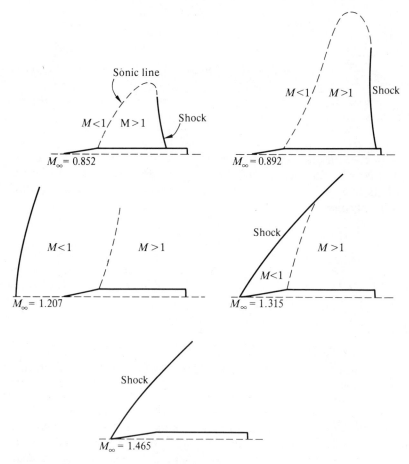

Figure 6-12 Flow around a wedge for various free stream mach numbers. After Liepmann and Roshko [1956].

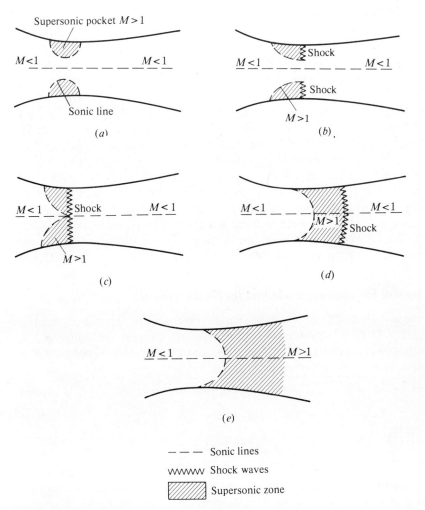

Figure 6-13 Two-dimensional flow through nozzle. (*a*) Lowest back pressure with a possible Venturi-type flow; (*b*) reduced back pressure toward downstream; (*c*) reduced back pressure toward downstream; (*d*) reduced back pressure downstream — similar to one-dimensional flow; (*e*) shock-free through entire nozzle.

necessity of shock fitting [Courant and Friedrichs, 1947] in which the incomplete information by the jump conditions must be supplemented by additional information coming to the shock from the fluid behind it. Since the shock-fitting technique often breaks down when shocks develop spontaneously within the fluid, von Neumann and Richtmyer avoided the direct use of the Hugoniot jump conditions but retained the basic conservation laws on which the Hugoniot conditions are based, in which the jump conditions still hold across the transition layer. Specifically, the dissipative mechanisms such as viscosity and heat conduction are included so that the surface of discontinuity is replaced by a thin transition layer in which quantities change rapidly, not discontinuously. The Hugoniot relations are satisfied such that the "smeared-out" shock travels with exactly the same speed as a discontinuous one would.

Alternative forms of the viscosity term have been proposed by a number of researchers such as Lax and Wendroff [1960], MacCormack [1969], and Murman and Cole [1971], among others. In the recent works of Jameson [1975], the conservation equations are centrally differenced for subsonic region and upwind differenced for the supersonic region for better representations of shock waves. Wellford and Oden [1975] presented the finite element applications to shock waves in solids. In these works, the derivative of the axial displacement is defined in terms of an unknown distance at which a shock front travels. The finite element with such discontinuity may be termed "moving shock element." In the flow field around a body, however, a shock is at least two-dimensional. Our attention is focused on whether there is a shock discontinuity at any given location and time. With this in mind, we introduce here a special finite element referred to as the "stationary shock element" [Chung and Hooks, 1976, 1977].

6-5-2 Element Discontinuity Method for Shock Waves

Let us assume that the shock wave passes through an element causing a pressure jump. Here we consider an explicit discontinuity at the center of the isoparametric element as depicted in Fig. 6-14. An independent interpolation of ϕ for each quadrant is given by (Fig. 6-15)

$$\phi^{(\alpha)} = \alpha_1 + \alpha_2\xi + \alpha_3\eta + \alpha_4\xi\eta + \alpha_5\xi^2 + \alpha_6\eta^2 \qquad (6\text{-}85a)$$

$$\phi^{(\beta)} = \beta_1 + \beta_2\xi + \beta_3\eta + \beta_4\xi\eta + \beta_5\xi^2 + \beta_6\eta^2 \qquad (6\text{-}85b)$$

$$\phi^{(\gamma)} = \gamma_1 + \gamma_2\xi + \gamma_3\eta + \gamma_4\xi\eta + \gamma_5\xi^2 + \gamma_6\eta^2 \qquad (6\text{-}85c)$$

$$\phi^{(\delta)} = \delta_1 + \delta_2\xi + \delta_3\eta + \delta_4\xi\eta + \delta_5\xi^2 + \delta_6\eta^2 \qquad (6\text{-}85d)$$

Here we introduce a total of 24 constants to be determined uniquely. Returning to equation (6-85), we may write 4 equations for ϕ at the corner nodes, 8 equations for ϕ at the midside nodes, and 4 equations for ϕ at the center node, resulting in 16 equations. We obtain 8 additional equations for the difference between the slopes of ϕ and their rates of change at the center node. They are

$$D_1 = \frac{\partial\phi^{(\alpha)}}{\partial\xi} - \frac{\partial\phi^{(\beta)}}{\partial\xi}, \qquad D_2 = \frac{\partial\phi^{(\beta)}}{\partial\eta} - \frac{\partial\phi^{(\gamma)}}{\partial\eta}$$

$$D_3 = \frac{\partial\phi^{(\gamma)}}{.\partial\xi} - \frac{\partial\phi^{(\delta)}}{\partial\xi}, \qquad D_4 = \frac{\partial\phi^{(\delta)}}{\partial\eta} - \frac{\partial\phi^{(\alpha)}}{\partial\eta}$$

$$E_1 = \frac{\partial^2\phi^{(\alpha)}}{\partial\xi^2} - \frac{\partial^2\phi^{(\beta)}}{\partial\xi^2}, \qquad E_2 = \frac{\partial^2\phi^{(\beta)}}{\partial\eta^2} - \frac{\partial^2\phi^{(\gamma)}}{\partial\eta^2}$$

$$E_3 = \frac{\partial^2\phi^{(\gamma)}}{\partial\xi^2} - \frac{\partial^2\phi^{(\delta)}}{\partial\xi^2}, \qquad E_4 = \frac{\partial^2\phi^{(\delta)}}{\partial\eta^2} - \frac{\partial^2\phi^{(\alpha)}}{\partial\eta^2} \qquad (6\text{-}86)$$

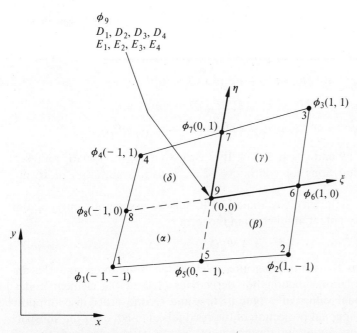

Figure 6-14 Element discontinuity method—nine-node isoparametric element.

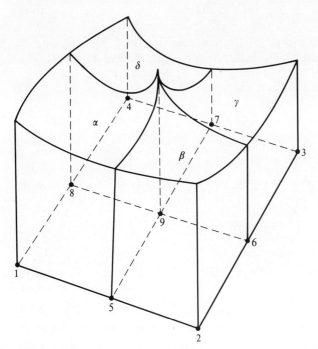

Figure 6-15 Discontinuous functional representation of velocity potential surface within finite element.

With these 24 equations now available, we solve for the unknown constants to be substituted back into (6-85). Thus we have

$$\phi^{(\alpha)} = \Phi_N^{(\alpha)}(\xi,\eta)\phi_N + F_r^{(\alpha)}(\xi,\eta)D_r + G_r^{(\alpha)}(\xi,\eta)E_r$$

$$\phi^{(\beta)} = \Phi_N^{(\beta)}(\xi,\eta)\phi_N + F_r^{(\beta)}(\xi,\eta)D_r + G_r^{(\beta)}(\xi,\eta)E_r$$

$$\phi^{(\gamma)} = \Phi_N^{(\gamma)}(\xi,\eta)\phi_N + F_r^{(\gamma)}(\xi,\eta)D_r + G_r^{(\gamma)}(\xi,\eta)E_r$$

$$\phi^{(\delta)} = \Phi_N^{(\delta)}(\xi,\eta)\phi_N + F_r^{(\delta)}(\xi,\eta)D_r + G_r^{(\delta)}(\xi,\eta)E_r$$

where $N = 1, 2, \ldots, 9$ and $r = 1, 2, 3, 4$. It is clear that ϕ_N, F_r, and G_r represent the interpolation functions for continuity of ϕ_N and discontinuities of D_r and E_r, respectively.

To obtain the finite element analog of (6-80), we assume an interpolation field for the velocity potential function in the form

$$\phi = \Psi_m Q_m \tag{6-87}$$

Here Ψ_m represents the continuous interpolation functions for ϕ and the discontinuous interpolation functions for derivatives of ϕ at the element nodes; Q_m denotes the nodal values of ϕ plus its first and second order discontinuous derivatives. An orthogonal projection of the residuals of (6-80) onto the subspace spanned by both continuous and discontinuous interpolation fields leads to

$$\int_\Omega (\phi_{,ii} - G)\Psi_m \, d\Omega = 0 \tag{6-88}$$

which is identical to (6-43) except for the new test functions Ψ_m replacing the standard interpolation functions Φ_N. Integrating by parts as in (6-44) yields

$$A_{mn}Q_n = G_m + F_m + S_m \tag{6-89}$$

where

$$A_{mn} = \int_\Omega \Psi_{m,i}\Psi_{n,i} \, d\Omega \tag{6-90}$$

$$G_m = \int_\Omega \frac{1}{a^2} \Psi_{n,i}\Psi_{p,j}\Psi_{q,i}\Psi_{m,j} \, d\Omega \, Q_n Q_p Q_q \tag{6-91}$$

$$F_m = \int_{\Gamma_1} \phi_{,i}n_i \overset{*}{\Psi}_m \, d\Gamma \tag{6-92}$$

$$S_m = -\int_{\Gamma_2} \beta_j n_j \overset{*}{\Psi}_m \, d\Gamma \tag{6-93}$$

with $\beta_j = (1/a^2)\phi_{,i}\phi_{,j}\phi_{,i}$. For the small perturbation theory, these matrices are of the form (6-45), now equipped with shock element interpolation functions instead of Φ_N. The boundary conditions are given by F_m representing the Neumann type on Γ_1, and S_m denoting the surfaces of pressure discontinuity on

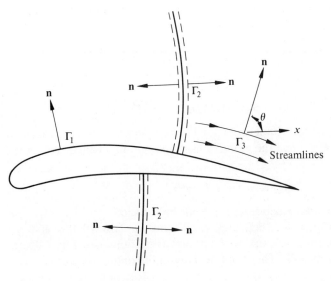

Figure 6-16 Boundary surfaces.

Γ_2 as shown in Fig. 6-16. We introduce a notation for discontinuity given by (6-93):

$$S_m = (S_m)_1 - (S_m)_2 \tag{6-94}$$

where the subscripts 1 and 2 refer to the upstream and downstream values of S_m, respectively, at the surface of pressure discontinuity.

From the element geometry (Fig. 6-14), we find that a typical 17×17 shock element influence coefficient matrix A_{mn} takes the form

$$\begin{aligned}
A_{mn} &= \int_\Omega \left(\frac{\partial \Psi_m}{\partial x} \frac{\partial \Psi_n}{\partial x} + \frac{\partial \Psi_m}{\partial y} \frac{\partial \Psi_n}{\partial y} \right) d\Omega \\
&= \int_{-1}^0 \int_{-1}^0 \frac{\partial \Psi_m^{(\alpha)}}{\partial x} \frac{\partial \Psi_n^{(\alpha)}}{\partial x} |\mathbf{J}| \, d\xi \, d\eta + \int_{-1}^0 \int_0^1 \frac{\partial \Psi_m^{(\beta)}}{\partial x} \frac{\partial \Psi_n^{(\beta)}}{\partial x} |\mathbf{J}| \, d\xi \, d\eta \\
&\quad + \int_0^1 \int_0^1 \frac{\partial \Psi_m^{(\gamma)}}{\partial x} \frac{\partial \Psi_n^{(\gamma)}}{\partial x} |\mathbf{J}| \, d\xi \, d\eta + \int_0^1 \int_{-1}^0 \frac{\partial \Psi_m^{(\delta)}}{\partial x} \frac{\partial \Psi_n^{(\delta)}}{\partial x} |\mathbf{J}| \, d\xi \, d\eta + \cdots
\end{aligned} \tag{6-95}$$

$$\frac{\partial \Psi_m^{(\alpha)}}{\partial x} = \frac{1}{|\mathbf{J}|} \left(\frac{\partial y}{\partial \eta} \frac{\partial \Psi_m^{(\alpha)}}{\partial \xi} - \frac{\partial y}{\partial \xi} \frac{\partial \Psi_m^{(\alpha)}}{\partial \eta} \right), \text{ etc., with } x = \Lambda_N x_N \text{ and } y = \Lambda_N y_N$$

Λ_N being the standard isoparametric interpolation function derived from

$$x, y = c_1 + c_2 \xi + c_3 \eta + c_4 \xi\eta + c_5 \xi^2 + c_6 \eta^2 + c_7 \xi^2\eta + c_8 \xi\eta^2 + c_9 \xi^2\eta^2$$

and

$$|\mathbf{J}| = \frac{\partial x}{\partial \xi} \frac{\partial y}{\partial \eta} - \frac{\partial x}{\partial \eta} \frac{\partial y}{\partial \xi} \tag{6-96}$$

$$\Psi_m^{(\alpha)} = [\Phi_1\Phi_2\cdots\Phi_9 F_1 F_2 F_3 F_4 G_1 G_2 G_3 G_4]^{(\alpha)}$$

$$\Psi_m^{(\beta)} = [\Phi_1\Phi_2\cdots\Phi_9 F_1 F_2 F_3 F_4 G_1 G_2 G_3 G_4]^{(\beta)}$$

$$\Psi_m^{(\gamma)} = [\Phi_1\Phi_2\cdots\Phi_9 F_1 F_2 F_3 F_4 G_1 G_2 G_3 G_4]^{(\gamma)}$$

$$\Psi_m^{(\delta)} = [\Phi_1\Phi_2\cdots\Phi_9 F_1 F_2 F_3 F_4 G_1 G_2 G_3 G_4]^{(\delta)}$$

$$Q_m = [\phi_1\phi_2\cdots\phi_9 D_1 D_2 D_3 D_4 E_1 E_2 E_3 E_4]^T$$

The Gaussian quadrature integration of the type (6-95) is also performed on the nonlinear term G_m. Thus the globally assembled finite element equations are written as

$$A_{\alpha\beta}Q_\beta = G_\alpha + F_\alpha + S_\alpha \tag{6-97}$$

Here the Neumann boundary conditions F_α can be satisfied automatically, but we require special treatment for the shock conditions S_α. Toward this end, we resort to direct applications of Rankine–Hugoniot conditions (see Fig. 6-17) to obtain an equivalent form of S_α. The Rankine–Hugoniot conditions are:

$$\rho_1 u_{n1} = \rho_2 u_{n2}$$

together with momentum normal and tangent to the wave,

$$\rho_1 u_{n1}^2 + P_1 = \rho_2 u_{n2}^2 + P_2$$

$$u_{t1} = u_{t2}$$

and the energy

$$H_1 + \tfrac{1}{2}u_1^2 = H_2 + \tfrac{1}{2}u_2^2 = \hat{H}$$

These equations, in view of the geometric relations and the appropriate equations of state, lead to [Brainerd, 1969],

$$1 - \varepsilon = A$$

with

$$A = (1 - \varepsilon_1)(1 - d)$$

$$\varepsilon = u_{n2}/u_{n1} \qquad \varepsilon_1 = \frac{\gamma - 1}{\gamma + 1} + \frac{2}{(\gamma + 1)M_1^2} = \frac{a_*^2}{u_1^2}$$

$$d = \tan^2\sigma/(M_1^2 - .1) \qquad a_*^2 = \frac{\gamma - 1}{\gamma + 1}\left(1 + \frac{2}{(\gamma - 1)M_\infty^2}\right)U^2$$

Thus the Rankine–Hugoniot condition results in

$$(1 - A)u_{n1} - u_{n2} = 0$$

or

$$\left[(1 - A)\cos\sigma - \frac{\sin\sigma}{\tan(\sigma + \delta)}\right]u_1 = 0 \tag{6-98a}$$

This can be easily recast in the form

$$q_{k\alpha}Q_\alpha = 0 \tag{6-98b}$$

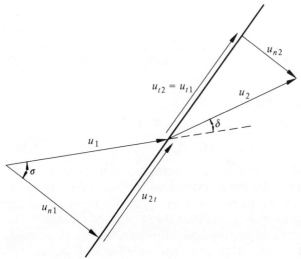

Figure 6-17 Oblique shock wave.

Here u_1 is the resultant local velocity upstream of the oblique shock. The shock angle σ is determined from the discontinuity values D_r at the element center. Note that the downstream velocity u_2 is eliminated through the deflection angle δ, which is determined between the tangential velocity u_{t2} and the downstream velocity u_2. The shock boundary condition matrix (6-98b) is now equivalent to (6-94). The quantity $q_{k\alpha}$ is called the shock boundary matrix with k denoting the number of shock elements.

To enforce the shock boundary condition (6-98b), we make use of Lagrange multipliers λ_k such that

$$\lambda_k q_{k\alpha} Q_\alpha = 0 \tag{6-99}$$

To replace S_α in (6-98) through the Lagrange multipliers, we construct an energy functional in quadratic form

$$\chi = \tfrac{1}{2} A_{\alpha\beta} Q_\alpha Q_\beta - R_\alpha Q_\alpha \tag{6-100}$$

where $R_\alpha = G_\alpha + F_\alpha$. Adding (6-99) to (6-100), we obtain

$$\chi = \tfrac{1}{2} A_{\alpha\beta} Q_\alpha Q_\beta - R_\alpha Q_\alpha + \lambda_k q_{k\alpha} Q_\alpha \tag{6-101}$$

Our objective is to seek an extremum of χ for the flow field satisfying the jump conditions with respect to every Q_α and λ_k

$$\delta\chi = \frac{\partial\chi}{\partial Q_\alpha} \delta Q_\alpha + \frac{\partial\chi}{\partial\lambda_k} \delta\lambda_k = 0$$

For all arbitrary values of δQ_α and $\delta\chi_k$, we must have $\partial\chi/\partial Q_\alpha = 0$ and $\partial\chi/\partial\lambda_k = 0$, leading to

$$\begin{bmatrix} A_{\alpha\beta} & q_{k\alpha} \\ q_{k\beta} & 0 \end{bmatrix} \begin{bmatrix} Q_\beta \\ \lambda_k \end{bmatrix} = \begin{bmatrix} R_\alpha \\ 0 \end{bmatrix} \tag{6-102}$$

Writing (6-102) in a compact form

$$B_{ij}X_j = Y_i \qquad (6\text{-}103)$$

and noting that Y_i contains R_α which is nonlinear, we may cast (6-103) in an iterative form

$$B_{ij}X_j^{(n+1)} = Y_i^{(n)}$$

Initially, we consider a shockless domain, with $R_\alpha = 0$:

$$A_{\alpha\beta}Q_\beta = 0 \qquad (6\text{-}104)$$

If the solution (6-104) yields Q_β indicating nonzero D_r, then (6-104) will be replaced by the expression containing the jump condition (6-102) in the subsequent iterations. In this iterative scheme R_α will be kept updated. The magnitude of D_r calculated at the center of an element signifies the strength of the shock. The direction of the shock is determined by connecting the centers of shock elements ($D_r \neq 0$). For multiple occurrences, interactions, and reflections of shocks, it is advisable to start with the shock element influence coefficients applied to the entire domain. At each iterative cycle, we remove the shock boundary matrix $q_{k\alpha}$ when D_r is found to be zero,

$$A_{\alpha\beta}Q_\beta = R_\alpha$$

In the interest of simplified iterative calculations, the expression (6-88) may be written in a global form as

$$\int_\Omega \left[\phi_{,ii} - \left(\frac{q}{a} \right)^2 \phi_{,ii} \right] \Psi_\alpha \, d\Omega = 0 \qquad (6\text{-}105a)$$

This leads to

$$A_{\alpha\beta}Q_\beta = G_\alpha + F_\alpha + S_\alpha \qquad (6\text{-}105b)$$

where

$$G_\alpha = K^2 \left[\int_\Omega \Psi_{\alpha,i}\Psi_{\beta,i} \, d\Omega \right] Q_\beta$$

$$S_\alpha = K^2 \int_\Gamma \Psi_{,i} n_i \overset{*}{\Psi}_\alpha \, d\Gamma$$

$$K^2 = \left(\frac{q}{a} \right)^2$$

with other quantities being the same as in (6-97). Notice that the quantity K^2 is to be held constant during each iterative cycle and updated for the following cycles. Had we used the Bateman principle with $\gamma = 2$, then we have $K^2 = (q/\hat{q})^2$ with

$$\hat{q} = \frac{2a_\infty^2}{\gamma - 1} + q_\infty^2$$

Once again, the shock surface boundary condition S_z is replaced by the Rankine-Hugoniot condition through the Lagrange multipliers (6-98). Otherwise, the finite element equations (6-105) are identical to those derived from (6-50) or (6-52).

If the entropy and enthalpy gradients are nonvanishing normal to the streamlines, then we need to modify (6-88) in the form

$$\int_\Omega (\nabla^2\phi - G - E)\Psi_m \, d\Omega = 0 \qquad (6\text{-}106)$$

with E given by (6-40)

$$
\begin{aligned}
E &= \frac{uv}{\varepsilon_{ij} V_j n_i} \left(\frac{1}{\gamma R} \eta_i n_i - \frac{1}{a^2} \hat{H}_{,i} n_i \right) \\
&= \frac{uv}{u\cos\theta - v\sin\theta} \left\{ \left(\frac{c_v}{\gamma RP} - \frac{\gamma}{a^2(\gamma-1)\rho} \right) \left(\frac{\partial P}{\partial x} \cos\theta + \frac{\partial P}{\partial y} \sin\theta \right) \right. \\
&\quad + \left(\frac{\gamma P}{a^2(\gamma-1)} - \frac{c_v}{R\rho} \right) \left(\frac{\partial \rho}{\partial x} \cos\theta + \frac{\partial \rho}{\partial y} \sin\theta \right) - \frac{1}{a^2} \left[u \left(\frac{\partial u}{\partial x} \cos\theta + \frac{\partial u}{\partial y} \sin\theta \right) \right. \\
&\quad \left. \left. + v \left(\frac{\partial v}{\partial x} \cos\theta + \frac{\partial v}{\partial y} \sin\theta \right) \right] \right\}
\end{aligned}
\qquad (6\text{-}107)
$$

where the angle θ is measured between the x axis and the direction normal to the streamlines. In this case, the velocity vector is given by

$$\mathbf{V} = \nabla\phi + \nabla \times \boldsymbol{\phi}$$

Here ϕ is the vector potential. This situation is represented by $E \neq 0$ or the entropy and enthalpy gradient vector

$$E_m = \int_{\Gamma_3} E\overset{*}{\Psi}_m \, d\Gamma \neq 0$$

which is added to (6-89)

$$A_{mn}Q_n = G_m + F_m + E_m + S_m \qquad (6\text{-}108)$$

The entropy and enthalpy gradient vector is calculated from the relation between the pressure-density and the local velocity distribution, lagging one step behind the iterative solution. It is observed that the enthalpy gradients in general are negligible. Direction cosines are determined from the angles between the x axis and the direction normal to the streamline which corresponds to the equipotential line.

The 17×17 element coefficient matrix is rather large in size and it may be advantageous to condense it to a smaller size via standard condensation schemes. We also note that the Gaussian quadrature integration must be performed within each element requiring the half interval. An alternative approach is to provide an independent interpolation field for each quadrant so that the Gaussian integration limits run -1 to 1. In this scheme, we provide at the center node two first derivatives of ϕ for each quadrant, $\partial\phi^{(\alpha)}/\partial\xi$ and $\partial\phi^{(\alpha)}/\partial\eta$, etc. Together with four nodal ϕ values, there are six generalized coordinates for each quadrant. Thus the interpolation fields are modified to

$$\phi^{(\alpha)} = \Phi_1^{(\alpha)}\phi_1 + \Phi_5^{(\alpha)}\phi_5 + \Phi_8^{(\alpha)}\phi_8 + \Phi_9^{(\alpha)}\phi_9 + F_\xi^{(\alpha)}\left(\frac{\partial\phi}{\partial\xi}\right)_9^{(\alpha)} + F_\eta^{(\alpha)}\left(\frac{\partial\phi}{\partial\eta}\right)_9^{(\alpha)}$$

$$\phi^{(\beta)} = \Phi_2^{(\beta)}\phi_2 + \Phi_5^{(\beta)}\phi_5 + \Phi_6^{(\beta)}\phi_6 + \Phi_9^{(\beta)}\phi_9 + F_\xi^{(\beta)}\left(\frac{\partial\phi}{\partial\xi}\right)_9^{(\beta)} + F_\eta^{(\beta)}\left(\frac{\partial\phi}{\partial\eta}\right)_9^{(\beta)}$$

$$\phi^{(\gamma)} = \Phi_3^{(\gamma)}\phi_3 + \Phi_6^{(\gamma)}\phi_6 + \Phi_7^{(\gamma)}\phi_7 + \Phi_9^{(\gamma)}\phi_9 + F_\xi^{(\gamma)}\left(\frac{\partial\phi}{\partial\xi}\right)_9^{(\gamma)} + F_\eta^{(\gamma)}\left(\frac{\partial\phi}{\partial\eta}\right)_9^{(\gamma)}$$

$$\phi^{(\delta)} = \Phi_4^{(\delta)}\phi_4 + \Phi_7^{(\delta)}\phi_7 + \Phi_8^{(\delta)}\phi_8 + \Phi_9^{(\delta)}\phi_9 + F_\xi^{(\delta)}\left(\frac{\partial\phi}{\partial\xi}\right)_9^{(\delta)} + F_\eta^{(\delta)}\left(\frac{\partial\phi}{\partial\eta}\right)_9^{(\delta)} \tag{6-109}$$

Now each quadrant can be treated as an independent element having the size 6×6. In the assembly, ϕ is common to all quadrants at the center node whereas all the second derivative terms remain as generalized coordinates, independent of adjacent quadrants. In this approach, although the shock strengths are not directly calculated, the definitions given in (6-86) enable us to examine the existence of the shocks and their strengths.

Other Approaches The first finite element applications to transonic flows are those by Chan, Brashears, and Young [1975] who employed the least squares in conjunction with the Galerkin approach using cubic triangular elements. No shock boundary treatment was implemented. Chan and Brashears [1975] further explored their method to include an unsteady state. Small perturbation theory is utilized in their studies.

Subsequently, Glowinski, Periaux, and Pironneau [1976] studied the transonic flow by optimal control [Polak, 1971] of distributed parameter systems, discretized in finite elements. The method of steepest descent was used to solve the system of governing finite element equations.

Wellford and Hafez [1976] examined a mixed variational principle for small disturbance transonic flow. The interpolation functions for the velocity potential and the velocity were introduced in the minimization of the mixed variational principle, resulting in two sets of equations, which may be solved iteratively.

Ecer and Akay [1976] studied the solution of full potential equation and reported difficulty locating a shock line but the accuracy of the results were otherwise good. In their formulation, the Bateman principle leading to (6-50b) was used. The resulting finite element equations are similar to (6-105b) but no shock element formulation (6-85, 6-98, 6-102) was considered in their study.

6-5-3 Unsteady Transonic Flow

For unsteady conditions, we invoke the equation of the form (6-82) and its finite element analog,

$$R_{NM}\ddot{\phi}_M + S_{NM}\dot{\phi}_M - A_{NM}\phi_M = F_N + G_N + S_N \qquad (6\text{-}110)$$

with

$$R_{NM} = \int_\Omega M_\infty^2 \Phi_N \Phi_M \, d\Omega$$

$$S_{NM} = \int_\Omega 2M_\infty^2 \Phi_N \frac{\partial \Phi_M}{\partial x} \, d\Omega$$

For oscillating airfoil problems, we return to (6-84) with an interpolation field of the form

$$\bar{\phi} = \Phi_N \bar{\phi}_N \qquad (6\text{-}111)$$

$$\bar{\phi}_N = \bar{\phi}_{N(R)} + i\bar{\phi}_{N(I)} \qquad (6\text{-}112)$$

where Φ_N is the interpolation function and $\bar{\phi}_{N(R)}$ and $\bar{\phi}_{N(I)}$ are the real and imaginary parts of the nodal values of $\bar{\phi}$. Substituting (6-111) into (6-84), we obtain two sets of equations similar to (6-110) but one for real part and another for imaginary part. These equations will be solved independently to obtain the nodal values of ϕ from (6-112).

In general, the imposition of boundary conditions for the finite element solutions is simpler than in the finite difference method. In the flight conditions of an airfoil, the exterior boundary is infinite. However, we must use a finite geometry along which the far field analytical solutions [Klunker, 1971; Murman and Cole, 1971] may be matched iteratively. On the solid boundary of a body, we require that the velocity normal to the surface be zero; that is

$$V_i n_i = 0$$

or

$$\frac{\partial \phi}{\partial n} = \phi_{,i} n_i = 0$$

Note also that at infinity, the perturbed velocity is zero,

$$\nabla \phi = \phi_{,i} = 0$$

These boundary conditions are applied in exactly the same manner as in incompressible flow discussed in Chap. 5. Refer to Fig. 5-1 for details of imposition of Dirichlet and Neumann boundary conditions.

For unsteady flows, we must satisfy additional boundary conditions, namely, the unsteady Kutter conditions; that is, the pressure across the trailing vortex sheet must be equal (Fig. 6-18),

$$C_p^+ = C_p^- \qquad (6\text{-}113)$$

where

$$C_p = -\frac{2}{U}\left[\frac{\partial \phi}{\partial t} + \frac{\partial \phi}{\partial x_1} + \frac{e}{2U}\left(\frac{\partial \phi}{\partial x_2}\right)^2\right] \qquad (6\text{-}114)$$

Thus, in terms of potential function, the unsteady Kutter conditions are

$$\left[\frac{\partial \phi}{\partial t} + \frac{\partial \phi}{\partial x_1} + \frac{e}{2U}\left(\frac{\partial \phi}{\partial x_2}\right)^2\right]^+ = \left[\frac{\partial \phi}{\partial t} + \frac{\partial \phi}{\partial x_1} + \frac{e}{2U}\left(\frac{\partial \phi}{\partial x_2}\right)^2\right]^- \qquad (6\text{-}115)$$

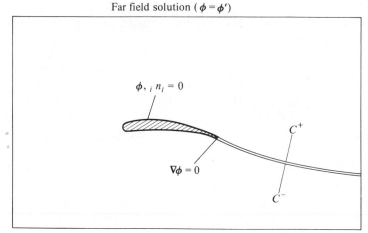

Figure 6-18 Boundary conditions.

The above conditions must be applied along the upper and lower surfaces of the mean wake position.

If the body is oscillating, then we require that velocity on the surface must be tangent. Let the position of the body geometry at any instant be defined as

$$B(x, y, z, t) = 0 \tag{6-116}$$

Then the surface tangency condition can be satisfied by requiring that the material derivative of the function B with respect to time be equal to zero [Landahl, 1961]

$$\frac{\partial B}{\partial t} + \frac{\partial B}{\partial x} + \frac{\partial \phi}{\partial x}\frac{\partial B}{\partial x} + \frac{\partial \phi}{\partial y}\frac{\partial B}{\partial y} + \frac{\partial \phi}{\partial z}\frac{\partial B}{\partial z} = 0 \quad \text{on } B = 0 \tag{6-117}$$

If the body is thin, we may approximate as

$$\frac{\partial B}{\partial t} + \frac{\partial B}{\partial x} + \frac{\partial \phi}{\partial y}\frac{\partial B}{\partial y} + \frac{\partial \phi}{\partial z}\frac{\partial B}{\partial z} = 0 \tag{6-118}$$

For a two-dimensional flow $(z = 0)$, if we let $B(x, y, t) = y - A(x, y, t)$, $A \ll 1$, $\partial A/\partial y \ll 1$

$$\frac{\partial \phi}{\partial y} = \frac{\partial A}{\partial x} + \frac{\partial A}{\partial t} \tag{6-119}$$

where the function A is taken as

$$A(x, y, t) = ah(x) + bm(x, t) \tag{6-120}$$

with a the airfoil thickness ratio $h(x)$ the thickness distribution along the axis, b the oscillation amplitude, and $m(x, t)$ the airfoil oscillation function.

In addition to these special boundary conditions (Kutter condition and oscillating body geometry), we must satisfy the standard boundary conditions of Dirichlet and Neumann types in the usual manner.

6-5-4 Example Problems

We consider here a missile consisting of 4 caliber tangent ogive and 9 caliber afterbody, which was studied experimentally in a wind tunnel by Spring [1973]. A typical finite element configuration is shown in Fig. 6-19. Note that the finite elements tangent to the body are installed with two sides constructed normal to the body. If the first layer is made very thin, then the satisfaction of Neumann boundary conditions will be demonstrated by approximately equal values of velocity potential between the two adjacent nodes in the direction normal to the body.

With the use of trial functions given by (6-109), $\gamma = 1.4$, angle of attack $= 0°$, the computed results (Fig. 6-20) for $M = 0.85$, 0.90, and 0.95 are compared with the wind tunnel data of Spring [Chung and Hooks, 1977]. In this study, an average of several cycles of iteration was required for convergence. For low Mach numbers, the shock boundary matrix vanishes, but tends to nonzero when the Mach number increases, indicating the formation of shock waves.

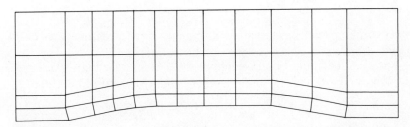

Figure 6-19 Finite element discretization of wind tunnel study (Spring, 1973): 4 caliber tangent ogive and 9 caliber afterbody.

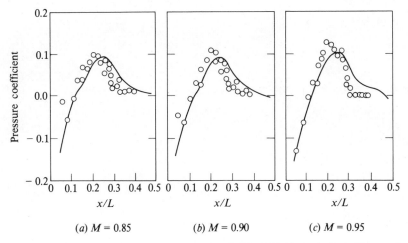

(a) $M = 0.85$ (b) $M = 0.90$ (c) $M = 0.95$

Figure 6-20 Pressure coefficients for the problem in Fig. 6-19: circles indicate the experimental results of Spring (1973).

In recent years, various investigators have reported finite element applications to transonic flow problems. These include Chan, et al. [1975], Wellford and Hafez [1976], Eccer and Akay [1976], and Glowinski, Periaux, and Pironneau [1976], among others, who used various approaches to transonic flow as briefly reviewed in Sec. 6-5-2. For the results of their studies, the reader should consult the literature cited above.

In conclusion, applications of finite elements to transonic flows, although still in the state of infancy, appear to be promising. Mathematical convergence rates as related to appropriate trial and test functions, treatment of shock interactions in both two- and three-dimensions, and various other features involved in transonic flows must await future studies.

MISCELLANEOUS FLOW PROBLEMS

7-1 GENERAL

In the foregoing chapters, we examined incompressible and compressible fluids in general in an attempt to establish generality rather than to show each specific detail in fluid dynamics applications.

Specific applications reported in the literature include seepage problems [Zienkiewicz, Mayer, and Cheung, 1966; Desai, 1975], lubrication [Argyris and Scharpf, 1969; Huebner, 1975], biomechanics problems [Tong and Fung, 1971], and liquid sloshing in elastic containers [Chung and Rush, 1976], among others. Additional applications are in the areas of diffusion, magnetohydrodynamics, and rarefied gas dynamics. These topics are detailed in the following sections.

7-2 DIFFUSION

7-2-1 Diffusion Equations

A process of diffusion refers to an equalization of concentrations within a single phase. We are concerned with the rate of flow of the diffusing substance in nonporous media. The concentration gradient is responsible for this flow. In general, there are two or more substances involved and, therefore, we will have more than one diffusion equation. Let J be defined as the quantity of substance per unit area per unit time. If x is the coordinate chosen perpendicular to the reference surface and c the concentration of the diffusing substance given as amount of substance per cubic centimeter, then Fick's first law of diffusion is of the form

$$J = -D\frac{\partial c}{\partial x} \tag{7-1}$$

where D is the coefficient of diffusion in square centimeters per second (cm^2/s). J may be considered as diffusion current of a vectorial quantity so that

$$\mathbf{J} = -D\nabla c \tag{7-2a}$$

or

$$J_i = -Dc_{,i} \tag{7-2b}$$

with $i = 1, 2, 3$ (in one-, two-, or three-dimensional space, respectively). In the case of transient state with time t, it can easily be shown that

$$\frac{\partial c}{\partial t} = D\frac{\partial^2 c}{\partial x^2} \tag{7-3}$$

which is called Fick's second law of diffusion. Here D is considered as constant. For multidimensional diffusion, (7-3) assumes the form

$$\frac{\partial c}{\partial t} = D\nabla^2 c = Dc_{,ii} \tag{7-4}$$

In case D is not constant, we revise (7-4) as

$$\frac{\partial c}{\partial t} = \nabla \cdot (D\nabla c) = (Dc_{,i})_{,i} \tag{7-5}$$

If D is not explicitly dependent on position, then we have

$$\frac{\partial c}{\partial t} = Dc_{,ii} + \frac{\partial D}{\partial c} c_{,i}c_{,i} \tag{7-6}$$

For anisotropic substance we may write

$$\frac{\partial c}{\partial t} = D_{xx}\frac{\partial^2 c}{\partial x^2} + D_{yy}\frac{\partial^2 c}{\partial y^2} + D_{zz}\frac{\partial^2 c}{\partial z^2} \tag{7-7}$$

where D_{xx}, D_{yy}, and D_{zz} refer to the diffusion coefficients in the x, y, and z directions, respectively.

Now let us consider concentration changes due to diffusion without the influence of external forces. In one-dimensional problems, we have

$$\frac{\partial c}{\partial t} = D\frac{\partial^2 c}{\partial x^2} - v\frac{\partial c}{\partial x} \tag{7-8}$$

with v the convection velocity. For multidimensional problems, we obtain

$$\frac{\partial c}{\partial t} = D\nabla^2 c - \mathbf{v} \cdot \nabla c \tag{7-9a}$$

or
$$\frac{\partial c}{\partial t} = Dc_{,ii} - v_i c_{,i} \tag{7-9b}$$

If there is an external force f acting in the x direction upon the dissolved molecules or colloidal particles or upon gas molecules of a certain kind, with m being the mobility of the particles under consideration, the resulting flux of matter becomes

$$cfm = cv$$

Thus the rate of concentration with time due to this current is

$$\frac{\partial c}{\partial t} = D \frac{\partial^2 c}{\partial x^2} - \frac{\partial}{\partial x}(cv) \tag{7-10}$$

For a multidimensional case we may write

$$\frac{\partial c}{\partial t} = Dc_{,ii} - (cv_i)_{,i} \tag{7-11}$$

If the rate of change of concentration of a single component is influenced partly due to diffusion and partly due to chemical reaction, the governing equation in one dimension assumes the form

$$\frac{\partial c}{\partial t} = D \frac{\partial^2 c}{\partial x^2} + f(c) \tag{7-12}$$

where $f(c)$ is the law of reaction rate and may depend on the concentrations of other components and on external influences such as illumination.

7-2-2 Finite Element Equations

The formulation of finite element equations may be based on equations (7-4), (7-6), (7-7), (7-9b), or (7-11) for multidimensional problems. For the purpose of illustration, let us consider the equation governed by (7-4)

$$\dot{c} - Dc_{,ii} = 0$$

with the concentration c approximated by

$$c = \Phi_N c_N$$

The local finite element equation is obtained as

$$\int_\Omega (\dot{c} - Dc_{,ii}) \Phi_N \, d\Omega = 0$$

or
$$A_{NM} \dot{c}_N + B_{NM} c_M = f_N \tag{7-13}$$

where
$$A_{NM} = \int_\Omega \Phi_N \Phi_M \, d\Omega \tag{7-14}$$

$$B_{NM} = \int_{\Omega} D\Phi_{N,i}\Phi_{M,i}\, d\Omega \tag{7-15}$$

$$f_N = \int_{\Gamma} Dc_{,i}n_i\Phi_N\, d\Gamma = \int_{\Gamma} J_i n_i \Phi_N\, d\Gamma \tag{7-16}$$

It is seen that the input flux at the boundaries is given by f_N (7-16). We may set $J_i n_i = J_0$ so that

$$f_N = \int_{\Omega} J_0 \Phi_N\, d\Omega$$

If the diffusion is governed by (7-9b), we have

$$\int_{\Omega} (\dot{c} - Dc_{,ii} + v_i c_{,i})\Phi_N\, d\Omega = 0 \tag{7-17a}$$

or

$$A_{NM}\dot{c}_M + B_{NM}c_M + E_{NM}c_M = f_N$$

where

$$E_{NM} = \int_{\Omega} v_i \Phi_N \Phi_{M,i}\, d\Omega$$

with other quantities the same as before. Here the steady velocity v_i is determined by the product of the external force f_i and the mobility of the particles under consideration.

Notice that in (7-17a), the standard test functions Φ_N identical to the trial functions are used. However, it is shown in Sec. 5-12 that the presence of the convective term contributes to an instability or oscillation in solutions with refinement of mesh sizes. It is also pointed out that the upwind finite elements can remedy this difficulty and produce a monotonic convergence. This can be accomplished by using special test functions of the order higher than the trial functions. In this case, we write

$$\int_{\Omega} (\dot{c} - Dc_{,ii} + v_i c_{,i})W_N\, d\Omega = 0 \tag{7-17b}$$

with $c = \Phi_N c_N$ where Φ_N may be considered as the standard linear interpolation functions. The test functions W_N then consist of

$$W_N = \Phi_N + \Psi_N \tag{7-18}$$

where Ψ_N for a one-dimensional case can be given by (5-203d),

$$\Psi_1 = \frac{3\alpha}{h^2} x(x - h) \qquad \Psi_2 = -\Psi_1$$

For two-dimensional isoparametric elements, Heinrich, et al. [1977], suggest that

$$W_N = [\Phi_N(\xi) + \Psi_N(\xi)][\Phi_N(\eta) + \Psi_N(\eta)] \tag{7-19}$$

$$\Psi_1(\xi) = \frac{3\alpha}{4} (\xi^2 - 1) = \Psi_4(\xi) = -\Psi_2(\xi) = -\Psi_3(\xi)$$

$$\Psi_1(\eta) = \frac{3\beta}{4} (\eta^2 - 1) = \Psi_2(\eta) = -\Psi_3(\eta) = -\Psi_4(\eta)$$

with signs of α and β coinciding with those of the velocity at a given node. The various matrices in (7-17b) are now given by

$$A_{NM} = \int_\Omega W_N \Phi_M \, d\Omega \qquad B_{NM} = \int_\Omega D W_{N,i} \Phi_{M,i} \, d\Omega \qquad E_{NM} = \int_\Omega v_i W_N \Phi_{M,i} \, d\Omega$$

7-2-3 Example Problems

Example 7-1 *Diffusion without convective term* For simplicity, let us first examine a one-dimensional diffusion problem with the convective term. The analytical solution for a cylinder of semi-infinite length and unit cross section [Jost, 1960] is

$$c = \frac{c_0}{\sqrt{4\pi Dt}} \exp\left(-\frac{x^2}{4Dt}\right) \tag{7-20}$$

where c_0 is the amount of substance placed at $x = 0$ that undergoes diffusion. In practical applications, however, we may be interested in the finite domain, or in holding the substance c_0 constant over a period of time, or taking the substance removed at a certain designated time. These features may be incorporated in the finite element solutions. To illustrate, we use one-dimensional domain of 10 elements with 11 nodes as shown in Fig. 7-1. The diffusion coefficient D is set equal to 0.1. The initial and boundary conditions are $c_0 = 10^{-5}$ at $x = 1$ and $\partial c/\partial x = 0$ at $x = 10$. The nodes adjacent to both ends are more closely spaced to better monitor the output data. The temporal operator of (3-54a) with $\theta = 0$ (implicit scheme) is used in this example. The optimum Δt is found to be $1 \text{ s} \le \Delta t \le 10 \text{ s}$ with oscillations occurring at $\Delta t < 1$ s. For $\Delta t > 10$ s, however, earlier histories become distorted. The input concentration is removed after the first time increment. The results shown in Fig. 7-1a indicate the general trend as seen in (7-20) for the infinite cylinder, but the effect of finite dimension is evident in this example. The steady state condition is reached at approximately $t = 500$ s. This is confirmed by several Δt values of 1, 10, and 100 s.

Figure 7-1b shows the results for the input concentration remaining indefinitely instead of being taken off after the first time increment. The steady state condition is reached at approximately 2,000 s. Solutions for $\Delta t = 10$ s lag slightly behind those for $\Delta t = 1$ s, possibly due to distortion of histories in the early stages, but they are quickly caught up as time progresses.

For a given type of temporal operator, the choice of Δt may be quite sensitive. The general guidelines for determining the limiting values of Δt

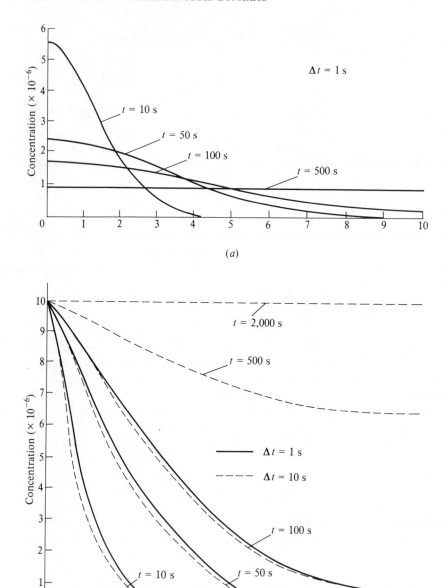

Figure 7-1 One-dimensional diffusion without convective term. Nodes at $x = 0$, 0.1, 0.5, 1, 2, 3, 4.5, 6.5, 8, 9.5, 10; $D = 0.1$, linear interpolation functions. (a) $c_0 = 10^{-5}$ removed after the first time increment; (b) $c_0 = 10^{-5}$ remaining indefinitely.

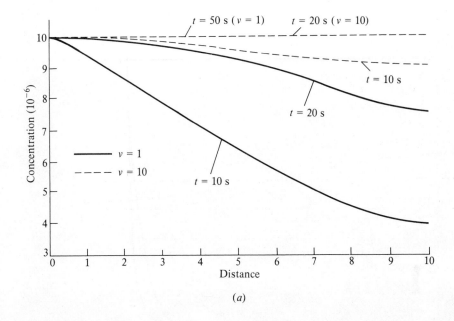

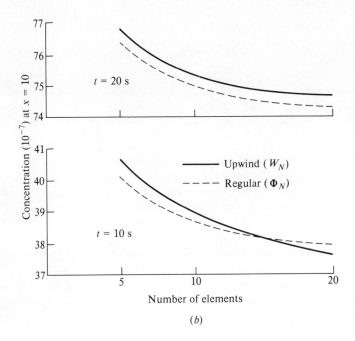

Figure 7-2 Convective-diffusion problem, 10 linear elements with equal length, $\Delta t = 10$ s, $D = 0.1$, implicit scheme (Eq. 3-56, $\theta = 0$). (a) Upwind finite elements with test function W_N; (b) convergence with mesh refinement and test functions Φ_N and W_N, $v = 1$.

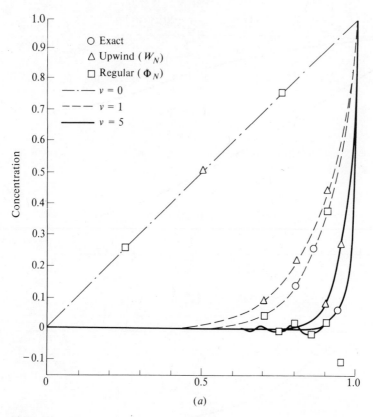

Figure 7-3 Convective-diffusion problem, $c(0) = 0$, $c(1) = 1$, $D = 0.1$, implicit scheme (Eq. 3-56, $\theta = 0$), $\Delta t = 10$ s. (*a*) Concentration at $t = 2000$ s, 20 equal elements; (*b*) convergence with mesh refinement $v = 5$ at $t = 2000$ s.

as given in Sec. 3-3-3 should be followed. It should be noted that in some cases, the accuracy deteriorates with larger values of Δt whereas the reverse is true for other cases. This is because the limiting values of Δt as they appear in the amplification matrix are influenced by the physical and geometrical parameters used in the finite element formulation.

Example 7-2 *Convective-diffusion* (*Neumann–Dirichlet problem*) The purpose of this example is to demonstrate the effect of the convective term characterized by (7-17a, b). Initial and boundary conditions are the same as in Example 7-1. With the linear trial functions Φ_N and the quadratic test functions W_N, the various matrices are determined†

$$A_{NM} = \frac{h}{12}\begin{bmatrix} 1 & -1 \\ 5 & 7 \end{bmatrix} \qquad B_{NM} = \frac{D}{h}\begin{bmatrix} 1 & -1 \\ -1 & 1 \end{bmatrix} \qquad C_{NM} = v\begin{bmatrix} 0 & 0 \\ -1 & 1 \end{bmatrix}$$

† For upwind difference we set $\alpha = 1$ in (7-19).

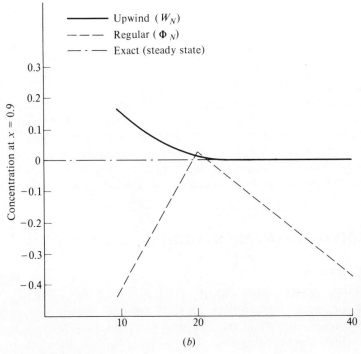

Figure 7-3 (continued)

The solution with 10 equal elements, $c_0 = 10^{-5}$ at $x = 0$ for $0 \leq t \leq \infty$, and $\partial c/\partial x = 0$ at $x = 10$ is shown in Fig. 7-2a. Notice that the higher convection velocity quickly pushes the response to the steady state condition. The convergence characteristics of solutions with mesh refinements using test functions Φ_N and W_N are shown in Fig. 7-2b. The effectiveness of the upwind finite elements using the test functions W_N as described in Sec. 5-12 is not obvious for the boundary conditions considered in this example. The merit of upwind finite elements is, however, associated with Dirichlet problems and examined below.

Example 7-3 *Convective-diffusion (two-point boundary value problem)* We consider the boundary conditions $c(0) = 0$ and $c(1) = 1$ in the solution of both (7-17a) and (7-17b). Although the steady state condition is reached approximately at $t = 50$ s, the time integration is carried out to 2,000 s (Fig. 7-3a) to assure its complete steadiness. The exact solution for the steady state is

$$c = \frac{\exp(Px) - 1}{\exp(P) - 1}$$

where $P = v/D$ is the Peclet number. For $v = 1$ ($P = 10$), the solution for the regular finite element (Φ_N) is almost identical to the exact steady state solution but oscillates violently when v is increased to $v = 5$ ($P = 50$). A glance at Fig. 7-3*b* reveals that the upwind finite element (W_N) yields a monotonic convergence to the exact solution whereas the regular finite element fails miserably.

Remarks The notion of upwind finite elements has been studied by Heinrich, et al. [1977], among others, and its effectiveness is demonstrated here for time-dependent convective diffusion problems associated with Dirichlet boundary conditions. A solution instability (or nonconvergence) occurs, in general, due to a combined effect of Δt, mesh refinement, and test functions.

An extension to two-dimensional problems can be made similarly, using the test functions (7-18) with (7-19). For details, see Heinrich, et al. [1977].

7-3 MAGNETOHYDRODYNAMICS (MHD)

7-3-1 Governing Equations

Magnetohydrodynamics (MHD) deals with the motion of a highly conducting fluid in the presence of a magnetic field. Such a motion generates electric currents which change the magnetic field, and the disturbed field in turn gives rise to mechanical forces which affect the flow field. This coupling between the electromagnetic and mechanical forces then characterizes hydromagnetic phenomena. Celestial bodies which contain large conducting masses are known to exhibit pronounced hydromagnetic phenomena.

The electromagnetic field is produced by a distribution of electric current and charge. The motion of charge constitutes a current which is determined by the magnitude of the charge and its velocity. The current density at a point is defined as the vector $\mathbf{J}$ by the equation

$$\mathbf{J} = \rho \mathbf{v} \tag{7-21}$$

where ρ is the charge density and $\mathbf{v}$ the velocity vector. It follows that in metals and valves, where the electricity is carried by electrons which are negatively charged, the direction of the current density vector is opposite to that of the moving electrons.

The current I across a surface is defined to be the rate at which a charge crosses that surface. Since a charge can cross S only by virtue of its velocity normal to S, we have

$$I = \mathbf{J} \cdot \mathbf{n} \, dS \tag{7-22}$$

where $\mathbf{n}$ is a unit vector normal to S.

Consider now two isolated charges q and q_1 moving in free space. The charge q is acted on by certain electrical forces due to q_1. If q is at rest, the electrical force here is $q\mathbf{E}$. The vector $\mathbf{E}$ is called the electric intensity. If q is moving with velocity $\mathbf{v}$, there is an additional force $q\mathbf{v} \times \mathbf{B}$ where the vector $\mathbf{B}$ is called the magnetic flux density. Two other vectors play a role in specifying the electro-magnetic field and they are related to the lines of force which emanate from charges and currents. The vector **D**, which is called the electric flux density, effectively measures the number of lines of force which originate from a charge. The vector **H**, which is called the magnetic intensity, is such that its value on a closed curve effectively measures the current which passes through the curve.

We shall assume that the vectors **E**, **B**, **D**, and **H** are continuous and possess continuous derivatives at ordinary points at which Maxwell's equations

$$\text{curl } \mathbf{E} + \frac{\partial \mathbf{B}}{\partial t} = 0 \tag{7-23a}$$

$$\text{curl } \mathbf{H} + \frac{\partial \mathbf{D}}{\partial t} = \mathbf{J} \tag{7-23b}$$

$$\text{div } \mathbf{D} = \rho \tag{7-23c}$$

$$\text{div } \mathbf{B} = 0 \tag{7-23d}$$

are satisfied.

Since the divergence of the curl of any vector vanishes identically, we obtain, by taking the divergence of (7-23b)

$$\text{div } \mathbf{J} = -\text{div } \frac{\partial \mathbf{D}}{\partial t} = -\frac{\partial}{\partial t}(\text{div } \mathbf{D}) \tag{7-24}$$

The interchange of operators div and $\partial/\partial t$ is permissible because **D** is assumed to be continuous. Substitution of (7-23b) into (7-24) gives

$$\text{div } \mathbf{J} + \frac{\partial \rho}{\partial t} = 0 \tag{7-25}$$

By analogy with a corresponding equation in hydrodynamics, (7-25) is called the equation of continuity.

In a field of infinite electrical conductivity, the fluid particles are tied to the lines of force of the magnetic field so that the lines of force may be thought of as possessing inertia, the mass per unit length being equal to the density of the fluid ρ.

To describe the magnetohydrodynamic behavior, we must have: 1. the mechanical equations embodying the effect of the electromagnetic forces as well as other forces on the motion, 2. the equation at continuity, 3. the equation of heat transport, and 4. the equation of state as well as the Maxwell equations given in Eq. (7-23).

Consider an incompressible viscous fluid in motion in which the only body forces are gravity and electromagnetic forces. The equation of motion can be written as

$$\rho \frac{\partial \mathbf{V}}{\partial t} + \rho(\mathbf{V} \cdot \nabla)\mathbf{V} = -\nabla \rho + \rho \mathbf{g} + \mathbf{J} \times \mathbf{B} + q\mathbf{E} + \rho\mu\{\nabla^2 \mathbf{V} + \tfrac{1}{3}\nabla(\nabla \cdot \mathbf{V})\} \tag{7-26}$$

in which

$$\mathbf{J} \times \mathbf{B} = (\nabla \times \mathbf{H}) \times \mathbf{B} \qquad (7\text{-}27)$$

and the term $q\mathbf{E}$ arising from the action of the electrical field on the resultant space charge is negligible compared with $\mathbf{J} \times \mathbf{B}$. The mechanical force $\mathbf{J} \times \mathbf{B}$ due to the action of the magnetic field on the electric current is perpendicular to the magnetic field and so will not affect the motion of the fluid along the lines of force.

The equations of continuity, heat conduction, and state are of the form

$$\frac{\partial \rho}{\partial t} + \rho(\nabla \cdot \mathbf{V}) = 0 \qquad (7\text{-}28)$$

$$T\frac{\partial \eta}{\partial t} = \frac{\partial \varepsilon}{\partial t} - \frac{P}{\rho}\frac{\partial \rho}{\partial t} = \frac{\kappa}{\rho}\nabla \cdot (\nabla T) + \frac{1}{\rho}\Phi \qquad (7\text{-}29)$$

and

$$P = \rho R T \qquad (7\text{-}30)$$

where η denotes the specific entropy, ε the internal energy, κ the thermal conductivity, and Φ the total dissipation function.

The condition to be satisfied at a fluid-fluid boundary or fluid-vacuum boundary are obtained by integration of the relevant equations across a thin stratum coinciding with the surface.

The motion of ionized gas belongs to the regime of plasma dynamics. The charged particles in a magnetic field are of interest in some astronomical problems. The characteristic feature of the motion of a charged particle in a magnetostatic field is its tendency to spiral around the magnetic lines of force; on this is superposed a slow drift normal to the magnetic field if this is not uniform. This drift will be in opposite sense for oppositely charged particles in a gravitational field or a field of force other than an electrical field. But in the case of crossed electrical and magnetic fields, the drift will be the same for the charges of opposite sign, irrespective of their masses and charges.

In an ionized gas we make certain assumptions in the classical theory of gases: 1. the assumption of molecular chaos in which it is supposed that particles having velocity resolutes lying within a certain range are, at any instant, distributed at random independently of the position and velocities of the other particles, and 2. it is assumed that only binary encounters between the particles may be sufficient for consideration.

The total number of particles in the element $d\mathbf{r}$ is defined as

$$n = f(\mathbf{V}, \mathbf{r}, t)\, d\mathbf{V} \qquad (7\text{-}31)$$

where f is called the velocity distribution function. The equation satisfied by the function f was first derived by Boltzmann in the form

$$\frac{\partial f}{\partial t} + (\mathbf{V} \cdot \nabla)f + (\mathbf{F} \cdot \nabla v)f = \left(\frac{\partial f_e}{\partial t}\right)_{\text{collision}} \qquad (7\text{-}32)$$

where $\mathbf{V}v$ stands for the gradient vector operator in the velocity space. Here $(\partial f_e/\partial t)$ collision is equal to the rate of change by encounters in the number of the class $\mathbf{V}$, $d\mathbf{V}$, per volume of real space in a fixed element of volume $d\mathbf{r}$ at $\mathbf{r}, t$. We assume that each particle is acted upon by a force $m\mathbf{F}$, independent of $\mathbf{V}$, where m is the mass. Boltzmann's equation for a gas consisting of ions of mass m and carrying a charge e is therefore

$$\frac{\partial f}{\partial t} + (\mathbf{V} \cdot \mathbf{V})f + \frac{e}{m}(\mathbf{E} + \mathbf{V} + \mathbf{B}) \cdot \mathbf{V}vf = \left(\frac{\partial f_e}{\partial t}\right)_{\text{collision}} \tag{7-33}$$

7-3-2 Finite Element Solutions for MHD Flow

In view of governing equations for MHD, it is evident that the finite element equations may be obtained as described in Chap. 5 or Chap. 6 with the exception of the electromagnetic forces $\mathbf{J} \times \mathbf{B}$ and $q\mathbf{E}$. In case of incompressible flow, the finite element equation for (7-26) is

$$\int_\Omega \left[\dot{V}_i + V_{i,j}V_j + \frac{1}{\rho}P_{,i} - g_i - F_i - v(V_{i,jj} + \tfrac{1}{3}V_{j,ji})\right]\Phi_N \, d\Omega = 0 \tag{7-34}$$

in which the electromagnetic force F_i is given by

$$F_i = \frac{1}{\rho}(\varepsilon_{ijk}J_jB_k + qE_i) \tag{7-35}$$

In terms of Hartmann number R_h, the electromagnetic force may be replaced by

$$F_i = -v\left(\frac{R_h}{L}\right)^2 V_i \tag{7-36}$$

where

$$R_h = \mu_0 H_0 L\left(\frac{\sigma_0}{\rho v}\right)^{1/2} \tag{7-37}$$

with H_0 the externally applied magnetic field, σ_0 the electric conductivity, v the kinematic viscosity, L the characteristic length, and of course, Φ_N the interpolation function for the finite element velocity distribution.

The finite element equations for continuity, energy, and state may also be derived similarly as described in earlier chapters. Wu [1973] presented the finite element solution for a special case of (7-34),

$$\int_\Omega \left[\dot{u} - vu_{,ii} + v\left(\frac{R_h}{L}\right)^2 u - \delta(t)\right]\Phi_N \, d\Omega = 0 \tag{7-38}$$

where $\delta(t)$ is the Dirac delta function. The finite element equation (7-38) takes the form

$$A_{NM}\dot{u}_M + B_{NM}u_M + C_{NM}u_M - E_N = 0 \tag{7-39}$$

Here the boundary term generated from the second term in (7-38) is set equal to zero, and

$$A_{NM} = \int_\Omega \Phi_N \Phi_M \, d\Omega$$

$$B_{NM} = \int_\Omega v\Phi_{N,i}\Phi_{M,i} \, d\Omega$$

$$C_{NM} = \int_\Omega v\left(\frac{R_h}{L}\right)^2 \Phi_N \Phi_M \, d\Omega$$

$$E_N = \int_\Omega \delta(t)\Phi_N \, d\Omega = \int_\Omega K_0\Delta(t)\Phi_N \, d\Omega$$

where K_0 is a constant and $\Delta(t)$ is the Heaviside unit function.

To solve (7-39), we consider a square duct of dimension $2a \times 2a$ discretized into 36 and 144 isoparametric elements for comparison (Fig. 7-4). The temporal operator of the type (3-54a, 3-56b) is used.

Figure 7-5 shows the dimensionless velocity profiles in the $y = 0$ plane for various Hartmann numbers at $\tau = vt/a^2 = 0.3$. The results demonstrate that an increase of Hartmann numbers brings the velocity profiles closer to the steady state as expected.

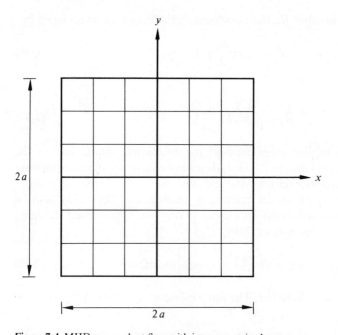

Figure 7-4 MHD square duct flow with isoparametric elements.

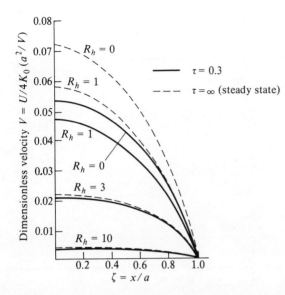

Figure 7-5 Velocity profiles for various Hartmann numbers at a specified time parameter with 144 elements after Wu [1973].

In Fig. 7-6, comparisons between the velocity profiles obtained by the point-matching method [Wu and Jeng, 1969] and the finite element method using 36 and 144 elements are given [Wu, 1973].

Figure 7-7 presents the transient velocity profiles in the $y = 0$ plane of the duct with 144 elements several time steps after the application of a constant pressure gradient for $R_h = 5.0$. Effects of the viscosity are small near the center of the duct in the early stages of time.

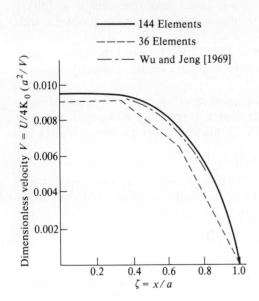

Figure 7-6 Velocity profiles for $R_h = 5$, $\tau = \infty$ after Wu [1973].

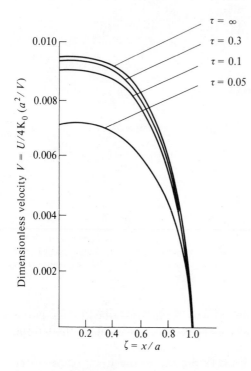

Figure 7-7 Velocity profiles for various time parameters and $R_h = 5$ with 144 elements— after Wu [1973].

7-3-3 Finite Element Analysis of MHD Instabilities

Magnetohydrodynamic instabilities of plasma are the important problems in thermonuclear fusion. In the past, the energy principle of Bernstein, et al. [1958] has been used to obtain analytical solutions in the instability problems with simple cases.

Takeda, et al. [1972] and Boyd, et al. [1973, 1975] studied the finite element applications to MHD instabilities. A brief summary on this subject is given in what follows.

Let us consider the cylindrical current-carrying plasma embedded in a longitudinal magnetic field and encased in a conducting wall separated by a vacuum. Following Bernstein, et al. [1958], the total potential energy is given by

$$W = W_p + W_v \tag{7-40}$$

where W_p and W_v are energy integrals of the plasma and the vacuum, respectively.

The magnetic field $\mathbf{B}(p, B_\theta, B_z)$ in the cylindrical coordinates and the plasma pressure P are related to the plasma current $\mathbf{j}(0, j_\theta, j_z)$ by

$$\mathbf{j} = \nabla \times \mathbf{B}$$

$$\nabla P = \mathbf{j} \times \mathbf{B}$$

Denoting also that $\xi = \xi_r$, $\eta = (im/r)\xi_\theta + ik\xi_z$, $\zeta = i\xi_\theta B_z - i\xi_z B_\theta$ with ξ_r, ξ_θ, and ξ_z being the components of the displacement vector ξ, we write the plasma energy integral in the form

$$W_p = \frac{\pi}{2} \int \left\{ \Lambda + \gamma P \left[\eta + \frac{1}{r} \frac{d}{dr} (r\xi) \right]^2 + \frac{k^2 r^2 + m^2}{r^2} (\zeta - \zeta_0)^2 \right\} r \, dr \quad (7\text{-}41)$$

with

$$\Lambda = \frac{1}{k^2 r^2 + m^2} \left[(krB_z + mB_\theta) \frac{d\xi}{dr} + (krB_z - mB_\theta) \frac{\xi}{r} \right]^2$$

$$+ \left[(krB_z + mB_\theta)^2 - 2B_\theta \frac{d}{dr} (rB_\theta) \right] \left(\frac{\xi}{r} \right)^2$$

$$\zeta_0 = \frac{r}{k^2 r^2 + m^2} \left[(krB_\theta - mB_z) \frac{d\xi}{dr} - (krB_\theta + mB_z) \frac{\xi}{r} \right]$$

The energy contained in the vacuum is

$$W_v = \frac{\pi}{2} \int |\nabla \times \mathbf{A}|^2 r \, dr \quad (7\text{-}42)$$

where $\mathbf{A}$ is the vector potential which must satisfy the boundary conditions

$$\mathbf{n} \times \mathbf{A} = -(\mathbf{n} \cdot \boldsymbol{\xi}) \mathbf{B} \quad (7\text{-}43)$$

at the plasma vacuum interface $r = a$, and

$$A_\theta = A_z = 0$$

at the surface of the conducting wall. For the linear pinch with fixed boundary of perfectly conducting wall without vacuum region, we have

$$\xi_r = 0 \qquad r = a$$

In this case, the energy of the vacuum should be eliminated from (7-40).

The kinetic energy of the plasma is given by

$$T = \frac{\pi}{2} \int_0^a \rho(\xi_r^2 + \xi_\theta^2 + \xi_z^2) r \, dr$$

where ρ is the plasma density.

Assuming that the displacement varies as $\exp(i\omega t)$, it can be shown that the Lagrangian of the system is

$$L = W - \omega^2 T \quad (7\text{-}44)$$

The finite element discretization is made along the plasma radius in one dimension with the interpolation relation given by

$$\xi_i = \Phi_N \xi_i^N \quad (7\text{-}45)$$

We observe that substitution of (7-45) into (7-44) leads to

$$L = \tfrac{1}{2} K_{NM}^{ij} \xi_i^N \xi_j^M - \omega^2 N_{NM}^{ij} \xi_i^N \xi_j^M \quad (7\text{-}46)$$

Minimizing the Lagrangian with respect to the nodal displacements yields

$$\frac{\partial L}{\partial \xi_i^N} = (K_{NM}^{ij} - \omega^2 M_{NM}^{ij})\xi_j^M = 0$$

$$|K_{NM}^{ij} - \omega^2 M_{NM}^{ij}| = 0 \qquad (7\text{-}47)$$

This is now a standard eigenvalue problem. The magnetohydrodynamic instability is determined by the presence of a negative eigenvalue in (7-47). The growth rate of the most unstable mode is given by

$$\Gamma = \sqrt{|\omega_{\min}^2|} \qquad (7\text{-}48)$$

The finite element solutions of (7-47) obtained by Takeda, et al. [1972] and Boyd, et al. [1973, 1975] are discussed below. The work of the latter confirms that of the former and the results of both studies agree reasonably well with the analytical solutions carried out by Shafranov [1970].

The first example deals with the pinch with fixed boundary. The analytical solution for the growth rate [Shafranov, 1970] is given by

$$\Gamma^2 = \frac{B_\theta^2(a)}{4\pi\rho a^2}\left[2\left|m - nq_a\right|\frac{Ka}{Z_m} - (m - nq_a)^2\right] \qquad (7\text{-}49)$$

where $q_a = 2\pi a B_z(a)/L B_\theta(a)$ is the safety factor, Z_m is the first zero of the mth order Bessel function $J_m(Z)$, and $K = 2\pi n/L$ is the longitudinal wave number with

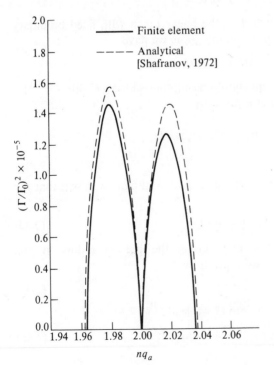

Figure 7-8 Growth rate of the $m = 2$ mode for an incompressible plasma with a fixed boundary, $K = 0.2$—after Takeda, et al. [1972].

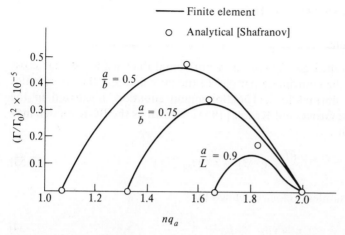

Figure 7-9 Growth rate for $m = 2$, $K = 0.2$ in a linear pinch with free boundary—after Boyd, et al. [1975].

L the pitch length. The results of the growth rate are shown in Fig. 7-8. Boyd, et al. [1975] noted that seven or more elements are adequate for a convergent solution. Boyd, et al. suggests that the slight disagreements between the analytical and finite element solutions may be due to the governing equations used for modal analysis of Shafranov that did not entirely agree with the energy principles of Bernstein. Despite this discrepancy, the region of instability does coincide with the analytical prediction [Shafranov, 1970]

$$m - \frac{2Ka}{Z_m} < nq < m < nq < m + \frac{2Ka}{Z_m} \qquad (7\text{-}50)$$

For the pinch with free boundary, the radius of the plasma–vacuum system is discretized in line elements such that a node occurs at the plasma–vacuum interface. In Fig. 7-9, we note that the ranges of instability agree well with the analytical solution [Shafranov]

$$m - 1 + \left(\frac{a}{b}\right)^{2m} < nq < m \qquad (7\text{-}51)$$

For $a/b = 0.5, 0.75$, the growth rates are in good agreement with the analytical solution

$$\Gamma^2 = \frac{B_\theta^2}{4\pi\rho a^2}\left[2\,|m - nq| + \frac{2(m - nq)^2}{1 - (a/b)^{2m}}\right] \qquad (7\text{-}52)$$

It was also shown that as K is increased, the maximum growth increases in magnitude and the range of instability increases with a limiting situation for which the instability range is approximately twice as large as for the case of small k. Accuracy tends to deteriorate for $a/b \geq 0.9$. A comparable dependence of growth rates on the position of the conducting wall was well documented by Takeda, et al.

7-4 RAREFIED GAS DYNAMICS

7-4-1 Basic Equations

Let us consider a rarefied gas flowing in a horizontal duct with irregular cross section with z being the coordinate parallel to the flow and x, y the coordinates of the cross section normal to z. The Boltzmann equation, linearized in the manner of Bhatnagar, Gross, and Krook [1954], known as the BGK model, may be written in the form

$$\frac{\partial f}{\partial t} + \mathbf{c} \cdot \nabla f = \frac{\beta_0}{\sigma} (f_{eq} - f) \tag{7-53}$$

where $\mathbf{c}$ is the dimensionless velocity defined by

$$\mathbf{c} = \mathbf{v}\beta_0 \qquad \beta_0 = \sqrt{\frac{m}{2kT_0}} \tag{7-54}$$

f is the singlet local distribution function, $\mathbf{v}$ the molecular velocity, σ the collision frequency, k the Boltzmann constant, m the mass of the gas per molecule, and T the absolute temperature. The subscript 0 refers to the properties at $z = 0$. Applying the usual Chapman and Ensko method of successive approximation, we may write

$$f = f_0(1 + \Phi) \tag{7-55}$$

$$n = n_0(1 + v) \tag{7-56}$$

$$T = T_0(1 + \tau) \tag{7-57}$$

with

$$f_0 = n\beta_0^3 \pi^{-3/2} \exp\left(-\mathbf{c}^2\right) \tag{7-58}$$

and

$$f_{eq} = n\beta_0^3 \pi^{-3/2} \exp\left[\frac{1}{(1 + \tau)} (\mathbf{c} - \mathbf{q}^*)^2\right] \tag{7-59}$$

where $\mathbf{q}^*$ is the dimensionless flow velocity which is defined as $\mathbf{q}^* = \mathbf{q}\beta_0$ and n is the number density. For the problem we consider here, the flow velocity can be expressed as

$$q_x^* = q_y^* = 0 \tag{7-60}$$

and

$$q_z^* = \frac{1}{n\beta_0^3} \int_{-\infty}^{\infty} \int_{-\infty}^{\infty} \int_{-\infty}^{\infty} f c_z \Phi \, dc_x \, dc_y \, dc_x \tag{7-61}$$

For small Mach number flow where $|q_z| \ll 1$, $v \ll 1$, $\tau \ll 1$, and $\Phi \ll 1$, the equilibrium local distribution function can be linearized as

$$f_{eq} = f_0\{1 + 2c_z q_z^* + (c^2 - \tfrac{3}{2})\tau\} \tag{7-62}$$

For simplicity, we consider a pure shear flow without heat transfer; namely, an isothermal flow at T_0. Thus, the temperature change due to compression will be ignored; that is, $\tau = 0$. This implies that the gas flows are resulting from a density gradient along the z direction, which in turn is caused by a pressure gradient. Therefore, the Boltzmann equation is linearized as

$$\frac{\partial \Phi}{\partial t} + c_x \frac{\partial \Phi}{\partial x} + c_y \frac{\partial \Phi}{\partial y} + c_z \bar{K} = \frac{1}{\theta} \{2c_z q_z^* - \Phi\} \tag{7-63}$$

where the nonlinear terms $(\Phi/n)(dn/dz)$ are neglected, and furthermore, we have restricted our attention to a fully developed flow, with $\partial \Phi / \partial z$ and $(d/dz)(1/p)(dp/dz)$ being zero. Finally, $\bar{K} = (1/p)(dp/dz)$, and $\theta = (\sigma/\beta_0)$.

On the walls of the duct, we shall assume that it reflects diffusely the molecules impinged on it. Thus, the boundary conditions for the distribution function will be characterized by Maxwellian; namely, on the boundary (F_0)

$$f[-\operatorname{sgn} c_x; \mathbf{r}_0, \mathbf{c}] = f[-\operatorname{sgn} c_y, \mathbf{r}_0, \mathbf{c}] = f_0 \tag{7-64}$$

with

$$\begin{bmatrix} \operatorname{sgn} c_x \\ \operatorname{sgn} c_y \end{bmatrix} = \begin{bmatrix} +1, c_x > 0 \text{ and } c_y > 0 \\ -1, c_y < 0 \text{ and } c_y < 0 \end{bmatrix} \tag{7-65}$$

The boundary condition for Φ then becomes

$$\Phi(-\operatorname{sgn} c_x; \mathbf{r}_0, \mathbf{c}) = \Phi(-\operatorname{sgn} c_y; \mathbf{r}_0, \mathbf{c}) = 0 \tag{7-66}$$

Introducing the dimensionless variables of the form

$$x^* = \frac{x}{r_0}, \quad y^* = \frac{y}{r_0}, \quad \Phi^* = \frac{\Phi}{Kr_0}, \quad \delta = \frac{r_0}{\theta} \tag{7-67}$$

and also assuming that

$$\Phi^* = c_z \psi(x^*, y^*, c_x, c_y) \tag{7-68}$$

we can rewrite (10-63) as

$$\frac{\partial \psi}{\partial t} + c_z \frac{\partial \psi}{\partial x} + c_y \frac{\partial \psi}{\partial y} + \bar{K} = \frac{1}{\theta} \left(\frac{1}{\pi} \int_{-\infty}^{\infty} \int_{-\infty}^{\infty} \psi \exp(-c_x^2 - c_y^2) \, dc_x \, dc_y - \psi \right) \tag{7-69}$$

where the superscript * is deleted for convenience.

Our objective is to determine ψ, called the perturbation function, and subsequently Φ from which we can calculate flow velocity by Eq. (7-61). In order to obtain the values for ψ, we shall apply the half range method [Gross, Jackson, and Zeiring, 1957] by dividing the ψ into four parts; namely,

$$\psi = \psi^{\pm \pm}(c_x, c_y) \tag{7-70}$$

and that $\psi^{\pm \pm}$ is only defined for $c_x > 0$ and $c_y > 0$, ψ^{+-} for $c_x > 0$ and $c_y < 0$, ψ^{-+} for $c_x < 0$ and $c_y > 0$, and ψ^{--} for $c_x < 0$ and $c_y < 0$. The integral in

Eq. (7-69) may be written as

$$\int_{-\infty}^{\infty} \int_{-\infty}^{\infty} \psi \exp\left(-c_x^2 - c_y^2\right) dc_x \, dc_y = \int_{0}^{\infty} \int_{0}^{\infty} \psi^{++} \exp\left(-c_z^2 - c_x^2 - c_y^2\right) dc_x \, dc_y$$

$$+ \int_{0}^{\infty} \int_{-\infty}^{0} \psi^{-+} \exp\left(-c_x^2 - c_y^2\right) dc_x \, dc_y + \int_{-\infty}^{0} \int_{0}^{\infty} \psi^{-+} \exp\left(-c_x^2 - c_y^2\right) dc_x \, dc_y$$

$$+ \int_{-\infty}^{0} \int_{-\infty}^{0} \psi^{--} \exp\left(-c_x^2 - c_y^2\right) dc_x \, dc_y$$

Finally, we calculate the volume flow rate Q_z

$$Q_z = \int \int q_z(x, y) \, dx \, dy$$

In the following sections, we shall demonstrate how these calculations are performed using the finite element technique.

7-4-2 Finite Element Solution of Boltzmann Equation

In the final form of the Boltzmann equation with BGK collision model (7-64), we shall assume that the perturbation function $\psi^{\pm\pm}(x, y, c_x, c_y)$ is given approximately by [Chung, Oden, and Wu, 1974]

$$\psi^{\pm\pm}(x, y, c_x, c_y) = \sum_{m=0}^{\infty} \sum_{n=0}^{\infty} \phi_{mn}(x, y) H_{mn}^{\pm\pm}(c_x, c_y) \tag{7-71}$$

wherein $H_{mn}^{\pm\pm} = h_m(c_x) h_n(c_y)$ and $h_m^{\pm}(c_x)$ and $h_n^{\pm}(c_y)$ are the Hermite polynomials of order m. Substituting (7-71) into (7-69), we obtain the residual function

$$R^{\pm\pm}(x, y, c_x, c_y) = \sum_{m=0}^{\infty} \sum_{n=0}^{\infty} \frac{\partial \psi_{mn}}{\partial t} H_{mn} + \sum_{m=0}^{\infty} \sum_{n=0}^{\infty} c_x \frac{\partial \psi_{mn}}{\partial x} H_{mn}^{\pm\pm}$$

$$+ \sum_{m=0}^{\infty} \sum_{n=0}^{\infty} c_y \frac{\partial \phi_{mn}}{\partial y} H_{mn}^{\pm\pm} + \bar{K} + \frac{1}{\theta} \left[\sum_{m=0}^{\infty} \sum_{n=0}^{\infty} \phi_{mn} H_{mn}^{\pm\pm} \right.$$

$$\left. - \frac{1}{\pi} \int_{-\infty}^{\infty} \int_{-\infty}^{\infty} \sum_{n=0}^{\infty} \sum_{m=0}^{\infty} \phi_{mn} H_{mn}^{\pm\pm} \exp\left(-c_x^2 - c_y^2\right) dc_x \, dc_y \right]$$

In the approximation (7-71), we choose ϕ_{mn} in such a manner that the averages of R with respect to $H_{ij}^{\pm\pm}$

$$\bar{R}_{ij}(x, y) = \int_{\infty}^{\infty} \int_{\infty}^{\infty} R H_{ij}^{\pm\pm} \, dc_x \, dc_y$$

vanish in velocity domain so as to obtain systems of partial differential equations in ϕ_{mn} of the form

$$\sum_{m=0}^{\infty} \sum_{n=0}^{\infty} \frac{\partial \phi_{mn}}{\partial t} W_{mnij} + \sum_{m=0}^{\infty} \sum_{n=0}^{\infty} \frac{\partial \phi_{mn}}{\partial x} A_{mnij} + \sum_{m=0}^{\infty} \sum_{n=0}^{\infty} \frac{\partial \phi_{mn}}{\partial y} B_{mnij}$$

$$+ \bar{K} E_{ij} + \frac{1}{\theta} \left[\sum_{m=0}^{\infty} \sum_{n=0}^{\infty} \phi_{mn} C_{mnij} - \sum_{m=0}^{\infty} \sum_{n=0}^{\infty} \phi_{mn} D_{mn} E_{ij} \right] = 0 \quad (7\text{-}72)$$

where

$$W_{mnij} = \int_{-\infty}^{\infty} \int_{-\infty}^{\infty} H_{mn}^{\pm\pm}(c_x, c_y) H_{ij}^{\pm\pm}(c_x, c_y) \, dc_x \, dc_y$$

$$A_{mnij} = \int_{-\infty}^{\infty} \int_{-\infty}^{\infty} c_x H_{mn}^{\pm\pm}(c_x, c_y) H_{ij}^{\pm\pm}(c_x, c_y) \, dc_x \, dc_y$$

$$B_{mnij} = \int_{-\infty}^{\infty} \int_{-\infty}^{\infty} c_y H_{mn}^{\pm\pm}(c_x, c_y) H_{ij}^{\pm\pm}(c_x, c_y) \, dc_x \, dc_y$$

$$C_{mnij} = \int_{-\infty}^{\infty} \int_{-\infty}^{\infty} H_{mn}^{\pm\pm} H_{ij}^{\pm\pm} \, dc_x \, dc_y$$

$$D_{mn} = \frac{1}{\pi} \int_{-\infty}^{\infty} \int_{-\infty}^{\infty} H_{mn}^{\pm\pm} \exp(-c_x^2 - c_y^2) \, dc_x \, dc_y$$

$$E_{ij} = \int_{-\infty}^{\infty} \int_{-\infty}^{\infty} H_{ij}^{\pm\pm} \, dc_x \, dc_y$$

Thus we have reduced the problem of solving (7-53) to that of solving an infinite system of partial differential equations (7-72) in the functions $\phi_{mn} = \phi_{mn}(x, y)$. We shall proceed to obtain approximate solutions of a truncated version of (7-72) by the finite element method.

Introducing the functional relationship in the form,

$$\phi_{mn}(x, y) = f_N(x, y) \phi_{mn}^N \quad (7\text{-}73)$$

and substituting (7-73) into (7-72), we obtain the new local residual

$$\hat{R}_{ij}(x, y) = \sum_{m=0}^{\infty} \sum_{n=0}^{\infty} \left\{ W_{mnij} f_N \frac{\partial \phi_{mn}^N}{\partial t} + A_{mnij} \frac{\partial f_N}{\partial x} \phi_{mn}^N \right.$$

$$\left. + B_{mnij} \frac{\partial f_N}{\partial y} \phi_{mn}^N + \frac{1}{\theta} (C_{mnij} f_N \phi_{mn}^N - D_{mn} E_{ij} f_N \phi_{mn}^N) \right\} + \bar{K} E_{ij}$$

where N is again summed from 1 to N_e and $f_N = f_N(x, y)$.

We shall now choose the local nodal values of ϕ_{mn}^N in such a manner that the local residual $\hat{R}_{ij}(x, y)$ is orthogonal to the subspace spanned by the functions $f_N(x, y)$ for each finite element; that is

$$\int \int \hat{R}_{ij} f_M \, dx \, dy = 0 \quad (M = 1, 2, \ldots, N_e)$$

As a consequence of this requirement, which is essentially the Galerkin method, we obtain for the finite element model of (7-72)

$$\sum_{m=0}^{\infty}\sum_{n=0}^{\infty} W_{mnij} W_{NM}\,\frac{\partial \phi_{mn}^N}{\partial t} + \sum_{m=0}^{\infty}\sum_{n=0}^{\infty}\left\{\left[A_{mnij}a_{NM} + B_{mnij}b_{mn}\,\frac{1}{\theta}\,(C_{mnij}\right.\right.$$

$$\left.\left. - D_{mn}E_{ij})c_{NM}\,\phi_{mn}^N\right]\right\} + E_{ij}\bar{K}d_M = 0 \qquad (7\text{-}74)$$

where

$$w_{NM} = \int\int f_N f_M\,dx\,dy \qquad (7\text{-}75a)$$

$$a_{NM} = \int\int \frac{\partial f_N}{\partial x} f_M\,dx\,dy \qquad (7\text{-}75b)$$

$$b_{NM} = \int\int \frac{\partial f_N}{\partial x} f_M\,dx\,dy \qquad (7\text{-}75c)$$

$$c_{NM} = \int\int f_N f_M\,dx\,dy \qquad (7\text{-}75d)$$

$$d_M = \int\int f_M\,dx\,dy \qquad (7\text{-}75e)$$

Equation (7-74) represents the general local finite element model of (7-72). Upon connecting elements together, (7-74) leads to a system of linear equations governing the global behavior of the discrete model of each function $\phi_{mn}(x, y)$. Boundary conditions amount to simply prescribing nodal values of $\phi_{mn}(x, y)$ at boundary nodes. In specific applications, appropriate forms of the interpolation functions $f_N(x, y)$ must be chosen and, of course, only a finite number of terms of the series in (7-73) can be used.

Having completed all integrations in (7-75), we obtain the finite element equation of the form

$$J_{NMij}^{mn}\dot{\phi}_{mn}^N + K_{NMij}^{mn}\phi_{mn} = F_{Mij} \qquad (7\text{-}76)$$

in which

$$J_{NMij}^{mn} = \sum_{m=0}^{\infty}\sum_{n=0}^{\infty} W_{mnij} w_{NM}$$

$$K_{NMij}^{mn} = \sum_{m=0}^{\infty}\sum_{n=0}^{\infty}\left\{A_{mnij}a_{NM} + B_{mnij}b_{NM} + \frac{1}{\theta}\,(C_{mnij} - J_{mn}E_{ij})c_{NM}\right\}$$

$$F_{Mij} = E_{ij}\bar{K}d_M$$

The number of equations generated in (7-76) depends on the order of Hermite polynomials approximations and the number of nodes in an element. Consider the mth Hermite polynomial defined by

$$H_m(\zeta) = (-1)^m \exp(\zeta^2) \frac{d^m}{d\zeta^m} \exp(-\zeta^2) \qquad (7\text{-}77)$$

The number of local finite element equations is determined by

$$r = (1 + m)^2 N$$

where N is the number of nodes in an element; for example, if we choose $m = 3$ and $N = 4$, then the number of local finite element equations becomes 64 with 16 equations at each node. The total number of equations for the entire cross section is 16 times the total number of nodes.

The global equation of (7-74) written in matrix form is

$$\mathbf{J}\dot{\phi} + \mathbf{K}\phi = \mathbf{F} \qquad (7\text{-}78)$$

where $\mathbf{J}$ and $\mathbf{K}$ are $s \times s$ matrices with $s = (1 + m)^2 N_T$, N_T is the total number of nodes in the cross section. The boundary conditions of (7-66) can be applied simply by deleting rows and columns of (7-78) corresponding to the boundary nodes. Once again, the solution of (7-78) is obtained by using the suitable temporal operators as discussed in Sec. 3-3-3.

The finite element solutions for the rarefied gas of one-dimensional Couette flow were carried out by Aguirre-Ramirez, Oden, and Wu [1973]. Similar works were also presented by Bramlette and Mallet [1970] and Bramlette [1971]. Chung, Oden, and Wu [1974] studied a square duct and solved the equations represented by (7-78). Brief discussions of their results are given below.

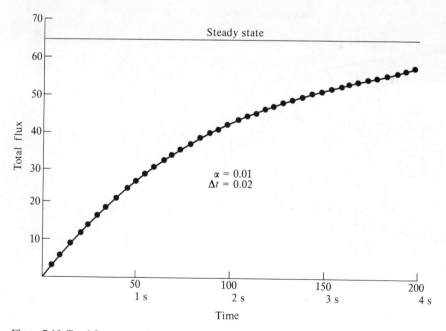

Figure 7-10 Total flux over unit square section versus time.

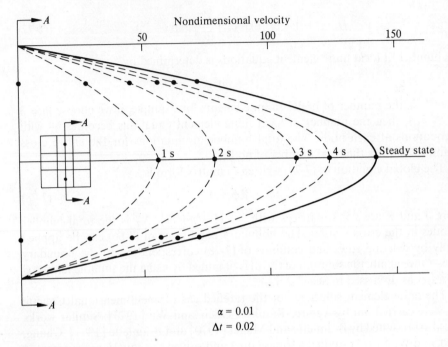

Figure 7-11 Transient velocity profile.

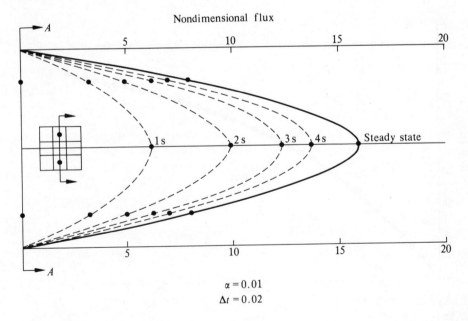

Figure 7-12 Transient flux profile.

Let us consider a square duct as shown in Fig. 7-10 [Chung, Oden, and Wu, 1974]. With the linear isoparametric elements of 3×3 and Gaussian quadrature integration of 6 points, the global equations are solved by using the temporal operator (3-55c) with $\Delta t = 0.02$ s, which provides the satisfactory stability and convergence.

Although various values of an inverse Knudsen number are chosen for study, the results of only $\alpha = 0.01$ are presented. Figure 7-10 shows the total nondimensional flux versus time. The computer calculations are cut off at 200 time steps. It is seen that approximately 80 percent of the steady state value of flux has been reached. The transient nondimensional velocity profiles are shown in Fig. 7-11 and the corresponding flux variations in Fig. 7-12.

The power of weighted residual approach for formulating the finite element equations is clearly demonstrated in (7-71) through (7-78). The orthogonal projection in the velocity space is superimposed on the subsequent orthogonal projection in the Euclidean space, a process which makes the finite element analysis extremely effective in the solution of the Boltzmann equation.

ABSCISSAE AND WEIGHT COEFFICIENTS OF THE GAUSSIAN QUADRATURE FORMULA

n	Weight coefficient w_k	Abscissae $\pm \xi_k, \pm \eta_k$
2	1.00000 00000	0.57735 02691
3	0.55555 55555	0.77459 66692
	0.88888 88888	0.00000 00000
4	0.34785 48451	0.86113 63115
	0.65214 51548	0.33998 10435
5	0.23692 68850	0.90617 98459
	0.47862 86704	0.53846 93101
	0.56888 88888	0.00000 00000
6	0.17132 44923	0.93246 95142
	0.36076 15730	0.66120 93864
	0.46791 39345	0.23861 91860
7	0.12948 49661	0.94910 79123
	0.27970 53914	0.74153 11855
	0.38183 00505	0.40584 51513
	0.41795 91836	0.00000 00000
8	0.10122 85362	0.96028 98564
	0.22238 10344	0.79666 64774
	0.31370 66458	0.52553 24099
	0.36268 37833	0.18343 46424
9	0.08127 43883	0.96816 02395
	0.18064 81606	0.83603 11073
	0.26061 06964	0.61337 14327
	0.31234 70770	0.32425 34234
	0.33023 93550	0.00000 00000
10	0.06667 13443	0.97390 65285
	0.14945 13491	0.86506 33666
	0.21908 63625	0.67940 95682
	0.26926 67193	0.43339 53941
	0.29552 42247	0.14887 43389

COMPUTER PROGRAMS

B-1

```
 1*      C  **************************************************************
 2*      C  **                                                          **
 3*      C  **    TWO DIMENSIONAL FLOW AROUND CYLINDER WITH TRIANGULAR ELEMENTS
 4*      C  **    BY STREAMFUNCTION FORMULATION -
 5*      C  **    (REFERENCE, LECTURE NOTES, T. J. CHUNG)
 6*      C  **                                                          **
 7*      C  **************************************************************
 8*      C
 9*      C  NDE  = NR OF NODES
10*      C  NEL  = NR OF ELEMENTS
11*      C  NBC  = NR OF DIRICHLET BOUNDARY CONDITIONS (WITH PSI = 0 OR ?)
12*      C  NVN  = NP OF NODES ALONG VERTICAL LINE ABOVE CREST
13*      C  NENN = TABLE OF NODE NRS BY ELEMENT
14*      C  ANM  = COEFFICIENT MATRIX FOR AN ELEMENT
15*      C  R    = INPUT VECTOR
16*      C  SK   = GLOBAL COEFFICIENT MATRIX
17*      C  XNODE = X COORD FOR GLOBAL NODE
18*      C  YNODE = Y COORD FOR GLOBAL NODE
19*      C
20*         PARAMETER NEL=137, NDE=96, NBC=30
21*         PARAMETER NVN = 5
22*      C
23*      C  **WARNING - ALL NODES ALONG THE VERTICAL LINE ABOVE THE CREST
24*      C             OF THE CYLINDER SHOULD BE NUMBERED CONSECUTIVELY
25*      C             FROM TOP TO BOTTOM IN ORDER TO BE COMPATIBLE
26*      C             WITH VELOCITY CALCULATIONS USED IN THIS PROGRAM
27*      C
28*         COMMON/TRNG/ ANM(3,3)
29*         COMMON/GEOM/XNODE(NDE),YNODE(NDE),NENN(NEL,3)
30*         DIMENSION SK(NDE,NDE),R(NDE),PSI(NDE)
```

```
31*          *          ,NODE(NPC),NODE(NBC),VAL(NBC),JC(NDE)
32*    C
33*    C     CLEAR GLOBAL MATRIX
34*          DO 3 I= 1,NDE
35*          DO 3 J= 1,NDE
36*          SK(I,J) = 0.0
37*    3     CONTINUE
38*    C
39*    C     READ COORDINATES
40*          READ(5,501) (XNODE(I),YNODE(I),I=1,NDE)
41*          WRITE (6,502)
42*          WRITE (6,503)(I,XNODE(I),YNODE(I),I=1,NDE)
43*    C
44*    C     READ ELEMENT NODE NRS
45*          READ(5,504)((NENN(I,J),J=1,3),I=1,NEL)
46*          WRITE(6,505)
47*          WRITE(6,506)(I,(NENN(I,J),J=1,3),I=1,NEL)
48*    C
49*    C     COMPUTE COEFFICIENT MATPIX FOR EACH ELEMENT
50*          DO 5 I=1,NEL
51*          CALL ELMT(I)
52*    C
53*    C     ASSEMBLE  GLOBALLY
54*          DO 10 J = 1,3
55*          JJ = NENN(I,J)
56*          DO 10 K = 1,3
57*          KK = NENN(I,K)
58*          SK(JJ,KK) = SK(JJ,KK) + ANM(J,K)
59*    10    CONTINUE
60*    5     CONTINUE
61*    C
```

```
62*          C          READ DIRICHLET BOUNDARY CONDITION INFORMATION
63*                     WRITE (6,107)
64*                     DO 13 I=1,NBC
65*                     READ (5,105) MODE(I),NODE(I),VAL(I)
66*                     WRITE (6,108) MODE(I),NODE(I),VAL(I)
67*          13         CONTINUE
68*                     DO 81 I = 1,NDE
69*          81         R(I) = 0.0
70*          C
71*          C          INTRODUCE BOUNDARY CONDITIONS
72*                     DO 82 I=1,NBC
73*                     NX=NODE(I)
74*                     IF (MODE(I).NE.0) GO TO 84
75*                     DO 85 J = 1,NDE
76*                     SK(NX,J)=0.0
77*                     SK(J,NX)=0.0
78*          85         CONTINUE
79*                     SK(NX,NX)=1.0
80*                     GO TO 82
81*          C
82*          C          COMPUTE INPUT VECTOR
83*          84         CONTINUE
84*                     DO 86 K = 1,NDE
85*                     R(K)=R(K)-SK(K,NX)*VAL(I)
86*          86         CONTINUE
87*                     DO 87 K = 1,NDE
88*                     SK(K,NX)=0.0
89*                     SK(NX,K)=0.0
90*          87         CONTINUE
91*                     SK(NX,NX)=1.0
92*                     R(NX)=VAL(I)
93*          82         CONTINUE
94*          C
95*          C          INVERT GLOBAL MATRIX AND MULTIPLY BY VECTOR
                        CALL IVMTX(SK,JC,NDE,NDE)
```

```
 96*         DO 97 I = 1,NDE
 97*         PSI(I) = 0.0
 98*         DO 97 J = 1,NDE
 99*         PSI(I) = PSI(I) + SK(I,J)*R(J)
100*      97 CONTINUE
101*   C
102*   C     WRITE NODAL STREAMLINE FUNCTION DATA
103*         WRITE(6,998)
104*         WRITE(6,999) (I,PSI(I),I=1,NDE)
105*   C
106*   C     VELOCITY PROFILE ON CREST OF CYLINDER
107*         NNN = NDE - NVN + 1
108*         MMM = NDE - 1
109*         WRITE(6,590)
110*         DO 555 I = NNN,MMM
111*         J = I + 1
112*         RD = YNODE(I) - YNODE(J)
113*         RM = (YNODE(I) + YNODE(J))/2.0
114*         SID = PSI(I) - PSI(J)
115*         VEL = SID/RD
116*         WRITE(6,591) RM,VEL
117*     555 CONTINUE
118*     808 STOP
119*     105 FORMAT (2I5,F10.0)
120*     107 FORMAT ('1 DIRICHLET BOUNDARY CONDITIONS',//6X,'NODE',5X,'NODE',
121*        *        5X,'VALUE'/)
122*     108 FORMAT(' ',2I9,F10.4)
123*     501 FORMAT(2F10.2)
124*     502 FORMAT('1NODE   X-COORD      Y-COORD')
125*     503 FORMAT(' ',I3,2F12.5)
126*     504 FORMAT(6I5)
127*     505 FORMAT('0ELEMENT NODE NUMBERS',/,'  ELMT  1  2  3')
128*     506 FORMAT(4I5)
129*     590 FORMAT (1H1,' VELOCITY PROFILE ON CREST OF CYLINDER'///)
```

```
130*        *     FORMAT(      6X,'Y',16X,'VEL'///)
131*      591 FORMAT (F10.5,F20.8/ )
132*      998 FORMAT( 1H1, 'NODAL STREAMLINE FUNCTION DATA'///)
133*      999 FORMAT (I5, F15.8)
134*          END

END OF COMPILATION:       NO  DIAGNOSTICS.

 1*   C     SUBROUTINE TO SOLVE ELEMENT COEFFICIENT MATRIX
 2*   C
 3*         SUBROUTINE ELMT(N)
 4*   C     X = X COORD FOR LOCAL NODE
 5*   C     Y = Y COORD FOR LOCAL NODE
 6*   C
 7*         PARAMETER NEL=137, NDE=86, NBC=30
 8*   C
 9*         COMMON/TRNG/ ANM(3,3)
10*         COMMON/GEOM/XNODE(NDE),YNODE(NDE),NENN(NEL,3)
11*         DIMENSION X(3),Y(3)
12*         DIMENSION B(3),C(3)
13*   C
14*   C     LOCAL NODE X AND Y COORDINATES
15*         DO 5 L = 1,3
16*         I = NENN(N,L)
17*         X(L) = XNODE(I)
18*         Y(L) = YNODE(I)
19*   5     CONTINUE
20*   C
21*   C     COMPUTE   2 X AREA
22*         A2 = X(2)*Y(3) + X(3)*Y(1) + X(1)*Y(2)
23*        1   -X(2)*Y(1) - X(3)*Y(2) - X(1)*Y(3)
24*         A = A2 / 2.0
25*         B(1) = (Y(2) - Y(3))/A2
26*         B(2) = (Y(3) - Y(1))/A2
```

```
27*         B(3) = (Y(1) - Y(2))/A2
28*         C(1) = (X(3) - X(2))/A2
29*         C(2) = (X(1) - X(3))/A2
30*         C(3) = (X(2) - X(1))/A2
31*         DO 6 I = 1,3
32*         DO 6 J = 1,3
33*         ANM(I,J) = (B(J) * B(I) + C(J) * C(I)) * A
34*       6 CONTINUE
35*         RETURN
36*         END
```

END OF COMPILATION: NO DIAGNOSTICS.

```
 1*         SUBROUTINE IVMTX (ATX,IRX,N,NR)
 2*   C *************************************************************
 3*   C       MATRIX INVERSION
 4*   C       SUBROUTINE
 5*   C *************************************************************
 6*   C
 7*         DIMENSION ATX(NR,1),IRX(NR)
 8*         DIMENSION TMP(3)
 9*   C
10*         DO 10 I = 1,N
11*         IRX(I) = I
12*      10 CONTINUE
13*         DET = 1.0
14*         DO 160 J = 1,N
15*         IF (J-N) 20,70,20
16*      20 CONTINUE
17*         TMP(1) = ABS(ATX(J,J))
18*         IMX = J
19*         JP = J+1
20*         DO 40 I = JP,N
21*         IF (TMP(1) - ABS(ATX(I,J)))30,40,40
```

```
22*         30 CONTINUE
23*            TMP(1) = ABS(ATX(I,J))
24*            IMX = I
25*         40 CONTINUE
26*            IF (IMX-J) 50,70,50
27*         50 CONTINUE
28*            DO 60 K = 1,N
29*            TMP(1) = ATX(IMX,K)
30*            ATX(IMX,K) = ATX(J,K)
31*            ATX(J,K) = TMP(1)
32*         60 CONTINUE
33*            I = IRX(IMX)
34*            IRX(IMX) = IRX(J)
35*            IRX(J) = I
36*            DET = -DET
37*         70 CONTINUE
38*            DET = DET*ATX(J,J)
39*            IF (DET) 80,300,80
40*         80 CONTINUE
41*            ATX(J,J) = 1. /ATX(J,J)
42*            DO 100 K = 1,N
43*            IF (K-J)90,100,90
44*         90 CONTINUE
45*            ATX(J,K) = ATX(J,K)*ATX(J,J)
46*        100 CONTINUE
47*            DO 160 I = 1,N
48*            IF (I-J) 110,160,110
49*        110 CONTINUE
50*            TMP(1) = ATX(I,J)
51*            ATX(I,J) = -ATX(I,J)*ATX(J,J)
52*            DO 150 K = 1,N
```

```
53*          IF (K-J) 120,150,120
54*      120 CONTINUE
55*          TMP(2) = ATX(I,K)
56*          TMP(3) = TMP(2) - TMP(1)*ATX(J,K)
57*          IF (ABS(TMP(3))- ABS(TMP(2))*1.E-06) 130,140,140
58*      130 CONTINUE
59*          TMP(3) = 0.0
60*      140 CONTINUE
61*          ATX(I,K) = TMP(3)
62*      150 CONTINUE
63*      160 CONTINUE
64*          NM = N-1
65*          DO 200 J = 1,NM
66*      170 CONTINUE
67*      180 IF (IRX(J) - J) 180,200,180
68*      180 CONTINUE
69*          K = IRX(J)
70*          IRX(J) = IRX(K)
71*          IRX(K) = K
72*          DO 190 I = 1,N
73*          TMP(1) = ATX(I,J)
74*          ATX(I,J) = ATX(I,K)
75*          ATX(I,K) = TMP(1)
76*      190 CONTINUE
77*          GO TO 170
78*      200 CONTINUE
79*          RETURN
80*   C
81*      300 CONTINUE
82*          WRITE (6,6100)
83*     6100 FORMAT('1 THE MATRIX IS SINGULAR')
84*          END
```

B-2

```
  2*     **
  3*     **   TWO DIMENSIONAL FLOW AROUND CYLINDER WITH ISOPARAMETRIC
  4*     **   ELEMENTS BY VELOCITY POTENTIAL -
  5*     **   (REFERENCE, LECTURE NOTES, T. J. CHUNG)
  6*     **
  7*     **
  8*  C
  9*  C   NEL  = NR OF ELEMENTS
 10*  C   NDE  = NR OF NODES
 11*  C   NBC  = NR OF DIRICHLET BOUNDARY CONDITIONS (WITH PHI = 0 OR ?)
 12*  C   NVN  = NR OF ELEMENTS ABOVE CREST
 13*  C   NSE  = NR OF ELEMENTS ALONG CYLINDER
 14*  C   NSN  = NR OF NODES ALONG SIDE OF CYLINDER
 15*  C   SK   = GLOBAL COEFFICIENT MATRIX
 16*  C   R    = INPUT VECTOR
 17*  C   AK   = ELEMENT COEFFICIENT MATRIX
 18*  C   XX   = X COORD FOR GLOBAL NODE
 19*  C   YY   = Y COORD FOR GLOBAL NODE
 20*  C   NENN = TABLE OF NODE NRS BY ELEMENT
 21*  C   CP   = PRESSURE ON CYLINDER
 22*  C
 23*        PARAMETER NEL=110,NDE=172,NBC=11,NVN=10
 24*        PARAMETER NSE=6, NSN=7
 25*  C
 26*        COMMON/BLK1/PSI(NDE)
 27*        COMMON/GEOM/XX(NDE), YY(NDE), NENN(NEL,4)
 28*        COMMON/SOLV/AK(4,4)
 29*        COMMON/INFO/W(6), ST(6)
 30*        DIMENSION SK(NDE,NDE)
```

```
31*       DIMENSION R(NDE), NE(NSE), CP(NSN), VX(2), VY(2)
32*      *,NODE(NBC),NODE(NBC),VAL(NBC),JC(NDE)
33*     C
34*     C CLEAR GLOBAL MATRIX
35*       DO 3 I = 1,NDE
36*       DO 3 J = 1,NDE
37*       SK(I,J) = 0.0
38*   3   CONTINUE
39*     C
40*     C READ COORDINATES
41*       READ (5,501) (XX(I),YY(I),I=1,NDE)
42*       WRITE (6,502)  (I,XX(I),YY(I),I=1,NDE)
43*       WRITE (6,503)  (I,XX(I),YY(I),I=1,NDE)
44*     C
45*     C READ ELEMENT NODE NRS
46*       READ(5,504)((NEN(I,J),J=1,4),I=1,NEL)
47*       WRITE(6,505)
48*       WRITE(6,506)(I,(NEN(I,J),J=1,4),I=1,NEL)
49*     C
50*     C WEIGHTS AND ABSCISSA FOR GAUSSIAN QUADRATURE
51*       W(1)=0.1713244923
52*       W(2)=0.3607615730
53*       W(3)=0.4679139345
54*       W(4)=W(3)
55*       W(5)=W(2)
56*       W(6)=W(1)
57*       ST(1)=-0.9324695142
58*       ST(2)=-0.6612093864
```

```
59*         ST(3)=-0.2386191860
60*         ST(4)= 0.2386191860
61*         ST(5)= 0.6612093864
62*         ST(6)= 0.9324695142
63*   C
64*   C     COMPUTE COEFFICIENT MATRIX FOR EACH ELEMENT
65*         DO 5 I=1,NEL
66*         CALL ELMT(I)
67*   C
68*   C     ASSEMBLE  GLOBALLY
69*         DO 10 J = 1,4
70*         JJ = NENN(I,J)
71*         DO 10 K = 1,4
72*         KK = NENN(I,K)
73*         SK(JJ,KK) = SK(JJ,KK) + AK(J,K)
74*   10    CONTINUE
75*   5     CONTINUE
76*   C
77*   C     READ DIRICHLET BOUNDARY CONDITION INFORMATION
78*         WRITE (6,107)
79*         DO 13 I=1,NBC
80*         READ (5,105) MODE(I),NODE(I),VAL(I)
81*         WRITE (6,108) MODE(I),NODE(I),VAL(I)
82*   13    CONTINUE
83*   C
84*         DO 81 I = 1,NDE
85*   81    R(I) = 0.0
86*   C     NONZERO NEUMANN BOUNDARY CONDITIONS
```

```
 87*          C          (DUE TO FREE STREAM VELOCITY INPUT)
 88*          C
 89*                     READ(5,111) THETA,XV,YV
 90*          C
 91*          C          NOTE - LOCAL NODE NRS BEGIN IN LOWER LEFT HAND CORNER
 92*          C                 THEREFORE NODES 1 AND 4 OF ELEMENT 1-10 WILL HAVE
 93*          C                 BOUNDARY CONDITIONS APPLIED
 94*          C
 95*                     DO 83 J = 1,10
 96*                     J1 = NENN(J,1)
 97*                     J4 = NENN(J,4)
 98*                     RL = YY(J4) - YY(J1)
 99*          C
100*          C          SINCE VALUES OF THETA, X VELOCITY, AND Y VELOCITY ARE EQUAL
101*          C          AT ALL NODES, F(1) AND F(2) ARE EQUAL
102*          C
103*                     FV = RL/3.0 *  (XV*COS(THETA) + YV*SIN(THETA) +
104*                    1              XV*COS(THETA)/2.0 + YV*SIN(THETA)/2.0)
105*                     R(J4) = R(J4) + FV
106*                     R(J1) = R(J1) + FV
107*          83         CONTINUE
108*                     WRITE(6,112)(I,R(I),I=1,11)
109*          C          INTRODUCE BOUNDARY CONDITIONS
110*                     DO 82 I=1,NBC
111*                     NX=NODE(I)
112*                     IF (MODE(I).NE.0) GO TO 84
113*                     DO 85 J = 1,NDE
114*                     SK(NX,J)=0.0
115*          85         SK(J,NX)=0.0
                         CONTINUE
```

```
116*              SK(NX,NX)=1.0
117*              GO TO 82
118*      C
119*      C       READJUST INPUT VECTOR FOR BOUNDARY CONDITIONS
120*      84      CONTINUE
121*              DO 86 K = 1,NDE
122*      86      R(K)=R(K)-SK(K,NX)*VAL(I)
123*              DO 87 K = 1,NDE
124*              SK(K,NX)=0.0
125*              SK(NX,K)=0.0
126*      87      CONTINUE
127*              SK(NX,NX)=1.0
128*              R(NX)=VAL(I)
129*      82      CONTINUE
130*      C
131*      C       INVERT GLOBAL MATRIX AND MULTIPLY BY VECTOR
132*              CALL IVMTX(SK,JC,NDE,NDE)
133*              DO 97 I = 1,NDE
134*              PSI(I) = 0.0
135*              DO 97 J = 1,NDE
136*              PSI(I) = PSI(I) + SK(I,J)*R(J)
137*      97      CONTINUE
138*      C
139*      C       WRITE NODAL VELOCITY POTENTIAL FUNCTION
140*              WRITE (6,998)
141*              WRITE(6,999) (I,PSI(I),I=1,NDE)
142*      C
143*      C       CALCULATE VELOCITIES ABOVE CREST OF CYLINDER
```

```
144*        NNN = NEL - NVN + 1
145*        WRITE (6,590)
146*        DO 66 I = NNN,NEL
147*        N1 = NENN(I,1)
148*        N2 = NENN(I,2)
149*        N3 = NENN(I,3)
150*        N4 = NENN(I,4)
151*        MM = 0
152*        XI = 1.0
153*        ETA = 1.0
154* 67     CONTINUE
155*        MM = MM + 1
156*        CALL POMEG( XI,ETA,VX,VY,MM,N1,N2,N3,N4)
157*        ETA = -1.0
158*        IF (MM .LT. 2) GO TO 67
159*        IF (I .EQ. NNN) GO TO 68
160*        VX(1) = (VX(1) + ADX) /2.0
161*        VY(1) = (VY(1) + ADY) /2.0
162* 68     CONTINUE
163*        ADX = VX(2)
164*        ADY = VY(2)
165*        WRITE (6,591)   N3,VX(1),VY(1)
166* 66     CONTINUE
167*        WRITE (6,591)   N2 ,VX(2),VY(2)
168* C
169* C      READ NUMBER OF ELEMENT  ALONG  SIDE
170* C      READ(5,595)(NE(J),J = 1,NSE)
171* C
172* C      CALCULATE VELOCITIES ALONG SIDE OF CYLINDER
```

```
173*       WRITE(6,592)
174*       DO 77 J = 1, NSE
175*       I = NE(J)
176*       N1 = NENN(I,1)
177*       N2 = NENN(I,2)
178*       N3 = NENN(I,3)
179*       N4 = NENN(I,4)
180*       MM = 0
181*       XI = -1.0
182*       ETA = -1.0
183* 78    CONTINUE
184*       MM = MM + 1
185*       CALL POMEG( XI, ETA, VX, VY, MM, N1, N2, N3, N4)
186*       XI = 1.0
187*       IF (MM .LT. 2) GO TO 78
188*       IF (J .EQ. 1) GO TO 79
189*       VX(1) = (VX(1) + ADX) /2.0
190*       VY(1) = (VY(1) + ADY) /2.0
191* 79    CONTINUE
192*       ADX = VX(2)
193*       ADY = VY(2)
194*       WRITE (6,591) N1, VX(1), VY(1)
195*  C    COMPUTE  PRESSURE ON CYLINDER
196*  C    CP(J) = VX(1)**2 + VY(1)**2
197* 77    CONTINUE
198*       CP(NSN) = VX(2)**2 + VY(2)**2
199*       WRITE (6,591) N2, VX(2), VY(2)
2CC*
```

```
C
C         CALCULATE VELOCITES OF EACH ELEMENT AT CENTROID
          WRITE (6,543)
          XI = 0.0
          ETA = 0.0
          DO 80 I = 1,NEL
          N1 = NENN(I,1)
          N2 = NENN(I,2)
          N3 = NENN(I,3)
          N4 = NENN(I,4)
          CALL FOMEG( XI,ETA,VX,VY,NN,N1,N2,N3,N4)
          WRITE (6,591) I,VX(1),VY(1)
       80 CONTINUE
C
C         PRINT PRESSURE
          WRITE(6,506)
          DO 90 J = 1,NSE
          I = NF(J)
          N1 = NENN(I,1)
          WRITE (6,594) N1,CP(J)
       90 CONTINUE
          N2 = NENN(I,2)
          WRITE (6,594) N2,CP(NSN)
          STOP
      868
      105 FORMAT (2I5,F10.0)
      107 FORMAT ('1 DIRICHLET BOUNDARY CONDITIONS',//6X,'MODE',5X,'NODE',
         * 5X,'VALUE'/)
      108 FORMAT (' ',2I9,F10.4)
```

```
230*      111 FORMAT(3F10.5)
231*      112 FORMAT('1 NONZERO NEUMANN BOUND. COND.'//5X,'NODE',5X,'VALUE',/
232*     *            ' ',I8,F10.4))
233*      501 FORMAT(2F10.2)
234*      502 FORMAT('1NODE    X-COORD     Y-COORD')
235*      503 FORMAT(' ',I3,2F12.5)
236*      504 FORMAT(8I5)
237*      505 FORMAT('0ELEMENT NODE NUMBERS',/,' ELMT   1   2   3   4')
238*      506 FORMAT(5I5)
239*      596 FORMAT(1H1,' VELOCITY PROFILE ON CREST OF CYLINDER'//
240*     *            6X,'NODE',10X,'X-VEL',15X,'Y-VEL'/)
241*      591 FORMAT(I10,2F20.8)
242*      592 FORMAT('0 VELOCITY PROFILE ALONG CYLINDER'//
243*     *            6X,'NODE',10X,'X-VEL',15X,'Y-VEL'/)
244*      593 FORMAT('1 VELOCITIES OF EACH ELEMENT AT CENTROID'//
245*     *            6X,'ELMT',10X,'X-VEL',15X,'Y-VEL',/)
246*      594 FORMAT(I10,F20.8)
247*      595 FORMAT(6I5)
248*      596 FORMAT(///' PRESSURE ON CYLINDER'//
249*     *            6X,'NODE',10X,'PRESSURE'/)
250*      998 FORMAT(1H1,'NODAL VELOCITY POTENTIAL FUNCTION'//)
251*      999 FORMAT(I5, F15.8)
252*          END

END OF COMPILATION:    NO DIAGNOSTICS.
```

```fortran
1*        SUBROUTINE ELMT(M)
2*  C
3*  C     SUBROUTINE TO COMPUTE COEFFICIENT MATRIX FOR ISOPARAMETRIC ELEMENT
4*  C
5*  C     X = X COORD FOR LOCAL NODE
6*  C     Y = Y COORD FOR LOCAL NODE
7*  C
8*        PARAMETER NEL=110,NDE=132,NBC=11,NVN=10
9*        COMMON/GEOM/XX(NDE), YY(NDE), NENN(NEL,4)
10*       COMMON/SOLV/AK(4,4)
11*       COMMON/INFO/W(6), ST(6)
12*       DIMENSION DTJ(3),AA(4,4),AB(2,4), AC(2,4), FK(6,4,4)
13*       DIMENSION X(4),Y(4),PS(6)
14* C     FIND LOCAL NODE COORDINATES
15*       DO 6 N=1,4
16*       NN = NENN(M,N)
17*       X(N)=XX(NN)
18*       Y(N)=YY(NN)
19*  6    CONTINUE
20* C
21* C     CALCULATE CONSTANTS IN DERIVATIVE OF INTERPOLATION FUNCTION
22*       AA(1,1) = X(4)-X(2)
23*       AA(1,2) = X(1)-X(3)
```

```
24*        AA(1,3) = X(2)-X(4)
25*        AA(1,4) = X(3)-X(1)
26*        AA(2,1) = Y(2)-Y(4)
27*        AA(2,2) = Y(3)-Y(1)
28*        AA(2,3) = Y(1)-Y(3)
29*        AA(2,4) = X(3)-X(4)
30*        AB(1,1) = X(4)-X(3)
31*        AB(1,2) = X(2)-X(1)
32*        AB(1,3) = X(1)-X(2)
33*        AB(1,4) = X(4)-Y(4)
34*        AB(2,2) = Y(3)-Y(4)
35*        AB(2,3) = Y(1)-Y(2)
36*        AB(2,4) = Y(2)-Y(1)
37*        AC(1,1) = X(4)-X(1)
38*        AC(1,2) = X(1)-X(4)
39*        AC(1,3) = X(1)-X(4)
40*        AC(1,4) = X(3)-X(2)
41*        AC(2,1) = Y(3)-Y(4)
42*        AC(2,2) = Y(1)-Y(4)
43*        AC(2,3) = Y(4)-Y(1)
44*        AC(2,4) = Y(2)-Y(3)
45*   C
46*   C
47*   C    COMPUTE JACOBIAN
48*        DTJ(1)=(X(4)-X(2))*(Y(1)-Y(3))-(X(1)-X(3))*(Y(4)-Y(2))
49*        DTJ(2)=(X(3)-X(4))*(Y(1)-Y(2))-(X(1)-X(2))*(Y(3)-Y(4))
50*        DTJ(3)=(X(4)-X(1))*(Y(2)-Y(3))-(X(2)-X(3))*(Y(4)-Y(1))
51*   C
```

```
C     COMPUTE ELEMENT COEFFICIENT MATRIX
      DO 71 I=1,4
      DO 72 J=1,4
      AK(I,J) = 0.0
      FK(1,I,J)=AA(1,I)*AA(1,J)+AA(2,I)*AA(2,J)
      FK(2,I,J)=AA(1,J)*AB(1,I)+AA(2,J)*AB(2,J)
     *+AA(1,J)*AB(1,I)+AA(2,J)*AB(2,I)
      FK(3,I,J)=AA(1,J)*AC(1,I)+AA(2,J)*AC(2,J)
     *+AA(1,J)*AC(1,I)+AA(2,J)*AC(2,I)
      FK(4,I,J)=AB(1,I)*AC(1,J)+AB(2,I)*AC(2,J)
     *+AB(1,J)*AB(1,I)+AB(2,J)*AB(2,I)
      FK(5,I,J)=AB(1,I)*AB(1,J)+AB(2,I)*AB(2,J)
      FK(6,I,J)=AC(1,I)*AC(1,J)+AC(2,I)*AC(2,J)
C
C     GAUSSIAN QUADRATURE INTEGRATION
      DO 35 K = 1,6
      DO 35 L = 1,6
C     ISOPARAMETRIC VARIABLES
      PS(1) = 1.0
      FS(2) = ST(K)
      PS(3) = ST(L)
      PS(4)=ST(K)*ST(L)
      PS(5)=ST(K)*ST(K)
      PS(6) = ST(L)*ST(L)
      DJA = 0.0
      DO 43 II = 1,3
   43 DJA = DJA + DTJ(II)*PS(II)/8.0
      ACOF=W(K)*W(L)    /(DJA*64.)
      DO 47 KK = 1,6
```

```
81*       AK(I,J)=AK(I,J) + FK(KK,I,J)*PS(KK)*ACOF
82*    47 CONTINUE
83*    35 CONTINUE
84*    72 CONTINUE
85*    71 CONTINUE
86*       RETURN
87*       END

 1*       SUBROUTINE POMEG(X,Y,VX,VY,M,N1,N2,N3,N4)
 2* C
 3* C     SUBROUTINE TO COMPUTE VELOCITY
 4* C
 5*       PARAMETER NEL=110,NDE=132,NBC=11,NVN=10
 6*       DIMENSION DTJ(3),POPZ(4),POPN(4),VX(2),VY(2)
 7*       COMMON/BLK1/PSI(NDE)
 8*       COMMON/GEOM/ XX(NDE), YY(NDE), NENN(NEL,4)
 9* C
10* C     CALCULATE DERIVITIVE OF FUNCTION
11*       POPZ(1) = (-1.0 + Y) /4.0
12*       POPZ(2) = (+1.0 - Y) /4.0
13*       POPZ(3) = (+1.0 + Y) /4.0
14*       POPZ(4) = (-1.0 - Y) /4.0
15*       POPN(1) = (-1.0 + X) /4.0
16*       POPN(2) = (-1.0 - X) /4.0
17*       POPN(3) = (+1.0 + X) /4.0
18*       POPN(4) = (+1.0 - X) /4.0
```

```
19*       PPPZ = POPZ(1) * PSI(N1) + POPZ(2) * PSI(N2)
20*      *     + POPZ(3) * PSI(N3) + POPZ(4) * PSI(N4)
21*       PPPN = POPN(1) * PSI(N1) + POPN(2) * PSI(N2)
22*      *     + POPN(3) * PSI(N3) + POPN(4) * PSI(N4)
23*       PYPZ = POPZ(1) * YY(N1) + POPZ(2) * YY(N2)
24*      *     + POPZ(3) * YY(N3) + POPZ(4) * YY(N4)
25*       PYPN = POPN(1) * YY(N1) + POPN(2) * YY(N2)
26*      *     + POPN(3) * YY(N3) + POPN(4) * YY(N4)
27*       PXPZ = POPZ(1) * XX(N1) + POPZ(2) * XX(N2)
28*      *     + POPZ(3) * XX(N3) + POPZ(4) * XX(N4)
29*       PXPN = POPN(1) * XX(N1) + POPN(2) * XX(N2)
30*      *     + POPN(3) * XX(N3) + POPN(4) * XX(N4)
31*       DTJ(1) = (XX(N4) - XX(N2))*(YY(N1) - YY(N3)) - (XX(N1) - XX(N3))*
32*      *         (YY(N4) - YY(N2))
33*       DTJ(2) = (XX(N3) - XX(N4))*(YY(N1) - YY(N2)) - (XX(N1) - XX(N2))*
34*      *         (YY(N3) - YY(N4))
35*       DTJ(3) = (XX(N4) - XX(N1))*(YY(N2) - YY(N3)) - (XX(N2) - XX(N3))*
36*      *         (YY(N4) - YY(N1))
37*       DET = DTJ(1) + DTJ(2) * X + DTJ(3) * Y
38*       DJ11 = 8.0 * (+PYPN) / DET
39*       DJ12 = 8.0 * (-PYPZ) / DET
40*       DJ21 = 8.0 * (-PXPN) / DET
41*       DJ22 = 8.0 * (+PXPZ) / DET
42*       VX(M) = -(DJ11*PPPZ + DJ12*PPPN)
43*       VY(M) = -(DJ21*PPPZ + DJ22*PPPN)
44**       RETURN
45*       END
```

END OF COMPILATION: NO DIAGNOSTICS.

SEE B-1 FOR SUBROUTINE IVMTX

A

B-3

```
1*    C ************************************************************
2*    C ************************************************************
3*    C **
4*    C **    AXISYMMETRIC IDEAL FLOW AROUND SPHERE WITH ISOPARAMETRIC
5*    C **    ELEMENTS BY STREAMFUNCTION FORMULATION -
6*    C **    (REFERENCE, LECTURE NOTES, T. J. CHUNG)
7*    C **
8*    C ************************************************************
9*    C ************************************************************
10*   C
11*   C    NOTATIONS USED IN CLASS NOTES
12*   C    AA(1,N)=AZ(N)                    AA(2,N)=AR(N)
13*   C    AB(1,N)=BZ(N)                    AB(2,N)=BR(N)
14*   C    AC(1,N)=CZ(N)                    AC(2,N)=CR(N)
15*   C    BB(N) = B(N)          BC(N) = C(N)          DD(N) = D(N)
16*   C    NELE = NR OF ELEMENTS
17*   C    NNODE = NR OF NODES
18*   C    NBC = NR OF DIRICHLET BOUNDARY CONDITIONS (WITH PSI = 0 OR ?)
19*   C    XX = X COORD FOR GLOBAL NODE
20*   C    YY = Y COORD FOR GLOBAL NODE
21*   C    X = X COORD FOR LOCAL NODE
22*   C    Y = Y COORD FOR LOCAL NODE
23*   C    IE = TABLE OF GLOBAL NODE NRS BY ELEMENT
24*   C    SK = GLOBAL MATRIX
25*   C    R = INPUT VECTOR
26*   C    AK = ELEMENT COEFFICIENT MATRIX
27*   C
28*   C
```

```
C **WARNING - ALL NODES ALONG THE VERTICAL LINE ABOVE THE CREST
C            OF THE SPHERE SHOULD BE NUMBERED CONSECUTIVELY
C            FROM TOP TO BOTTOM IN ORDER TO BE COMPATABLE
C            WITH VELOCITY CALCULATIONS USED IN THIS PROGRAM.
C
      PARAMETER NELE=96, NNODE=117, NBC=37
      PARAMETER NVN=9, NHN=13
C
      DIMENSION AA(2,4),AB(2,4),AC(2,4),BB(4),BC(4),BD(4),
     *          DTJ(3),XR(4),W(4),ST(6),PS(8),AK(4,4),PSI(NNODE)
     *         ,YF(4,4),YG(4,4),X(4),Y(4),FK(6,4,4),GK(8,4,4)
     *         ,IE(NELE,4),XX(NNODE),YY(NNODE),SK(NNODE,NNODE),R(NNODE)
     *         ,NODE(NBC),NODE(NBC),VAL(NBC),JC(NNODE)
C
C     READ NODE NRS AND COORDINATES
      WRITE (6,203)
   11 READ (5,102) N,XX(N),YY(N)
      WRITE(6,204) N,XX(N),YY(N)
      IF (N .LT. NNODE) GO TO 11
C
C     READ ELEMENT NR AND GLOBAL NODE NRS
      WRITE (6,200)
   12 READ (5,103) M,(IE(M,I),I=1,4)
      WRITE(6,206) M,(IE(M,I),I=1,4)
      IF (M .LT. NELE) GO TO 12
C
C     READ DIRICHLET BOUNDARY CONDITION INFORMATION
      WRITE (6,107)
```

```
      DO 13 I=1,NBC
      READ (5,105) NODE(I),NODF(I),VAL(I)
      WRITE (6,108) NODE(I),NODF(I),VAL(I)
   13 CONTINUE
C
C  WEIGHTS AND ABSCISSA FOR GAUSSIAN QUADRATURE
      W(1)=0.1713244923
      W(2)=0.3607615730
      W(3)=0.4679139345
      W(4)=W(3)
      W(5)=W(2)
      W(6)=W(1)
      ST(1)=-0.9324695142
      ST(2)=-0.6612093864
      ST(3)=-0.2386191860
      ST(4)= 0.2386191860
      ST(5)= 0.6612093864
      ST(6)= 0.9324695142
C
      BB(1) = -1.
      BB(2) = -1.
      BB(3) = -1.
      BB(4) = -1.
      BC(1) = -1.
      BC(2) = -1.
      BC(3) = -1.
      BC(4) = -1.
      BD(1) = -1.
      BD(2) = -1.
      FD(3) = -1.
```

```
                 RD(4) = -1.0
C
C     CLEAR GLOBAL MATRIX
      DO 38 I = 1,NNODE
      DO 38 J = 1,NNODE
38    SK(I,J)=0.0
C
C     CALCULATE COEFFICIENT MATRIX FOR EACH ELEMENT
      DO 5 M=1,NELE
C
C     FIND LOCAL NODE COORDINATES
      DO 6 N=1,4
      NN=IF(M,N)
      X(N)=XX(NN)
      Y(N)=YY(NN)
6     CONTINUE
C
C     CALCULATE CONSTANTS IN DERIVATIVE OF INTERPOLATION FUNCTION
      AA(1,1) = X(4)-X(2)
      AA(1,2) = X(1)-X(3)
      AA(1,3) = X(2)-X(4)
      AA(1,4) = X(3)-X(1)
      AA(2,1) = Y(2)-Y(4)
      AA(2,2) = Y(3)-Y(1)
      AA(2,3) = Y(4)-Y(2)
      AA(2,4) = Y(1)-Y(3)
      AB(1,1) = X(3)-X(4)
      AB(1,2) = X(4)-X(3)
      AB(1,3) = X(2)-X(1)
      AB(1,4) = X(1)-X(2)
      AB(2,1) = Y(4)-Y(3)
```

```
117*          AB(2,2) = Y(3)-Y(4)
118*          AB(2,3) = Y(1)-Y(2)
119*          AB(2,4) = Y(2)-Y(1)
120*          AC(1,1) = X(2)-X(3)
121*          AC(1,2) = X(4)-X(1)
122*          AC(1,3) = X(1)-X(4)
123*          AC(1,4) = X(2)-X(2)
124*          AC(2,1) = Y(3)-Y(2)
125*          AC(2,2) = Y(1)-Y(4)
126*          AC(2,3) = Y(4)-Y(1)
127*          AC(2,4) = Y(2)-Y(3)
128*    C
129*    C  COMPUTE JACOBIAN
130*          DTJ(1)=(X(2)-X(4))*(Y(3)-Y(1))-(X(3)-X(1))*(Y(2)-Y(4))
131*          DTJ(2)=(X(4)-X(3))*(Y(2)-Y(1))-(X(2)-X(1))*(Y(4)-Y(3))
132*          DTJ(3)=(X(1)-X(3))*(Y(2)-Y(2))-(X(3)-X(2))*(Y(1)-Y(4))
133*    C
134*    C  INTERPOLATION FIELD
135*          XXR(1)=Y(1)+Y(2)+Y(3)+Y(4)
136*          XXR(2)=-Y(1)+Y(2)+Y(3)+Y(4)
137*          XXR(3)=-Y(1)-Y(2)+Y(3)+Y(4)
138*          XXR(4)=Y(1)-Y(2)+Y(3)-Y(4)
139*    C
140*    C  COMPUTE ELEMENT COEFFICIENT MATRIX
141*          DO 71 I=1,4
142*          DO 72 J=1,4
143*          XF(I,J)=0.0
144*          YG(I,J)=0.0
145*          FK(1,I,J)=AA(1,I)*AA(1,J)+AA(2,I)*AA(2,J)
146*          FK(2,I,J)=AA(1,I)*AB(1,J)+AA(2,I)*AB(2,J)
```

```
147*          *            +AA(1,J)*AE(1,I)+AA(2,J)*AE(2,I)
148*       FK(3,I,J)=AA(1,I)*AC(1,J)+AA(2,I)*AC(2,J)
149*          *            +AA(1,J)*AC(1,I)+AA(2,J)*AC(2,I)
150*       FK(4,I,J)=AA(1,I)*AC(1,J)+AA(2,I)*AC(2,J)
151*          *            +AA(1,J)*AE(1,I)+AA(2,J)*AC(2,I)
152*       FK(5,I,J)=AC(1,I)*AC(1,J)+AE(2,I)*AC(2,J)
153*       FK(6,I,J)=AC(1,I)*AC(1,J)+AE(2,I)*AC(2,J)
154*       GK(1,I,J)=AA(1,J)+BD(I)+AE(1,J)
155*       GK(2,I,J)=AA(1,J)+BC(I)+AC(1,J)
156*       GK(3,I,J)=AA(1,J)+BC(I)+AC(1,J)
157*       GK(4,I,J)=BC(I)*AE(1,J)+BC(I)*AC(1,J)+BF(I)*AC(1,J)+ED(I)*AA(1,J)
158*       GK(5,I,J)=BD(I)*BC(I)*AC(1,J)
159*       GK(6,I,J)=BC(I)*AE(1,J)+AC(1,J)
160*       GK(7,I,J)=BD(I)*AC(1,J)
161*       GK(8,I,J)=BD(I)*AC(1,J)
162*    C
163*    C     GAUSSIAN QUADRATURE INTEGRATION
164*          DO 35 K = 1,6
165*          DO 35 L = 1,6
166*    C
167*    C     ISOPARAMETRIC VARIABLES
168*          PS(1) = 1.0
169*          PS(2) = ST(K)
170*          PS(3) = ST(L)
171*          PS(4)=ST(K)*ST(L)
172*          PS(5)=ST(K)*ST(K)
173*          PS(6) = ST(L)*ST(L)
174*          PS(7) = ST(K)*ST(K)*ST(L)
175*          PS(8) = ST(K)*ST(L)*ST(L)*ST(L)
176*          DJA = 0.0
```

```
      DO 43 II = 1,3
   43 DJA = DJA + DTJ(II)*PS(II)/8.0
      XR = 0.0
      DO 42 II = 1,4
   42 XR = XR + XXR(II)*PS(II)/4.0
      ACOF=W(K)*W(L)*XR/(DJA*64.0)
      DO 47 KK = 1,6
      XF(I,J)=XF(I,J) + FK(KK,I,J)*PS(KK)*ACOF
   47 CONTINUE
      APGF=W(K)*W(L)/16.
      DO 46 KK = 1,8
      XG(I,J) = XG(I,J) + GK(KK,I,J)*PS(KK)*APGF
   46 CONTINUE
   45 CONTINUE
      AK(I,J) = XF(I,J)+XG(I,J)
   72 CONTINUE
   71 CONTINUE
C
C     ASSEMBLE GLOBAL COEFFICIENT MATRIX
C
      DO 65 JJ = 1,4
      J = IE(N,JJ)
      DO 65 KK = 1,4
      K = IE(M,KK)
      SK(J,K) = SK(J,K) + AK(JJ,KK)
   65 CONTINUE
    5 CONTINUE
      DO 81 I = 1,NNODE
   81 R(I) = 0.0
C
```

```
207*  C     INTRODUCE BOUNDARY CONDITIONS
208*        DO 82 I=1,NFC
209*        NX=NODE(I)
210*        IF (NODE(I).NE.0) GO TO 84
211*        DO 85 J = 1,NNODE
212*        SK(NX,J)=0.0
213*        SK(J,NX)=0.0
214*  85    CONTINUE
215*        SK(NX,NX)=1.0
216*        GO TO 82
217*  C
218*  C     COMPUTE INPUT VECTOR
219*  84    CONTINUE
220*        DO 86 K = 1,NNODE
221*        R(K)=R(K)-SK(K,NX)*VAL(I)
222*        DO 87 K = 1,NNODE
223*        SK(K,NX)=0.0
224*        SK(NX,K)=0.0
225*  87    CONTINUE
226*        SK(NX,NX)=1.0
227*        R(NX)=VAL(I)
228*  82    CONTINUE
229*  C
230*  C     INVERT GLOBAL MATRIX AND MULTIPLY BY VECTOR
231*        CALL IVMTX(SK,JC,NNODE,NNODE)
232*        DO 97 I = 1,NNODE
233*        PSI(I) = 0.0
234*        DO 97 J = 1,NNODE
235*        PSI(I) = PSI(I) + SK(I,J)*R(J)
236*  97    CONTINUE
237*  C
```

```fortran
238*   C   WRITE NODAL STREAMFUNCTION DATA
239*       WRITE (6,998)
240*       WRITE(6,999) (I,PSI(I),I=1,NNODE)
241*   C
242*   C   VELOCITY PROFILE ON CREST OF SPHERE
243*       NNN = NNODE - NVN + 1
244*       MMM = NNODE - 1
245*       WRITE (6,590)
246*       DO 555 I = NNN,MMM
247*       J = I + 1
248*       RD= YY(I)**2 -YY(J)**2
249*       SID= (PSI(I) -PSI(J))*2
250*       VEL= SID/RD
251*       RM = (YY(I) + YY(J))/2.0
252*       WRITE(6,591) RM,VEL
253*   555 CONTINUE
254*   808 STOP
255*   102 FORMAT (I5,2F10.0)
256*   103 FORMAT (5I5)
257*   105 FORMAT (2I5,F10.0)
258*   106 FORMAT (2I5)
259*   107 FORMAT ('1 DIRICHLET BOUNDARY CONDITIONS',//6X,'MODE',5X,'NODE',
260*       * 5X,'VALUE'/)
261*   108 FORMAT (' ',2I9,F10.4)
```

```
263*      110 FORMAT (' ',2I9)
264*      203 FORMAT (1H1,20H NODAL POINT DATA    //
265*        1 5X,12H NODAL POINT,15X,1HX,20X,1HY )
266*      204 FORMAT (I12,6X,F16.4,4X,F16.4)
267*      206 FORMAT (5I10)
268*      400 FORMAT (1H1,'NODAL ASSEMBLY DATA'// 4X,'ELEMENT',5X,'1',9X,'2',
269*        * 9X,'3',9X,'4'//)
270*      590 FORMAT (1H1,' VELOCITY PROFILE ON CREST OF SPHERE'///
271*        * 6X,'Y',16X,'VEL'//)
272*      591 FORMAT (F10.5,F20.8/ )
273*      998 FORMAT(1H1,'NODAL STREAMFUNCTION DATA'//)
274*      999 FORMAT (I5, F15.8)
275*          END

END OF COMPILATION:      NO DIAGNOSTICS.


SEE B-1 FOR SUBROUTINE IVMTX
```

REFERENCES

Adams, R. A. [1975]: *Sobolev Spaces*, Academic Press, New York.

Aguirre-Ramerez, G., Oden, J. T., and Wu, S. T. [1973]: *A Numerical Solution of the Boltzmann Equation by the Finite Element Technique*, Proc., 7th Intl. Symp. Rarefied Gas Dyn., Universita Degli Studi Di Pisa, Pisa, Academic Press, New York.

Akin, J. E. [1973]: *A Least Squares Finite Element Solution of Nonlinear Operators*, The Mathematics of Finite Elements and Applications, Ed. Whiteman, Academic Press, London.

Arakawa, A. [1966]: Computational Design for Long-Term Numerical Integration of the Equations of Fluid Motion: Two-Dimensional Incompressible Flow, Part I, *J. comp. Physics*, I, pp. 119–143.

Argyris, J. H. [1963]: *Recent Advances in Matrix Methods of Structural Analysis*, Pergamon Press, Elmsford, New York.

Argyris, J. H. and Mareczek, G. [1972]: Potential Flow Analysis by Finite Elements, *Ingenieur-Archiv.*, **41**, pp. 1–25.

Argyris, J. H., and Scharpf, D. W. [1969]: The Incompressible Lubrication Problem, 12th Lanchester Memorial Lecture, Appendix IV, *The Aeronautical Journal of the Royal Aeronautical Soc.*, Vol. 73, pp. 1044–1046.

Aubin, J. P. [1967]: Behavior of the Error of the Approximate Solutions of Boundary-Value Problems for Linear Elliptic Operators by Galerkin's and Finite Difference Method, *Annal. della Scuola Normale di Pisa*, Series, Vol. 21, pp. 599–637.

Aubin, J. P. [1972]: *Approximation of Elliptic Boundary Value Problems*, Wiley-Interscience, New York.

Babuska, I. [1970]: Finite Element Method for Domains with Corners, *Computing 6*, pp. 264–273.

Babuska, I. and Aziz, A. K. [1972]: Lectures on the Mathematical Foundations of the Finite Element Method, Mathematical Foundations of the Finite Element Method with Applications to Partial Differential Equations, A. K. Aziz (Ed.), Academic Press, New York, pp. 1–345.

Baker, A. J. [1972]: Finite Element Computation Theory for Three-Dimensional Boundary Layer Flow, *AIAA*, Paper 72-108.

Baker, A. J. [1973]: Finite Element Solution Algorithm for Viscous Incompressible Fluid Dynamics, *Intl. J. numer. Meth. in Engng*, **6**, 1, pp. 89–101.

Baker, A. J. [1975]: *Finite Element Solution Algorithm for Incompressible Fluid Dynamics*, Finite Elements in Fluids, Vol. 2, Eds Gallagher, Oden, Taylor, and Zienkiewicz; John Wiley and Sons, New York.

Barnhill, R. E. and Whiteman, J. R. [1973]: *Error Analysis of Finite Element Methods with Triangles for Elliptic Boundary Value Problems, The Mathematics of Finite Elements and Applications*, J. R. Whiteman (Ed.), Academic Press, London, pp. 83–112.

Bateman, H. [1929]: Notes on a Differential Equation Which Occurs in the Two-Dimensional Motion of a Compressible Fluid and the Associated Variational Problems, *Proc. Roy. Soc. (London), Ser. A*, Vol. 125, No. 700, pp. 598–618.

Bateman, H. [1930]: Irrotational Motion of a Compressible Inviscid Fluid, *Oroc. National Acad. Sci. U.S.A.*, **16**, pp. 816–825.

Berkhoff, J. C. M. [1975]: *Linear Weave Propagation Problems and the Finite Element Method, Finite Elements in Fluids*, Eds R. H. Gallagher, J. T. Oden, C. Taylor, O. C. Zeinkiewicz; John Wiley and Sons, New York.

Bernstein, I. B., Frieman, E. A., Kruskal, M. D., and Kulsrud, R. M. [1958]: An Energy Principle for Hydromagnetic Stability Problems, *Proc. Roy. Soc.*, **A-244**, 17.

Bhatnagar, E. P., Gross, E. P., and Krook, M. [1954]: A Model for Collision Processes in Gases; I. Small Amplitude Processes in Charged and Neutral One-Component Systems, *Phys. Rev.*, **94**, p. 5111.

Birkhoff, G., Schultz, M. H., and Varga, R. [1968]: Piecewise Hermite Interpolation in One and Two Variables with Applications to Partial Differential Equations, *Numer. Meth.*, **11**, pp. 232–256.

Boyd, T. J. M., Gardner, G. A., and Gardner, L. R. T. [1973]: Numerical Study of MHD Stability Using the Finite Element Method, *Nuclear Fusion*, **13**, p. 764.

Boyd, T. J. M., Gardner, G. A., and Gardner, L. R. T. [1975]: *Hydromagnetic Stability Studies Using the Finite Element Method*, Finite Elements in Fluids, Vol. 2, Eds Gallagher, Oden, Taylor, and Zienkiewicz, pp. 255–274.

Bozeman, J. D., and Dalton, C. [1973]: Numerical Study of Viscous Flow in A Cavity, *J. comp. Phy.*, **12**, pp. 348–363.

Brainerd, J. J. [1969]: Shock Wave Similarity, Recent Advances in Engineering Sciences, Ed. A. C. Eringen, **12**, 4, pp. 173–193.

Bramble, J. H., and Hilbert, S. R. [1970]: Estimation of Linear Functionals on Sobolev Spaces with Application to Fourier Transforms and Spline Interpolation, *SIAM J. of Numerical Analysis*, **7**, pp. 112–124.

Bramble, J. H., and Hilbert, S. R. [1971]: Bounds for A Class of Linear Functionals with Applications to Hermite Interpolation, *Numer. Math.*, **16**, pp. 362–369.

Bramlette, T. T. [1971]: Plane Poiseuille Flow of A Rarefied Gas Based on the Finite Element Method, *Phys. of Fluids*, **14**, No. 2, pp. 288–293.

Bramlette, T. T., and Mallet, R. H. [1970]: A Finite Element Solution Technique for the Boltzmann Equation, *J. Fluid Mech.*, **42**, p. 177.

Bratanow, T., Ecer, A., and Kobiske, M. [1973]: Finite Element Analysis of Unsteady Incompressible Flow around An Oscillating Obstacle of Arbitrary Shape, *AIAA J.*, **11**, No. 11, pp. 1471–1477, Nov.

Carey, G. F. [1975]: *A Dual Perturbation Expansion and Variational Solution for Compressible Flows Using Finite Elements, Finite Elements in Fluids*, Vol. 2, Eds Gallagher, Oden, Taylor, and Zienkiewicz, John Wiley & Sons, pp. 159–177.

Carlson, G. A., and Hornbeck, R. W. [1973]: A Numerical Solution for Laminar Entrance Flow in A Square Duct, *J. Appl. Mech. T.A.S.M.E., Series E.*, pp. 25–30.

Chan, S. T. K., and Brashears, M. R. [1975]: Finite Element Analysis of Unsteady Transonic Flow, AIAA Paper No. 75-875, June.

Chan, S. T. K., Brashears, M. R., and Young, V. Y. C. [1975]: Finite Element Analysis of Transonic Flow by the Method of Weighted Residuals, *AIAA Paper* No. 75-79.

Chan, S. T., and Larock, B. E. [1972]: Potential Flow around A Cylinder Between Parallel Walls by Finite Element Methods, *J. Engng. mech. Div., ASCE*, **98**, No. EM 5, pp. 1317–1322.

Cheng, R. T. [1972a]: Numerical Investigation of Lake Circulation around Islands by the Finite Element Method, *Int. J. numer. Meth. in Engng.*, **5**, No. 1, pp. 103–112.

Cheng, R. T. [1972b]: Numerical Solution of the Navier–Stokes Equations by the Finite Element Method, *Physics of Fluids*, **15**, No. 12, pp. 2093–2105, Dec.

Chung, T. J. [1975]: Convergence and Stability of Nonlinear Finite Element Equations, *AIAA J.*, **13**, No. 7, pp. 963–966.

Chung, T. J., and Chiou, J. N. [1976]: Analysis of Unsteady Compressible Boundary Layer Flow via Finite Elements, *Intl. J. for Computers and Fluids*, **4**, No. 1A, pp. 1–12.

Chung, T. J., and Hooks, C. G. [1976]: Discontinuous Functions in Transonic Flow, *AIAA Paper* No. 76-329.

Chung, T. J., and Hooks, C. G. [1977]: Discontinuous Trial and Test Functions in Full Potential Equation, Proc., Intl. Symposium on Innovative Numerical Analysis in Engineering and Applied Sciences, Versailles, France, May 23–27, 1977.

Chung, T. J., Oden, J. T., and Wu, S. T. [1974]: The Finite Element Analysis of Transient Rarefied Gas Flow, *Proc., Intl. Symposium on Finite Element Methods in Flow Problems*, Swansea, England, Jan.

Chung, T. J., and Rush, R. H. [1976]: Dynamically Coupled Motion of Surface-Fluid-Shell Systems, *J. appl. Mech., T.A.S.M.E.*, **43**, Series E, No. 3, pp. 507–508.

Ciarlet, P. G., and Raviart, P. A. [1972]: General Lagrange and Hermite Interpolation in R^n with Applications to the Finite Element Method, *Arch. Rat. Mech. Anal.*, **46**, pp. 177–199.

Clough, R. W. [1960]: The Finite Element Method in Plane Stress Analysis, *Proceedings of 2nd Conf. on Electronic Computation, American Society of Civil Engineers*, Pittsburgh, Penn., pp. 345–378.

Cole, J. D. [1951]: On A Quasilinear Parabolic Equation Occurring in Aerodynamics, *Quart. Appl. Math.*, **9**, pp. 225–236.

Courant, R., Friedrichs, K. O., and Lewy, H. [1947]: The Partial Equations of Mathematical Physics, *IBM J.*, 1967, p. 215. (Trans. of original paper in *Math. Ann.*, **100**, p. 32.)

Courant, R., and Hilbert, D. [1953]: *Methods of Mathematical Physics*, Interscience Publishers, New York, Vols 1 and 2.

Cowper, G. R., Kosko, E., Lindberg, G. M., and Olson, M. D. [1969]: Static and Dynamic Applications of A High Precision Triangular Plate Bending Element, *AIAA J.*, **7**, pp. 1957–1965.

Crandall, S. H. [1956]: *Engineering Analysis,* McGraw-Hill, New York.

Crank, J., and Nicolson, P. [1947]: A Practical Method for Numerical Evaluation of Solutions of Partial Differential Equations of the Heat Conduction Type, Proc., *Cambridge Philosophical Soc.*, **43**, No. 50, pp. 50–67.

Dennis, S. C. R., and Chang, G. Z. [1970]: Numerical Solutions for Steady Flow Past a Circular Cylinder at Reynolds Numbers up to 100, *J. Fluid Mech.* **42**, pp. 471–489.

Desai, C. S. [1975]: *Finite Element Methods for Flow in Porous Media, Finite Elements in Fluids*, **1**, Eds Gallagher, Oden, Taylor, and Zienkiewicz, John Wiley & Sons, New York.

Douglas, J., and Dupont, T. [1970]: Galerkin Methods for Parabolic Equations, *SIAM J. Numer. Anal.*, **7**, No. 4, pp. 575–626.

Dupont, T. [1973]: L^2—Estimates for Galerkin Methods for Second Order Hyperbolic Equations, *SIAM, J. Numer. Anal.*, **10**, pp. 880–889.

Ecer, A., and Akay, H. U. [1976]: Application of Finite Element Method to the Solution of Transonic Flow, *Proc., 2nd Intl. Symposium on Finite Element Methods in Flow Problems*, Italy, June.

Eisenstat, S. C., and Schultz, M. H. [1972]: *Computational Aspects of the Finite Element Method, The Mathematical Foundations of the Finite Element Method with Applications to Partial Differential Equations*, Ed. A. K. Aziz, Academic Press, New York, pp. 505–524.

Eringen, A. C. [1967]: *Mechanics of Continua*, John Wiley & Sons, New York.

Evans, M. E., and Harlow, F. H. [1957]: The Particle-in-Cell Method for Hydrodynamic Calculations, Los Alamos Scientific Lab., Report No. LA-2139, Los Alamos, New Mexico.

Finlayson, B. A. [1972]: *The Method of Weighted Residuals and Variational Principles*, Academic Press, New York.

Fix, G., and Nassif, N. [1972]: On Finite Element Approximation to Time Dependent Problems, *Numer. Meth.*, **19**, pp. 127–135.

Fried, I. [1969]: Finite-Element Analysis of Time-Dependent Phenomena, *AIAA J.*, **7**, No. 6, pp. 1170–1172.

Fried, I. [1971]: Discretization and Computational Error in Higher-Order Finite Elements, *AIAA*, **9**, pp. 2071–2073.

Fujii, H. [1972]: *Finite Element Schemes: Stability and Convergence. Advances in Computational Methods in Structural Mechanics and Design*, Eds J. T. Oden, R. W. Clough, and Y. Yamamoto, UAH Press, Huntsville.

Galerkin, B. G. [1915]: Rods and Plates, Series Occurring in Various Questions Concerning the Elastic Equilibrium of Rods and Plates (in Russian), *Vestn. Inghenevov*, **19**, pp. 897–908.

Gallagher, R. H. [1975]: *Finite Element Lake Circulation and Thermal Analysis, Finite Elements in Fluids*, Eds Gallagher, Oden, Taylor, and Zienkiewicz, Vol. 1, John Wiley and Sons, New York.

Gallagher, R. H., and Chan, S. T. K. [1973b]: *Higher-Order Finite Element Analysis of Lake Circulation, Computers and Fluids*, Vol. I, Pergamon Press, pp. 119–132.

Gallagher, R. H., Liggett, J., and Chan, S. [1973a]: Finite Element Analysis of Circulation in Variable-Depth Shallow Homogenous Lakes, *J. Hyd. Div., ASCE*, 99.

Gelder, D. [1971]: Solution of the Compressible Flow Equations, *Intl. J. num. Meth. in Engng*, **3**, pp. 35–43.

Glowinski, R., Periaux, J., and Pironneau, O. [1976]: Transonic Flow Simulation by the Finite Element Method via Optimal Control, *Proc., 2nd Intl. Symp. Finite Element Methods in Flow Problems*, ICCAD, Italy.

Gross, E. P., Jackson, E. A., and Ziering, S. [1957]: Boundary Value Problems in Kinetic Theory of Gases, *Annals of Phys.*, **1**, pp. 141–167.

Habashi, W. G. [1976]: Compressible Potential Flows by the Finite Element Method, *Proc., 2nd Intl. Symp. Finite Element Methods in Flow Problems*, ICCAD, Italy.

Heinrich, J. C., Huyakorn, P. S., Zienkiewicz, O. C., and Mitchell, A. R. [1977]: An Upwind Finite Element Scheme for Two-Dimensional Convective Transport Equation, *Intl. J. num. Meth. Engng*, **11**, No. 1, pp. 131–144.

Herrivel, J. W. [1955]: The Derivation of the Equations of an Ideal Flow by Hamilton's Principle, *Proc. Camb. Phil. Soc.*, **51**, pp. 344–349.

Hildebrand, F. B. [1956]: *Introduction to Numerical Analysis*, McGraw-Hill, New York.

Huebner, K. H. [1975]: *The Finite Element Method for Engineers*, John Wiley and Sons, New York.

Imai, I. [1941]: On the Flow of a Compressible Fluid Past a Circular Cylinder, II, *Proc. Phys. Math. Soc. Japan*, **23**, pp. 180–193.

Jameson, A. [1975]: Transonic Potential Flow Calculations Using Conservation Form, *AIAA Paper 75-000*, June.

Jamet, P., and Raviart, P. A. [1974]: Numerical Solution of Stationary Navier–Stokes Equation by Finite Element Method, Computing Methods in Applied Science Lecture Notes on Computer Science, Dec. (3), Springer-Verlag.

Jespersen, D. C. [1974]: Arakawa's Method Is A Finite Element Method, *J. Comput. Phys.*, **16**, pp. 383–390.

Jost, W. [1960]: *Diffusion in Solids, Liquids, Gasses*, Academic Press, New York.

Kawahara, M., and Okamota, T. [1976b]: Finite Element Analysis of Steady Flow of Viscous Fluid Using Stream Function, *Proc. JSCE*, No. 247, pp. 123–135.

Kawahara, M., Yoshimura, K., and Ohsaka, H. [1976a]: Steady and Unsteady Finite Element Analysis of Incompressible Viscous Fluid, *Intl. J. Num. Mech. Engng*, **10**, pp. 437–456.

Kellogg, B. [1971]: Singularities in Interface Problems, *SYNSPADE*, pp. 351–400.

Klunker, E. B. [1971]: Contribution to Methods for Calculating the Flow about Thin Lifting Wings at Transonic Speeds—Analytical Expressions for the Far Field, *NASA Technical Notes, NASA TN D-6530*.

Kolmogorov, A. N., and Fomin, S. V. [1957]: *Elements of the Theory of Functions and Functional Analysis*, Vol. 1, *Metric and Normed Spaces*, translated from the 1954 Russian edn by L. F. Baron, Graylock Press, Rochester, New York.

Kolmogorov, A. N., and Fomin, S. V. [1961]: *Elements of the Theory of Functions and Functional Analysis*, Vol. 2. *Measure: The Lebesgue Integral, Hilbert Space*, translated from the 1960 Russian edn by H. Kamel and H. Komin, Graylock Press, Rochester, New York.

Landahl, M. T. [1961]: *Unsteady Transonic Flow*, Pergamon Press, New York.

Lax, P. D. and Richtmyer, R. D. [1956]: Survey of the Stability of Finite Difference Equations, *Comm. pure and appl. Math.*, **9**, pp. 267–293.

Lax, P. D., and Wendroff, B. [1960]: Systems of Conservation Laws, *Comm. pure and appl. Math.*, **15**, p. 363.

Lee, J. K. [1976]: Convergence of Mixed-Hybrid Finite Element Methods, Ph.D. Dissertation, The University of Texas, Austin.

Lehman, R. S. [1959]: Development at an Analytic Corner of Solutions of Elliptic Partial Differential Equations, *J. math. Mech.*, **8**, pp. 727–760.

Leonard, J., and Melfi, D. [1972]: 3-D Finite Element Model for Lake Circulation, Proc., 3rd Conf. on Matrix Methods in Structural Mech., Dayton, Ohio.

Liepmann, H. W., and Roshko, A. [1957]: *Elements of Gas-dynamics*, John Wiley and Sons, London.

Lions, J. L., and Magenes, E. [1972]: Nonhomogeneous Boundary-Value Problems and Applications, Vol. I (Trans. from 1968 French edition by P. Kenneth), Springer-Verlag.

Lynn, P. O. [1974]: Least Squares Finite Element Analysis of Laminar Boundary Layers, *Intl. J. Num. mech. Engng*, **8**, p. 865.

Lynn, P. O., and Boppana, R. R. [1976]: Least Squares Collocation Finite Element Numerical Method for the Navier–Stokes Equations, *Proc., 2nd Intl. Symp. Finite Element Methods in Flow Problems*, ICCAD, Italy.

MacCormack, R. W. [1969]: The Effect of Viscosity in Hypervelocity Impact Cratering, AIAA 66-354, pp. 1–7.

Mikhlin, S. G. [1964]: *Variational Methods in Mathematical Physics*, Pergamon Press, Elmsford, New York.

Mirels, H. [1956]: Boundary Layer Behind Shock or Thin Expansion Wave Moving Stationary Fluid, NASA TN 3712.

Morse, P. M., and Feshback, H. [1953]: *Methods of Theoretical Physics*, McGraw-Hill, New York, p. 1001.

Murman, E. M., and Cole, J. D. [1971]: Calculations of Plane Steady Transonic Flows, *AIAA J.*, **9(1)**, pp. 114–121, Jan.

Napolitano, L. G. [1976]: Functional Analysis Derivation of Hybrid Variational Functionals for Fourth Order Elliptical Operators, *Proc., 2nd Intl. Symp. Finite Element Methods in Flow Problems*, S. Margherita Ligure, Italy, June 14–18.

Neumann, J. von, and Richtmyer, R. D. [1950]: A Method for the Numerical Calculation of Hydro-dynamic Shock, *J. appl. Phys.*, **21**, p. 232, March.

Nitche, J. [1968]: Ein Kriterium fur die Quasi-Optimalitat des Ritzchen Verfahrens, *Num. Math.*, Vol. 11, pp. 346–348.

Norrie, D. H., and deVries, G. [1972]: Applications of Finite Element Methods in Fluid Dynamics, Ed. Smolderen, J. J., *AGARD Lecture Series*, LS-48, May.

Norrie, D. H., and deVries, G. [1975]: *Application of the Pseudo-Functional Finite Element Method to Nonlinear Problems in Finite Elements in Fluids*, Vol. 2, Ed. Gallagher, John Wiley and Sons, New York.

Oden, J. T. [1972]: *Finite Elements of Nonlinear Continua*, McGraw-Hill, New York.

Oden, J. T. [1973]: *The Finite Element Method in Fluid Dynamics, in Lectures on Finite Element Methods in Continuum Mechanics*, Eds J. T. Oden and E. R. A. Oliveira, UAH Press.

Oden, J. T., and Reddy, J. N. [1975]: *Variational Methods in Theoretical Mechanics*, Springer-Verlag.

Oden, J. T., and Reddy, J. N. [1976]: *Introduction to Mathematical Theory of Finite Elements*, John Wiley and Sons, New York.

Oden, J. T., and Wellford, L. C., Jr. [1972]: Analysis of Flow of Viscous Fluids by the Finite Element Method, *AIAA J.*, **10**, No. 12, pp. 1590–1599.

Olson, M. D. [1975]: *Variational Finite Element Methods for Two-Dimensional and Axisymmetric Navier–Stokes Equations in Finite Elements in Fluids*, Vol. 1, Eds Gallagher, Oden, Taylor, and Zienkiewicz, John Wiley and Sons, New York.

Ortega, J. M., and Rheinboldt, W. C. [1970]: *Iterative Solution of Nonlinear Equations in Several Variables*, Academic Press, New York.

Periaux, J. [1975]: Three-Dimensional Analysis of Compressible Potential Flows with the Finite Element Method, *Intl. J. num. Meth. in Engng*, **9**, No. 4, pp. 775–831.

Pian, T. H. H., and Tong, P. [1969]: Basis of Finite Element Methods for Solid Continua, *Intl. J. num. Meth. in Engng*, **1**, pp. 3–28.

Polak, E. [1971]: *Computational Methods in Optimization*, Academic Press.

Rachford, H. H., and Wheeler, M. F. [1975]: *An H^{-1} Galerkin Procedure for the Two-Point Boundary-Value Problems*, in deBoor, C., Ed., *Mathematical Aspects of Finite Element Methods in Partial Differential Equations*, Academic Press, New York, pp. 353–382.

Ralston, A., and Wilf, H. S. [Ed. 1967]: *Mathematical Methods for Digital Computers*, Vol. 2, John Wiley and Sons, New York.

Rayleigh, J. W. S. [1877]: *Theory of Sound*, 1st edn revised, Dover Publications, Inc., New York, 1945.

Reiner, M. [1945]: *A Mathematical Theory of Dilatancy*, American, Vol. 67, pp. 305–362.

Richardson, L. F. [1910]: The Approximate Arithmetical Solution by Finite Differences of Physical Problems Involving Differential Equations with An Application to the Stresses in Masonry Dam, *Trans. Roy. Soc. Lond., Ser. A*, **210**, pp. 307–357.

Richtmyer, R. D., and Morton, K. W. [1967]: *Difference Methods of Initial-Value Problems*, 2nd edn, Interscience Publishers, New York.

Ritz, W. [1909]: Uber Eine Neue Methods zur Losung Gewisser Variations-probleme der Mathematischen Physik, *J. Reine Angew. Math.*, **135**, No. 1, p. 1.

Rivlin, R. S. [1948]: The Hydrodynamics of Non-Newtonian Fluids, *Proc. Roy. Soc.*, **A193**, pp. 260–281.

Rivlin, R. S., and Ericksen, J. L. [1955]: Stress-Deformation Relations for Isotropic Materials, *J. rational Mech. and Anal.*, **4**, pp. 323–425.

Roache, P. J. [1972]: *Computational Fluid Dynamics*, Hermosa Publishers, Albuquerque, N. M.

Schlichting, H. [1968]: *Boundary-Layer Theory*, McGraw-Hill, New York.

Schwartz, L. [1966]: *Mathematics for the Physical Sciences*, Addison-Wesley, Reading, Mass.

Serrin, J. [1959]: Mathematical Principles of Classical Fluid Mechanics, *Handbuch der Physik*, Vol. 8, pp. 125–262.

Shafranov, V. D. [1970]: Hydromagnetic Stability of a Current Carrying Pinch in a Strong Longitudinal Magnetic Field, *Sov. Phy. Tech. Phys.*, **15**, p. 175.

Shen, S. F. [1975]: An *Aerodynamicist Looks at the Finite Element Method*, *Finite Elements in Fluids*, Vol. 2, Eds Gallagher, Oden, Taylor, and Zienkiewicz, John Wiley and Sons, New York.

Shen, S. F., and Habashi, W. G. [1974]: Local Linearization of the Finite Element Method and Its Applications to Compressible Flows, 1st Intl. Conf. on Computational Methods in Nonlinear Mechanics, The Univ. of Texas in Austin.

Sobolev, S. L. [1950]: Applications of Functional Analysis in Mathematical Physics, Translated from the 1950 Russian edition by F. Browder, *Translations of Mathematical Monographs*, Vol. 7, American Mathematical Society, Providence, R. I., 1963.

Sokolnikoff, I. S. [1967]: *Tensor Analysis, Theory and Applications to Geometry and Mechanics of Continua*, John Wiley and Sons, New York.

Southwell, R. V. [1956]: *Relaxation Methods in Theoretical Physics*, Vol. 2, The Clarendon Press, Oxford.

Spalding, D. B. [1972]: A Novel Finite Difference Formulation for Differential Equations Involving Both First and Second Derivatives, *Int. J. num. Meth. Engng*, **4**, pp. 551–559.

Spring, D. J. [1973]: Comparisons between experiment and an approximate transonic calculative method, *Tech. Report RD-73-36*, U.S. Army Missile Command, Huntsville, Ala.

Stoker, J. J. [1950]: *Nonlinear Vibrations*, Interscience Publishers, New York, p. 223.

Strang, G., and Fix, G. [1972]: *An Analysis of the Finite Element Method*, Prentice Hall, Englewood Cliffs, N.J.

Swartz, B., and Wendroff, B. [1974]: The Relative Efficiency of Finite Difference and Finite Element Methods. I: Hyperbolic Problems and Splines, *SIAM J. Num. Anal.*, **11**, No. 5, pp. 979–993.

Takeda, T., Simomura, Y., Ohta, M., and Yoshikawa, M. [1972]: Numerical Analysis of MHD Instabilities by the Finite Element Method, *Phys. of Fluids*, **15**, p. 2193.

Taylor, A. E. [1958]: *Introduction to Functional Analysis*, John Wiley and Sons, New York.

Taylor, C., and Davis, J. M. [1975]: *Tidal Propagations and Dispersion in Estuaries, Finite Elements in Fluids*, Eds Gallagher, Oden, Taylor, and Zienkiewicz, John Wiley and Sons, New York.

Taylor, C., and Hood, P. [1973]: A Numerical Solution of the Navier–Stokes Equations Using the Finite Element Technique, *Intl J. Computers and Fluids*, **1**, No. 1, pp. 73–100.

Thompson, J. F., Thames, F. L., and Mastin, C. W. [1974]: Automatic Numerical Generation of Body-Fitted Curvilinear Coordinate System for Field Containing any Number of Arbitrary Two-Dimensional Bodies, *J. comp. Phys.*, **15**, No. 3, July, pp. 299–319.

Tong, P. [1971]: *The Finite Element Method in Fluid Flow Analysis, Recent Advances in Matrix Methods of Structural Analysis and Design*, Gallagher, R. H., Yamada, Y., and Oden, J. T., Univ. of Ala. Press, Tuscaloosa, pp. 787–808.

Tong, P., and Fung, Y. C. [1971]: Slow Viscous Flow and Its Application to Biomechanics, *J. appl. Mech.*, **38**, Series E, No. 4, pp. 721–728.

Tonti, E. [1969]: Variational Formulation of Nonlinear Differential Equations I, II, *Bull. Acad. Roy. Belg.* (Classe des sci) (5), 55, pp. 137–165, pp. 262–278.

Tricomi, F. [1923]: Sulle equazion: lineari alle derivate Parzial: di secondo ordine, di tipo misto, Rendiconti, Atti dell Academia Nazionale deri Linceri, Series 5, 14, pp. 134–247.

Truesdell, C., and Toupin, R. [1960]: *The Classical Field Theories, Encyclopedia of Mechanics*, Vol. III/I, Ed. Flügge, S., Springer-Verlag, New York.

Turner, M. J., Clough, R. W., Martin, H. C., and Topp, L. P. [1956]: Stiffness and Deflection Analysis of Complex Structures, *J. aeron. Sci.*, **23**, No. 9, pp. 805–823.

Vainberg, M. M. [1964]: *Variational Methods for the Study of Nonlinear Operators*, Holden-Day, San Francisco, Calif.

Varga, R. S. [1967]: *Hermite Interpolation-Type Ritz Methods for Two-Point Boundary Value Problems*, Ed. Bramble, J. H., Numerical Solutions of Partial Differential Equations, pp. 365–373, Academic Press, New York.

Wellford, L. C., Jr., and Hafez, M. M. [1976]: Finite Element Analysis of Linear Elliptic-Hyperbolic Systems Using a Local Parabolic Region, *Proc., 2nd Intl. Symp. on Finite Element Methods in Flow Problems*, ICCAD, Italy.

Wellford, L. C., and Oden, J. T. [1975]: Discontinuous Finite Element Approximations for the Analysis of Shock Waves in Nonlinear Elastic Solids, *J. comp. Phys.*, **19**, p. 179.

Whiteman, J. R. [1975]: *Numerical Solution of Steady State Diffusion Problems Containing Singularities, Finite Element in Fluids*, Eds Gallagher, Oden, Taylor, and Zienkiewicz, John Wiley, New York.

Wilkinson, J. [1965]: *The Algebraic-Eigenvalue Problems*, Oxford Univ. Press (Clarendon), London and New York.

Wu, S. T. [1973]: Unsteady MHD Duct Flow by the Finite Element Method, *Numer. Meth. Engng*, **6**, No. 1.

Wu, S. T., and Jeng, D. R. [1969]: Unsteady MHD Flow in Arbitrary Ducts by Point Matching Method, *AIAA J.*, 7, pp. 1612–1614.

Zienkiewicz, O. C. [1971]: *The Finite Element Method in Engineering Science*, 2nd edn, McGraw-Hill, London.

Zienkiewicz, O. C., and Cheung, Y. K. [1965]: Finite Elements in the Solution of Field Problems, *The Engineer*, pp. 507–510.

Zienkiewicz, O. C., and Godbole, P. N. [1975]: *Viscous, Incompressible Flow with Special Reference to Non-Newtonian (Plastic) Fluids, Finite Elements in Fluids*, Vol. 1, Eds Gallagher, Oden, Taylor, and Zienkiewicz, John Wiley, New York.

Zienkiewicz, O. C., Mayer, P., and Cheung, Y. K. [1966]: Solution of Anisotropic Seepage Problems by Finite Elements, *Proc., Amer. Soc. civ. Eng.*, **92**, EM1, pp. 111–120.

Zlamal, M. [1973]: The Finite Element Method in Domains with Curved Boundaries, *Intl. J. numer. Meth. Engng*, **5**, pp. 367–373.

INDEX